North Woods River

WISCONSIN LAND AND LIFE

ARNOLD ALANEN
Series Editor

North Woods River

The St. Croix River
in Upper Midwest History

Eileen M. McMahon

and

Theodore J. Karamanski

THE UNIVERSITY OF WISCONSIN PRESS

Publication of this volume has been made possible, in part, through support from LOYOLA UNIVERSITY CHICAGO, XCEL ENERGY, ANDERSON CORPORATION, the ST. CROIX NATIONAL SCENIC RIVERWAY, and the ST. CROIX VALLEY FOUNDATION.

The University of Wisconsin Press
1930 Monroe Street, 3rd Floor
Madison, Wisconsin 53711-2059

uwpress.wisc.edu

3 Henrietta Street
London WC2E 8LU, England

5   4   3   2   1   .

Printed in the United States of America

Library of Congress Cataloging-in-Publication Data
McMahon, Eileen M., 1954–
North woods river : the St. Croix River in Upper Midwest history / Eileen M. McMahon and
    Theodore J. Karamanski.
        p.   cm.—(Wisconsin land and life)
        Includes bibliographical references and index.
ISBN 978-0-299-23424-9 (pbk.: alk. paper)
ISBN 978-0-299-23423-2 (e-book)
1. Saint Croix River (Wis. and Minn.)—History.   2. Saint Croix River Valley
    (Wis. and Minn.)—History.   I. Karamanski, Theodore J., 1953–   II. Title.
    III. Series: Wisconsin land and life.
F587.S14M35   2009
977.5′1—dc22
2009009307

To
TED
and
JOE

Contents

Contents

Illustrations

Figures

Illustrations

Acknowledgments

It has been a pleasure to get to know this rolling old river. And like anyone else who has had the pleasure of canoeing its serene waters, we would like to thank those people who are responsible for the protection and care of the St. Croix, from the Minnesota and Wisconsin Departments of Natural Resources to the private citizens of the St. Croix River Association to the staff of the Minnesota–Wisconsin Boundary Area Commission, but most importantly, the women and men of the National Park Service.

This book began as a historic resource study for the National Park Service. Among the staff of the St. Croix National Scenic Riverway, we would particularly like to thank Jean M. Schaeppi and later Julie Galonska, supervisory park ranger for interpretation; Michael Lindquist, Marshland District maintenance employee; and Tony Andersen, superintendent of the riverway. Schaeppi in particular was a valuable partner in gathering historical site information regarding the riverway. We gratefully acknowledge the careful research and insightful writing of Rachel Franklin-Weekley of the National Park Service, who first began work on the recreational history of the area. We are also grateful to Don Stevens of the Midwest Regional Office for helping to facilitate this project in many ways, most especially through his helpful and astute comments on the draft of this project submitted to the National Park Service.

The research staffs of the Minnesota Historical Society and the Wisconsin Historical Society were extremely helpful, especially Harry Miller in Madison, who likely knows more about the sources of Wisconsin history than any other living person. Joseph DeRose of the State Historic Preservation Office in Wisconsin played an important role in helping us to develop site-specific information on the St. Croix Valley. With marvelous state historical societies in both Wisconsin and Minnesota, it was a pleasure to engage in this research project.

Another wonderful resource for historical research was the numerous local historical societies and public libraries within the St. Croix Valley. Let us single out the Taylors Falls Historical Society, the Washington County Historical Society, the History Network of Washington County, and the Burnett County Historical Society. The collections and publications of these wonderful grassroots organizations have kept alive the stories of the people who built the towns, mills, and farms of the St. Croix. We also would be remiss if we did not acknowledge the dean of all those who write about St. Croix history, James Taylor Dunn, whose love of the river lives on in his fine volume *The St. Croix: Midwest Border River.*

We are grateful to Loyola University Chicago for their generous subvention to the University of Wisconsin Press to assist in the publication of this manuscript. We are also appreciative of our friends and colleagues at Loyola University Chicago and Lewis University for the interest and support for this research effort.

On a more personal level we would like to thank Thomas McMahon for generously volunteering his service as an unpaid research assistant and the use of his mobile research vehicle (a.k.a. Tom's camper) during our first research trip to the St. Croix Valley. Lastly, we would like to thank our sons, Ted and Joe, for making it all worthwhile.

North Woods River

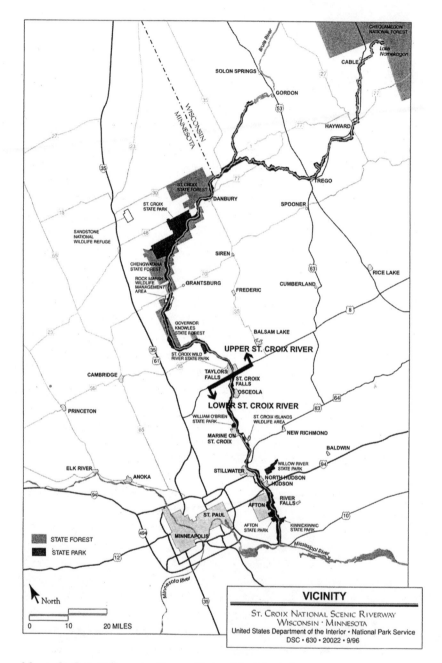

Map 1. St. Croix Valley today (Source: National Park Service).

Introduction

In 1907 Ray Stannard Baker, a son of St. Croix Falls, Wisconsin, penned an essay titled "The Burden of the Valley of Vision." At the time, Baker was one of Progressive Era America's leading "muckraking" journalists. The essay, however, was an elegiac call for urban America to return to nature. In the piece, he described life in a picturesque valley of neat farms, "acres and contentment," yet Baker's rural vision included links to the refinements of the city via inter-urban trains and daily newspaper delivery. Baker argued happiness lay in a middle ground between the simple and the civilized. The St. Croix Valley has flourished in the tension between the rural and the urban. Through a combination of conscious decision and regional evolution, the St. Croix Valley has become the twenty-first-century embodiment of Baker's "valley of vision" for urbanites who boat on its water, build on its bluffs, and bike on its pathways. The St. Croix River occupies a great heart-shaped drainage in Minnesota and Wisconsin between Lake Superior and the Mississippi River. The valley rests in the ever-lengthening shadow of the Twin Cities of Minneapolis and St. Paul. While in art and literature rivers often serve as symbols of hope, historically they have always been agents of change. The St. Croix may be a "valley of vision" today, but history shows that the river's image is as mutable as its past identification as a farmer's valley of plenty or a lumberman's river of pine.[1]

The ancient philosopher Heraclites demonstrated he understood the historical character of rivers when he observed, "No man ever steps into the same river twice, for it is not the same river and he is not the same man." Historically change is the way of the river. Rivers are highways that bring together people from distant places. Rivers also serve as barriers and boundaries. In the seventeenth century, the St. Croix River brought the Ojibwe invaders who, after a century of bloodshed, drove the Dakota from the land of their fathers. In more recent times the river has brought— thanks to the U.S. Army Corps of Engineers, lock and dam projects, and global trade patterns—invaders from the Baltic Sea, in the form of zebra mussels. For this exotic species, like human immigrants from Europe before

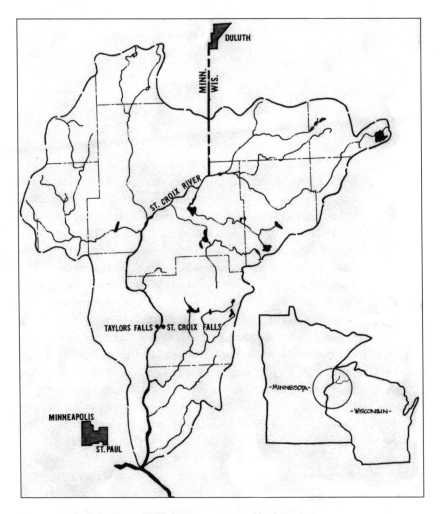

Map 2. St. Croix Valley watershed (Source: National Park Service).

them, the St. Croix has been a river of opportunity as new colonies flourish and indigenous populations are vanquished from a transformed ecosystem. As a highway of change the St. Croix has consistently exerted its stiffest penalties on those populations—human, animal, and plant—that reside closest to the river; and it was the property owners along the Upper St. Croix who lost their homes when the federal government declared it a wild and scenic river.

The St. Croix has been the vehicle by which a significant portion of the Minnesota and Wisconsin North Woods has been transformed; yet the river has also served as a barrier to travel, people, and change. It has not been as bloody or decisive a boundary as Europe's Rhine River or even America's Potomac; still, the St. Croix has been a border river. Where nature created one valley and one watershed, politicians saw the St. Croix as a logical dividing line between the states of Wisconsin and Minnesota. Rivers are natural barriers, better at impeding the spread of wildfires than preventing the migration of plant or animal species. For human populations the St. Croix sped the movement of people and goods along its north-south axis, while at the same time the river was an obstacle to east-west movement, necessitating bridges and ferries. These improvements were unknown before the nineteenth century and rare before the twentieth century, making possible the river's function as a frontier, first between the Ojibwe and the Sioux, and later between two states. The political boundary hardened the natural division of the valley between east and west bank. School districts, local governments, and road commissions defied the logic of propinquity and excluded people living on the other side of the river. A river valley shared in common by two Indian peoples and then two states too often became a resource for both to exploit and neither to protect.

In his 1965 contribution to the Rivers of America series, James Taylor Dunn dubbed the St. Croix a "Midwest Border River." The theme of the border river embraces more than the political division between Minnesota and Wisconsin. The St. Croix Valley is also divided among ten different counties whose jurisdiction overlaps with eleven different municipalities. Hundreds of township and thousands of individual property lines further subdivide the valley. Quite accurately one of the earliest European American settlers in the region, William H. C. Folsom, described the establishment of St. Croix county as beginning the "dismemberment of the St. Croix valley."[2] The Wild and Scenic Rivers Act of 1968 and the Lower St. Croix National Scenic Riverway Act in 1972 created a new set of boundaries along the river. Upriver from Stillwater, Minnesota, the St. Croix River was divided between the narrow wild waters of the Upper St. Croix and the broad, lakelike reaches of the Lower St. Croix. The former was a North Woods river, its valley dominated by wilderness, timber extraction, cut-over farms, and, more recently, tourism. The lower river is marked by farms rooted in deep black soil, prosperous river ports, the scene a century ago of belching smokestacks from mills and steamboats. The upper river has been the hinterland, the resource-rich frontier, of the more heavily populated and urbanized lower river.

The upper river is a remote corner of the American Midwest. In contrast, the lower river is on the border of the dynamic Minneapolis–St. Paul metropolitan area. Originally, St. Croix towns such as Stillwater and Prescott, Wisconsin, vied with the Twin Cities on the Mississippi for metropolitan status. Having decisively lost that competition, the Lower St. Croix gradually fell under the ever-lengthening suburban shadow of the Twin Cities. During the early twentieth century, the metropolitan traffic flows became realities for the Lower St. Croix Valley.

In recent years, the St. Croix's location on the fringe of a growing urban center has overshadowed the crucial historical position of the river valley on the border between the two great inland waterways in North America, the Great Lakes and the Mississippi River. Before there was a Minneapolis or a St. Paul, before the sources of the Mississippi were known, the headwaters of the St. Croix were accessed via portage trail by Indians and fur traders on the move from Lake Superior to the Mississippi. The portage between Lake Superior waters and the St. Croix was crucial to making the region the cockpit of the conflict between the Ojibwe and Dakota and the scene of intense fur trade rivalry between the North West Company, the XY Company, and later the American Fur Company. The unsuitability of the rapid, rock-strewn Upper St. Croix to offer navigation to more than birch bark canoes brought an end to the river's strategic role as a regional link, although the desire to maintain and later revive the waterway through the building of a modern canal was sustained throughout the nineteenth and twentieth centuries. In spite of the fact that the upper river itself was bypassed by commerce, the valley corridor of the St. Croix continued to link the Upper Mississippi with Lake Superior through railroads and later highways.

The St. Croix is not a large river in terms of its size, nor a great river in terms of its impact on the development of the United States. It flows for 165 miles from its source, a long narrow finger of water known as Upper Lake St. Croix, to the Mississippi River at Point Douglas, where the river, once again placid and lakelike, ends its journey. Between the river's mouth and its course, the St. Croix drains 7,760 square miles.[3] Much smaller rivers, such as the Chicago River and Buffalo Creek, have given rise to great cities. Other larger rivers became crucial pathways to the interior, such as the Columbia and the Hudson rivers. The St. Croix was selected in 1968 to be one of the first wild and scenic rivers within the National Park System, yet for most of its history the significance of the St. Croix has been within the framework of the Upper Midwest region. The St. Croix's national significance is its experience with unrestrained private enterprise in developing the frontier and later the desire of local, state, and national institutions to impose a comprehensive structure for future economic, recreational, and residential development.

Like the river itself, the St. Croix's history is tributary to that of the Mississippi valley. Waterways have played a large role in defining this great heartland. Rivers together with portage and canal links to the Great Lakes were a framework promoting regional unity. People, produce, products, and ideas circulated via the waterways. Free movement shaped the Upper Midwest, which in turn shaped the nation. More than a century ago historian Frederick Jackson Turner saw the waterways as a definitive factor in the development of the Midwest. He called it the "typically American region," because it had "no barriers to shut out its frontier from its settled region, and with a system of connecting waterways, the Middle region mediated between East and West as well as between North and South." The label that it is the "typically American region" constitutes the particular "burden," if we may paraphrase the great southern historian C. Vann Woodward, of midwestern history. While the South or New England has a very clear identity, the Midwest, because of its "typical" image, has no distinct profile. This is particularly true for the Upper Midwest, where the farm belt meets the North Woods. In historiographic terms, the Upper Midwest lies hidden in plain sight.[4]

Yet, this history of the St. Croix River's place in Upper Midwest history highlights two critical themes that are unique to the region: the abundance of water and the influence of seasonality. Desert-born Patricia Nelson Limerick rejected Turner's attempt to generalize midwestern experience into a national frontier narrative in part because the native of Portage, Wisconsin, "grew up in a place with plenty of water."[5] Easy access to water for transportation, agriculture, and individual consumption is universal in the Upper Midwest. The states of this section, Michigan, Wisconsin, and Minnesota, all derive their names from defining waterways. Taking the Badger State as an example, the state is unified by the central drainage of the Wisconsin River, yet on its fringes is a quilt of land and life made up of a patchwork of river valleys: the Menominee in the north; the Fox and Milwaukee in the east; the Fever, Kickapoo, Black, Chippewa, and Mississippi in the west; and in the northwest, the St. Croix. The character of each of these rivers has to a large extent defined what is unique in the valleys, while at the same time each illustrates the pervasive influence of abundant water.

Seasonality modifies abundance. It freezes rivers and lakes, shortens growing seasons, and alters animal behavior. Birds migrate; livestock bellow for fodder; the pelts of fur-bearing animals become thicker and more valuable. Today life on the St. Croix alternates with seasonal rhythms from boating season to fall color to deer hunting to skiing season. In the past, seasonality dictated by necessity, not fancy, the movement of peoples across the landscape. For the Dakota, the St. Croix was most vital during the autumn harvest of wild rice. For nineteenth-century European Americans, each turn of the season set in motion a stream of young males. In the late fall, they surged north to the logging camps where winter snows made possible the movement of bulky pine from the cuttings to the riverbank. In spring, these same men might turn their hands to spring plowing in the lower valley, to the dangerous life of the lumber raftsman, Great Lakes sailor, or riverboat deckhand. Work and profits on the St. Croix followed the dictates of the season, and that experience was largely shared by most of the river valleys of the Upper Midwest region.

This study is an environmental and social history. It began as a public history project in cooperation with the St. Croix National Scenic Riverway. The book is published as part of the Wisconsin Land and Life series, so it necessarily seeks to explore the interactions between the land and waters of the St. Croix and the people who have called the valley home. It is a history of the land, how the land was perceived, and altered, and how people adjusted to the reconfigured environment. The dynamic interplay between the cultural and the natural landscape is fundamental to historical understanding. "The valley was considered too far north and the soil too sterile for cultivation," recalled William Folsom, one of the first European American settlers in Taylors Falls, Minnesota. Climate and soil quality are examples of how the environment shapes historical development. Yet, an environmental history seeks to do more than describe the immutable. Just a few years after the St. Croix was dismissed as an agricultural region, it became the seat of hundreds of new frontier farms. Folsom noted, "Many of those who came here in 1838 found out their mistake and made choice of the valley for their permanent home."[6] What had changed was not the length of the growing season or the quality of the soil, but the settlers' perception of the valley. Environmental imagination, the interplay between land and culture, is a critical ingredient of this North Woods story. The St. Croix Valley offered opportunities and imposed constraints on all the plant and animal communities within its riverine corridor. Just as important, however, were the ideas and institutions of the people who came to live there, what anthropologists refer to as a people's "cultural script." The stories about the St. Croix that follow are about the dialectic between a natural blueprint and a cultural script.

The St. Croix Valley encapsulates the history of the Upper Midwest, from its role as a voyageur's highway to its Bunyanesque contribution to the logging frontier. Its painful transition from a countryside patchwork of ethnically diverse cutover farms to a thinly inhabited tourist haven in the "land of sky-blue waters" mirrors a transformation forced upon much of the region. The Upper Midwest, the area bordering the Upper Great Lakes of Huron, Michigan, and Superior and the Upper Mississippi River, does not have a firm place in the regional history of the United States. Yet, the people of much of Michigan, Wisconsin, and Minnesota share a common experience of land, labor, and cultural heritage; that experience must be understood before it can be used to bind separate states into a self-conscious region. The St. Croix River Valley, on the frontier between the Mississippi and Lake Superior drainage, on the border between Wisconsin and Minnesota, with its rich lower valley and more rugged up-country, has a story that reflects the experience of a region and illuminates in part the nature of a nation. As writer James Grey observed a half century ago, "The Upper Midwest contains within itself the memory of everything that America has been and the knowledge of what it may become."[7]

I | Valley of Plenty, River of Conflict

Moving almost silently through the forest, Little Crow approached the place where he had set one of his steel beaver traps. Through the morning mist the Mdewakanton Sioux leader saw that someone had preceded him to the site. The stranger lifted the trap, heavy with a fine, fresh beaver carcass, and was about to remove the valuable catch when he suddenly looked up to see Little Crow. With "a loaded rifle in his hands," Little Crow "stood maturely surveying him." The stranger was not a Sioux, or as Little Crow himself would have referred to his people, a Dakota. The man was dressed in the manner of the Ojibwe. For two generations the Dakota and the Ojibwe had been at war for control of the Upper Mississippi and St. Croix river valleys. This Ojibwe had been caught not only deep in Dakota Territory, but also in the act of committing the worst type of thievery, robbing another hunter's trap. As an act of war and self-defense, Little Crow "would have been justified in killing him on the spot, and the thief looked for nothing else, on finding himself detected."[1]

"Take no alarm at my approach," said Little Crow. Instead of raising his rifle, the Dakota chief spoke gently. "I only come to present to you the trap of which I see you stand in need. You are entirely welcome to it." The wary Ojibwe was further taken back when Little Crow held out his rifle. "Take my gun also, as I perceive you have none of your own." The chief capped this unlikely encounter by offering the stunned Ojibwe a healthy piece of advice: "depart . . . to the land of your countrymen, but linger not here, lest some of my young men who are panting for the blood of their enemies, should discover your foot steps in our country, and fall on you." With that, Little Crow turned his back on the re-armed enemy and traced his steps back to his village.

The story of Little Crow's gesture was recorded by the United States Indian agent Henry Rowe Schoolcraft in his narrative of an 1820 journey to the Upper Mississippi country. Schoolcraft included the story because it

Figure 1. Little Crow's village on the Mississippi near the mouth of the St. Croix. Henry Lewis, *Das illustrirte Mississippithal* (1848).

illustrated the contradictory perception held by European Americans of the Dakota people. Lieutenant Zebulon M. Pike, who had visited Little Crow's village in 1805, had described the Dakota as "the most warlike and independent nation of Indians within the boundaries of the United States, their every passion being subservient to that of war." Yet Schoolcraft also noted that they were "a brave, spirited, and generous people."[2] Little Crow's gesture was magnanimous, but it also was an exercise of supreme self-confidence by a warrior whose mastery over his opponent did not depend upon his ownership of a mere firearm. Through his exaggerated generosity, Little Crow counted a notable coup, as the Ojibwe thief was reduced in status from that of an invader to a mere beggar. The encounter also underscores an important historical point. For the Dakota and the Ojibwe, the most important event on the St. Croix between the mid-eighteenth and mid-nineteenth centuries was neither the expansion of the fur trade nor the arrival of European American settlers, but a terrible and persistent intertribal war. It was the interests and actions of the Indians, not those of a handful of fur traders or Indian agents, that shaped the early history of the valley.

The Dakota and Their Neighbors

The Dakota's ability as warriors, their generosity, and their pride as a nation were all defining characteristics of the first historic inhabitants of the St. Croix Valley. The Dakota could afford to be generous because they occupied one of the largest and richest regions of the North American interior. The early French fur trader Nicholas Perrot called it "a happy land, on account of the great numbers of animals of all kinds that they have about them, and the grains, fruits, and roots which the soil there produces in abundance."[3] The St. Croix was the northeastern border of a Dakota homeland that extended along the Mississippi River and its tributaries, from the mouth of the Wisconsin River on the south to the headwater lakes in the north, and along the Minnesota River westward to the Great Plains. Not just its vast extent made this homeland rich. The diversity of landscape at the disposal of the Dakota offered a cornucopia of resources to the nation's hunters and gatherers. The Dakota lands straddled the northern woodlands and transitional prairie ecosystems and were united by rich riverine corridors and pockmarked by countless lacustrine clusters. Only the long, hard winters of north-central America tempered the possibilities of an otherwise lavish and diverse environment.

"Places are defined," observed historian Elliott West, "in part when people infuse them with imagination." The Upper Mississippi landscape found by European American explorers such as Pike and Schoolcraft was shaped by the choices made by its Dakota inhabitants. Other Indian peoples, such as the Shawnee or the Huron, would have looked upon the rich bottom lands along the Mississippi and envisioned fields of maize, or later white settlers would have seen commercial lumber in white pine thickly arrayed in ranks along the margins of the northern lakes. The Dakota, however, imagined their homeland as a grand hunting preserve. Like most Native American peoples of the Upper Midwest, the Dakota structured their lives around a seasonal subsistence cycle. In the case of the Mdewakanton, this cycle was based on hunting, not the gardening of maize or beans that played an important role in the lives of the Algonquin Indians who dominated the Great Lakes region. Dakota men were hunters and warriors. Fittingly, they approached hunting as they approached war, cooperating with other Dakota to overwhelm their prey, yet always alert to the possibilities of individual recognition.[4]

The Dakota began their year amid the thousand lakes of northern Minnesota and Wisconsin. Large lakes of the St. Croix Valley, such as Chisago, Pokegama, and Upper St. Croix, became the sites of villages of a hundred or more deerskin lodges. Men were active throughout the winter hunting white-tailed deer. Generally able to structure their hunts to suit their palates, the Dakota hunters would alternate the taking of deer with the hunting of winter bears. In winter, deer and elk were a bit too lean and therefore dry when cooked to suit the taste of the Dakota. Bear, on the other hand, were heavy with fat in the winter, and rendered bear fat added savor to other meat. Women prepared meals and treated hides. During the late February and March days, when the winter sun formed a crust of ice upon the deep snowdrifts of the forest, the Dakota hunters stalked herds of elk. These graceful grazing animals favored the open prairies during most of the year but retreated to the fringes of the forest when winter was at its worst. Moving swiftly over the frozen snow with their snowshoes, the Dakota could take large numbers of elk, as they broke through the surface snow and struggled in the drifts.[5]

The proud hunters were greeted with the cry "Kous! Kous!" as young boys saw the men return to the village burdened with heavy loads of meat. Soon every lodge was empty as the entire community, young and old, rushed out to honor the hunters. The shouts continued to fill the evening air until the men laid down the meat at the door of their lodges. A successful late winter elk hunt became the occasion for a great round of feasting among the Dakota lodges. A hunter established his status in part by forcing upon his guests more food than could be consumed. Eating to the point of nausea was the mark of a true Dakota. When elk hunting failed, as it occasionally did because of a lack of snow, the Dakota relied on fish taken in the adjacent lakes. Like true hunters, the Dakota favored spearing fish to the use of nets or hooks, and if their efforts failed or yielded meager results, they accepted a shortage of food as a natural part of the season. Wild plants helped to bridge the rare seasons of want and the more common seasons of plenty. In 1767 Jonathan Carver witnessed the Dakota chewing the soft, inner fibers of "a shrub," perhaps the red willow, which he said tasted "not unlike the turnip."[6]

When the sap of the maple tree began to run, in March or April, the specter of a season of want disappeared. Women took the lead organizing the work of tapping maple trees, gathering sap, and boiling the liquid into sugar. Besides a few old men or boys who might help tend the fires, the sugar camps were composed entirely of women. Most of the men were off trapping or hunting waterfowl. Women united by kinship ties often came together to share the work and fun of making sugar. The sugar camp might be occupied for as long as a month and as many as a hundred trees could be tapped. The hardest part of the sugar making was the preparation of wooden troughs used for boiling. Although the bottoms of these hallowed logs were smeared with mud to retard their burning, exposure to the direct flames of the rendering fires meant that troughs had to be continuously replaced. Such work was well rewarded when the finished sugar was gathered in birch bark containers and the women of the family held feasts in which bark pans of sugar were passed around for all to enjoy. Amid the laughter and stories that were shared, the women and children joined in jokes and dares. A frequent dare was to see who could drink the most of what one anthropologist called "a revolting concoction," liquid tallow. The tallow was used in small amounts to help process the sugar. Around the sugar campfire some women responded to their challengers by drinking cupfuls. Then everyone awaited the resulting effect on the winner, who often became sick or sleepy.[7]

In summer, whole villages of Dakota took to their canoes and journeyed down the St. Croix to its junction with the Mississippi. Amid the hills and river terrace prairies just west of the great river roamed herds of buffalo. Before the Europeans came, the buffalo ranged throughout the domain of the Dakota, and their herds accounted for the abundance that normally marked the life of the Mdewakanton. The Dakota held their summer buffalo hunts on both banks of the St. Croix. Bison ranged throughout western Wisconsin, and small herds were even known to graze in the marshy Pine Barrens of the St. Croix's headwaters region. The most popular place to hunt the buffalo, however, was on the Lower St. Croix and along the Upper Mississippi. In 1680 the missionary-explorer Father Louis Hennepin accompanied the members of a village of Mille Lacs Dakota on a buffalo hunt as far south as Lake Pepin, on the Mississippi. There they killed more than 120 bison.[8]

The summer buffalo hunt was a defining cultural experience for the Dakota of the St. Croix Valley. The buffalo provided the means and the rationale for the Dakota community. In contrast to many of their Algonquin neighbors who lived much of the year in small groups of only several families, the Dakota lived in villages composed of hundreds of people. The village functioned as a unit, not as a congregation of individual hunters. This discipline was established by the requirements of the buffalo hunt. "They assemble at nightfall on the eve of their departure," the fur trader Nicholas Perrot observed, "and choose among their number the man whom they consider most capable of being the director of the expedition." This master of the hunt and his adjutants assigned each man his role in the coming endeavor: scout, shooter, or policeman enforcing tribal discipline. Unlike the popular image of a Sioux buffalo hunt, with hunters racing over the plains on horseback, shooting their prey, the Dakota approached the hunting grounds via birch bark canoes. Upon the receipt of reports from the scouts, the leader would quietly dispatch the hunters, sometimes with the use of smoke signals, and the hunters would drive the herd toward its destroyers. Hennepin witnessed two hundred men converge on a buffalo herd from opposite slopes of a large hill. The two groups of hunters "shut in the buffalo whom they killed in great confusion." Sometimes the bison could be driven by means of prairie fires over a high riverbank and dispatched in that way. The traditional technique of the Dakota buffalo hunt was a group effort leading to a massive slaughter of game. The aftermath of such a hunt, the ground packed with bleeding animals in their death throes, might strike modern readers, as it did the nineteenth-century artist Paul Kane, as, "more painful than pleasing," but such a sentiment would have been foreign to a hunting people like the Dakota.[9]

The excitement of the hunt slowly gave way to the drudgery of processing the harvest of meat and hides. In the disciplined structure of the Dakota buffalo camp, much of this work fell to the women. Some were given the task of quartering and butchering the bison. Others may have been regarded as specialists preparing hides that would become blankets, clothing, and, crucial to the Dakota's mobility, tents. The most laborious task of the women was the drying of thousands of pounds of meat. This was done over slow burning fires with heat, smoke, and sun joining to preserve thin strips of buffalo for up to a year. So important was this task that the prudent hunt leader never selected a kill site far removed from a large supply of firewood. It often took weeks to properly dry and store the meat of a single large kill.[10]

A successful buffalo hunt provided the Dakota with security from want for the remainder of the year. Hunters continued to pursue game throughout the year, including buffalo. The July hunt, however, was designed to produce not fresh meat for the moment, but an insurance policy for the rest of the year. The Mille Lacs Dakota with whom Hennepin lived in 1680 were particularly scrupulous to husband their harvest for the future. The Frenchman observed that "The women buucanned [dried] the meat in the sun, eating only the poorest, in order to carry the best to their villages, more than two hundred leagues from this great butchery." So fundamental was this hunt to the prosperity of the Dakota that hunt leaders were given extraordinary powers to ensure that nothing or no one endangered the community endeavor. Hennepin encountered one group of Dakota celebrating an early buffalo hunt. They arrived ahead of the rest of the village and, rather than wait, made a large killing on their own. The hunters of the main party were furious and destroyed the early arrivals' lodges and took all of their meat. One of them explained to the priest, "Having gone to the buffalo-hunt before the rest, contrary to the maxims of the country, any one had the right to plunder them, because they put the buffaloes to flight before the arrival of the mass of the nation."[11]

In late summer the Dakota would return to the northern lakes. Men would hunt waterfowl and deer, while the women prepared for the vital harvest of wild rice. The rice harvest was second in importance to the buffalo for the prosperity of the Dakota. The Upper St. Croix country excelled as a habitat for the tall aquatic grass known as wild rice. Early French explorers, such as Nicholas Perrot, described the plant as "wild oats," which was actually more accurate because it is not a rice at all, but an annual cereal grass. In later years the European American fur traders labeled the St. Croix Valley as the "Folle Avoine country," using the French term for "wild rice" to characterize the region. The dam on the St. Croix River at Gordon, Wisconsin, destroyed one of the finest wild rice habitats in the region when it flooded the marshy shores of the natural river to create a large recreational lake known as the St. Croix Flowage. For generations before the dam was built, Indian women relied on this rich stretch of river. Dakota women would sometimes seed lake or stream shores to increase their future harvests, but the majority of the wild rice crop grew naturally. Harvesting the crop so as to ensure its return the next year and processing it as a food source required considerable ingenuity and long hours of work. Women gathered the rice in a canoe in which they carefully shook the grain from the tops of the grass, often by means of a wooden stick, so as not to damage the plant. Once a canoe load was brought ashore, "The rice was then separated from the chaff by scorching it in a kettle," recalled an early Minnesota settler, "and then beating it in a mortar made by digging a circular hole in the ground and lining it with deer skin."[12]

Prior to the eighteenth century, agriculture seems to have played a very small part in Dakota life. The amount of wild rice available in the homeland of the Mdewakanton Dakota assured a steady source of natural cereal. Small plots of corn were sometimes planted near village sites, but the amount was never enough for maize to serve as a sustaining element in their diet. Its role seems to have been as a source of menu variety. Similarly they would establish plots of tobacco near their villages. Most of what the Dakota desired was obtained by hunting or through the gathering of wild plants.

The final stage of the Dakota's annual subsistence cycle began in October or November, "the moon of the deer." For one or two moons the large villages would break up into smaller bands that would then cooperate in a communal hunt of white-tailed deer and occasionally elk or woodland caribou. Often they would employ tactics reminiscent of the buffalo hunt, coordinating the movement of large numbers of hunters to drive the deer toward designated shooters. Before the hunt ended in January, with a return to their large semi-permanent lakeside villages, the deer hunters often succeeded in bagging large numbers of deer. Samuel Pond, an early settler who knew the Sioux well, estimated that two Dakota bands combined to kill two thousand deer during the year.[13]

The abundance of food resources that was a manifest part of Dakota life in the St. Croix Valley could give the impression, as historian Gary Anderson has observed, that their lives were "rather idyllic." But the abundance came at a cost. The toll was levied, in part, by the Dakota's frequent movement across the Upper Mississippi landscape. "They have no fixed abode," declared Pierre de Charlevoix somewhat erroneously, "but travel in great companies like the Tartars, never stopping in any place longer than they are detained by the chase." Early French geographers referred to the Dakota as the "wandering Sioux." The French did not see that the abundance of Dakota life was based on their movement, the ability to exploit each segment of their varied homeland at its peak for hunting and gathering. Their large, semi-permanent lakeside villages were an exception to this movement, and it was while in residence in these villages that the Dakota were susceptible to a shortage of resources. Because these sites were occupied repeatedly over the years, they suffered from a shortage of firewood and reduced game populations in their vicinity. The very young and, especially, the very old bore the burden of the seasonal cycle. Those who could not keep up with the group risked the health and well-being of the other family members.[14]

"Hunting is the principal occupation of the Indians," declared Jonathan Carver, who lived among the Dakota during the 1760s. Typical of European observers, he sneered at what he perceived as the "indolence peculiar to their nature," but he did not see the contradiction when he described Dakota hunters as "active, persevering, and indefatigable." The fact was that successful hunting cultures such as the Dakota required considerable energy and sacrifice from their hunters. Bringing down an enraged buffalo or elk at close quarters with a compound bow was fraught with risk, as was the pursuit of game across a frozen boreal landscape. Dakota women also bore a heavy burden. On the move the burden was not merely metaphorical but might entail a pack and tumpline of well over one hundred pounds.

The abundance of the Dakota excited the envy of their Indian neighbors and attracted the notice of the French. Pierre Esprit Radisson and

WAA-NA-TAA OR THE FOREMOST IN BATTLE

Chief of the Sioux tribe.

Figure 2. The Dakota warrior Waa-na-taa (Foremost in Battle). Chief of the Sioux tribe, painted by James Lewis Otto at the Treaty of Prairie du Chien, 1825. The American government used the treaty council to unsuccessfully mediate the Dakota–Ojibwe dispute. Men like Waa-na-taa, however, continued to defend the Dakota's right to lands east of the Mississippi and St. Croix rivers. Wisconsin Historical Society, WHi-1867.

Médard Chouart, Sieur des Groseilliers, the first Europeans to penetrate the interior of northern Wisconsin, first encountered the Dakota during the winter of 1659. Unusual climatic conditions had ruined the winter hunts of the Menominee among whom the French were staying. The village inhabitants were forced to eat their dogs, and starvation gradually consumed its inmates. As the horrible winter came to a close, representatives of the Dakota arrived in the village. The eight well-fed Dakota men, each accompanied by two wives bearing baskets of wild rice, strongly impressed the haggard Europeans. Without hesitation they accepted the invitation of the Dakota to visit their lands.[15]

The Dakota demonstrated considerable forbearance, even generosity, toward their Algonquin neighbors to the east. But like Little Crow's decision to spare the life of an Ojibwe robbing his trap, the Dakota peoples' generosity was calculated. As early as 1650, the Dakota allowed remnants of the Huron, who had been driven from their homelands near Georgian Bay by Iroquois invaders, and a small group of their Ottawa allies to settle in Dakota Territory near Lake Pepin. The Dakota were at least in part motivated by the desire to obtain French trade goods from the two tribes that had been the backbone of the early western fur trade. But Dakota generosity seems to have been misinterpreted as a sign of weakness by the Huron and Ottawa, who tried to drive the Dakota away from the Mississippi. This major miscalculation of Dakota capability and intent led to the complete expulsion of the Huron and Ottawa from Dakota lands. For the next hundred years the Dakota were intermittently at war with the eastern Algonquin bands, who, like the Huron, had been driven west by the Iroquois. Traditional enemies such as the Cree to the north and the Illinois to the south continued to be the focus of annual Dakota war parties, but warfare with the Huron, the Ottawa, and, especially, the Fox also became common. The Fox arrived in Wisconsin in the seventeenth century, driven from the Michigan peninsula first by the Iroquois and later by the Ojibwe. Their arrival in what is now Wisconsin brought them into collision with the Eastern Dakota bands. Ojibwe oral tradition holds that for a brief time the Fox actually occupied the Upper St. Croix Valley and the region around Rice Lake, Wisconsin. The Dakota and the Ojibwe began their relationship, which would later stain the waters of the St. Croix with much blood, as allies against the Fox. In 1680 a joint Dakota–Ojibwe war party, perhaps as many as eight hundred men, fell upon the Fox villages in east central Wisconsin. Only after suffering severe losses were the Fox able to repulse the attack.[16]

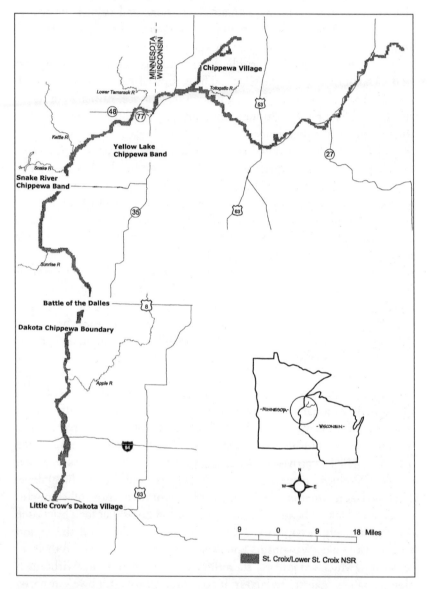

Map 3. Dakota–Ojibwe conflict sites, St. Croix National Scenic Riverway (Source: National Park Service).

As the enemy of the Dakota's enemy, the Ojibwe became neighbors with whom the eastern bands shared hunting grounds, trade, and brides. The Ojibwe originally entered the river and lake country of the Wisconsin border as Dakota guests, not as invaders. As allies, the Ojibwe were allowed to hunt and trap in the St. Croix and Chippewa river valleys. It was an alliance of two of the most numerous and expansionistic Native peoples of the North American interior sealed with Fox blood and sustained by substantial and mutual benefits. The Dakota shared in the Ojibwe's regular access to French trading goods. The Ojibwe won access to lands rich in white-tailed deer, beaver, and wild rice. The Dakota secured metal tools and firearms to improve subsistence activities and their military efficiency. According to Ojibwe oral tradition, the Dakota first encountered firearms when an Ojibwe peace delegation arrived in a Mdewakanton encampment on the St. Croix River. The incident ended badly when a proud Dakota warrior denied the power of a musket and dared the Ojibwe to shoot at him. One fearful crack of the gun led to the death of the man. This incident damaged the alliance. Nonetheless, relations were patched, and early in the eighteenth century the Ojibwe were allowed to establish a village in Dakota Territory, near Spooner, Wisconsin.[17]

French Fur Traders on the St. Croix

By the beginning of the eighteenth century, regular access to firearms had become vital to the Dakota. French fur traders pushing up the Mississippi River had armed their traditional enemies in the Illinois country, while the *coureurs de bois* of the Great Lakes region had provided enemies to the east, such as the Fox, with muskets and powder. Those enemies and the remoteness of the Dakota lands frustrated the efforts of early French traders to establish themselves among the Mdewakanton. The first European to enter the Dakota lands along the St. Croix was Daniel Greysolon, Sieur du Luth (Duluth). In 1679 the explorer claimed the region for France and offered gifts to the Dakota. At the mouth of the St. Croix he secured the release of Father Louis Hennepin, a Recollet missionary whom they had taken prisoner on the Mississippi. Duluth made the first recorded passage of the Bois Brule–St. Croix portage that allowed access from Lake Superior to the Mississippi valley, a route he likely learned about from the Ojibwe at La Pointe.

The name of the St. Croix River dates from this early period of French activity in the Upper Mississippi region. Hennepin tried to fix on it the name *Riviere du Tombeau,* or River of the Grave, after the Dakota buried a snakebite victim on its bank. That name did not take, nor did one that appeared on several early maps, *Riviere de la Madeleine.* Many stories concerning the name *St. Croix* link it to the early missionaries. One story credits the name to French priests, who saw the shape of a holy cross in the river's right angle junction with the Mississippi. Another version attributes the name to a rock formation on the bank of the river in the Dalles that appeared to have the shape of a cross. Other than Hennepin, who tried to put a morbid moniker on the river, the only other early missionary to see the river was the Jesuit Gabriel Marest, who was a missionary to the Dakota and witness to Nicholas Perrot's vainglorious ceremony of "taking possession" of the region in 1689. In that ceremony Perrot claimed "in his Majesty's name" a vast track of land stretching westward from Green Bay, including "the rivers St. Croix and St. Peter, and other places more remote." The use of the name *St. Croix* at so early a date suggests the river may indeed have been named, like the St. Peter's River, by missionaries. Another story dates from Jean Baptiste Bénard de La Harpe's 1700 account of Pierre Charles Le Suer's journey a year earlier up the Mississippi to the Minnesota River valley. "He left on the east of the Mississippi, a great river," La Harpe wrote, "called St. Croix, because a Frenchman of that name was wrecked at its mouth." While the exact origin will never be known for sure, the name *St. Croix* is one of the earliest European place-names in the upper Midwest region.[18]

Other missionaries and traders followed, and a series of trading posts were temporarily established in the late 1600s and early 1700s along the Upper Mississippi or Lower Minnesota rivers. Each venture, however, failed due to distance and the opposition of the Fox, who attacked posts and hindered resupply efforts. In 1695 Tiyoskate, a Dakota emissary to the governor general of New France, with carefully staged tears in his eyes, implored, "All the nations had a father who afforded them protection; all of them have iron." He concluded by describing himself as a "bastard in quest of a father." Although the governor general was stirred to grandiloquently promise Tiyoskate "iron" tools and weapons, Dakota trade contacts remained intermittent—all of which served to make the alliance with the Ojibwe more important to the Dakota.[19]

The Mdewakanton made French efforts to extend trade to them more difficult by refusing to forgo their wide-ranging military campaigns. With the firearms they were able to obtain, the Dakota ravaged the Illinois country. In 1700, French officials encountered Piankashaw Indians in Illinois who were attacked by recent Dakota raids. They had been so devastated by this event that they were unwilling to risk retaliation. This aggressive approach to their southern frontier even extended to French traders moving north from Illinois, whom the Dakota routinely robbed, as they were regarded as allies of the Illini, Miami, or Piankashaw. This vigorous approach to war, together with the large number of warriors the Sioux were traditionally able to marshal for war, inclined the French to refer to the Dakota as the "Iroquois of the West."[20]

The power of the Dakota, however, began to ebb in the eighteenth century. Over the years the Sioux divided into two distinct groups. In 1680 Father Louis Hennepin described one subdivision of the Sioux as the "Tinthonha (which means prairie-men)." Pierre Charles Le Sueur, who attempted to establish a trading post (just below the mouth of the St. Croix) among the Dakota in 1694, distinguished between the "Sioux of the West," who resided along the Minnesota River, and the "Sioux of the East," who dwelled along the Upper Mississippi. During the late 1600s, the western Sioux or Lakota began to expand from the Upper Minnesota River over the plains to the Upper Missouri River country. This historic migration would in the course of the next century bring the bulk of the Sioux from the forest to the grasslands. The birch bark canoe was gradually forsaken in favor of the horse, which the Lakota began obtaining by trade around 1707. Buffalo hunting went from being one part of the annual seasonal cycle, dictated by the arrival of bison herds along the Upper Mississippi, to the sustaining food source available year-round to hunting villages made mobile by the horse.[21]

Like all movements of peoples, this one resulted from both a perception of opportunity and the dictates of necessity. As warriors, the Dakota were engaged in conflict on all of their frontiers. Opponents to the north, south, and east were all as well or better armed than the Dakota. To the west, however, were Indian peoples whose trade contacts with European fur traders were inferior to those of the Dakota. Nor was the population size of the Upper Missouri peoples as formidable as in Wisconsin, which in the seventeenth century became densely populated with eastern Indians fleeing the Iroquois. Although the Dakota appear to have been generally successful in defending their large, rich homeland in the late seventeenth and early eighteenth centuries, the soft frontier to the west drew aggressive Sioux warriors out onto the plains. Another factor was the opportunity to secure buffalo seemingly at will. The buffalo hunt was a defining cultural element among all the Sioux, their grand group experience. To build a lifestyle around it must have been alluring, particularly because hunting options in Minnesota may have declined at the critical time of the move onto the plains. In 1717 and 1718, French traders noted that animal populations south and west of Lake Superior declined due to an unknown disease. "All the elk were attacked by a sort of plague, and were found dead," according to one report. Indians who ate the flesh of the infected animals also died. There is evidence that by the 1730s even the eastern-most Dakota, the Mdewakanton, were forced to substantially increase their reliance on buffalo hunting, perhaps due to a decline in the elk herds. If buffalo was becoming the dietary backbone of many Sioux, the movement onto the plains made economic sense.[22]

The Origins of the Dakota–Ojibwe War

As the Sioux nation as a whole expanded westward, the Dakota, manning as it were the eastern gateway to their lands, were placed in an untenable position. The Sioux in the west were no longer available for, nor interested in, fighting battles against traditional enemies on the eastern border. As the Sioux frontier expanded dramatically to the west, the ability of the nation to maintain its control over the large and rich hunting grounds along the Upper Mississippi was necessarily compromised. For all their military prowess and impressive numbers the Sioux could not dominate both the northern plains and the Upper Mississippi valley; as the western bands moved to accomplish the former, the eastern Dakota were thrown into a desperate attempt to maintain the land of their fathers. New diseases introduced by the Europeans, especially smallpox and malaria, and the accelerated pace of intertribal warfare also contributed to the decline of the eastern Dakota. Between 1680 and 1805, the number of Dakota in the Mississippi valley may have declined by as much as one-third. These factors, together with the migration of their western kinsmen, made the Dakota vulnerable to the equally expansionistic Ojibwe.[23]

The Ojibwe pursued their own manifest destiny in the seventeenth and eighteenth centuries. In the ancient traditions of the tribe, the Ojibwe once inhabited lands near the Atlantic Ocean. Upon migrating to the Great Lakes region, they split into three groups, the modern Ottawa, Potawatomi, and Ojibwe. The latter group occupied a vast arc of lands stretching from the shores of Lake Erie to rivers and lakes west of Lake Superior. This vast homeland was much larger than even the domain of the Dakota, but it was much less diverse in its landforms and, therefore, less abundant in resources, notably lacking the sustaining herds of elk and buffalo. Fish from the Great Lakes filled the protein gap for the Ojibwe, but they did not produce the residuals of leather and robes. Together with their Potawatomi and Ottawa cousins, the Ojibwe had been among the earliest interior tribes to become engaged with the fur trade. As traders and trappers, they were always on the lookout for new peoples with whom to exchange or new lands to trap. Through the seventeenth century the Dakota both provided the Ojibwe with furs and made their hinterland available to pioneer Ojibwe bands. This economic alliance became frayed when the Dakota were gradually able to obtain European weapons and goods directly from the French. In turn, the Dakota came to resent the Ojibwe's middleman commerce with the Cree and Assiniboine, hereditary enemies of the Sioux.[24]

Violence between the Ojibwe and the Dakota began slowly in the 1720s and escalated to a full, prolonged war in 1736. Conflict between the tribes forced the Dakota to abandon traditional wintering villages at Leech Lake and Mille Lacs, sites that were soon colonized by the Ojibwe. In the oral tradition of the latter, this process was rendered heroic by tales of an epic battle of three days that left the former Dakota villages littered with the bodies of the slain. More plausible was a calculated withdrawal closer to the increasingly more important buffalo herds along the Mississippi and away from regions exposed to both Ojibwe and Cree attack. Far from being completely routed from the region, the Dakota continued to occupy a village on the Rum River (which drains Mille Lacs) for another generation. The bulk of the eastern Dakota, however, concentrated along the mainstream of the Mississippi River and the Lower St. Croix. During the seventeenth century the Dakota did not use the St. Croix as intensively as they did the large headwaters lakes in north-central Minnesota. As a result, in the 1700s the St. Croix was a much more reliable source of game. Not surprisingly, control of the hunting grounds and rice marshes of the St. Croix was hotly contested by both sides in the war.[25]

The Dakota–Ojibwe war was a tragedy for both peoples and a source of frustration to the Europeans who, during the 1700s, began to influence events in the Upper Mississippi region. Conflict made the already precarious existence of a frontier fur trader downright dangerous. Men supplying weapons to one side were naturally regarded as enemies by the other and were often dealt with violently. In 1741, following Ojibwe attacks on two Dakota camps, the latter retaliated against French traders in Wisconsin. Warfare also hurt trapping, as Indian hunters were reluctant to stray far from their family's winter camps. In the more than a century and a quarter that the Ojibwe and the Dakota were locked in warfare, European traders repeatedly tried to arrange truces. This desire for parlays on the part of the Europeans should not obscure the fact that fur traders sometimes, often unknowingly, encouraged hostilities between the Dakota and the Ojibwe. Traders based on the Upper Mississippi and dwelling with the Dakota naturally supported Sioux claims to the St. Croix while those based on Lake Superior and in Ojibwe territory naturally desired their hunters to dominate as large an area as possible. Most of all, traders wanted fur trapping to be pursued aggressively. During the early 1730s, when the conflict between the two tribes began to simmer, French fur traders exasperated the situation by encouraging the Ojibwe and the Winnebago to expand their trapping grounds along the Upper Mississippi. The fur traders did not think the Dakota were trapping anywhere near the full potential of their rich trapping grounds and impatiently encouraged interlopers to fill the perceived void. The entire history of the fur trade in the St. Croix Valley took place under the cloud of a bloody, often internecine, war.[26]

The experience of Paul and Joseph Marin, French fur traders in the region from 1750 to 1754, reveals the relationship between war, peace, and trade. By bringing the governor general of New France into partnership, Paul Marin, a veteran fur trader, was granted a lease to the fur trade of the Upper Mississippi. Marin's political connection initially assured that the vital access points to the region—Green Bay, which controlled the Fox–Wisconsin water route to the Mississippi, and La Pointe, which was critical to the Bois Brule–St. Croix route—were controlled by his friends. In fact, his son Joseph was granted command of the La Pointe garrison. The elder Marin astutely cultivated Dakota leaders. Canoe loads of gifts were lavished on them, and in turn they acted as trading liaisons, collecting furs from all of the hunters who had received goods in advance from Marin. The French extended their trading presence to most of the villages of the eastern Dakota and were rewarded with annual returns of 150,000 francs. Meanwhile, young Joseph Marin, based in Ojibwe territory, worked to ease tensions between his trading partners and the Dakota. Eventually the father and son managed to work out a division of the disputed territory. On the St. Croix this led to the Dakota recognizing Ojibwe rights to the valley from its headwaters to the mouth of the Snake River.[27]

The tenure of the Marins demonstrates that, properly managed, the fur trade could have been a means to control the level of violence between the Ojibwe and the Dakota. Yet the corrupt French colonial administration in Quebec had different priorities and would not long manage its western affairs in a consistent manner. Paul Marin's appointment to the Upper Mississippi had been made with the assurance that he share the profits with the Marquis de la Jonquiere. That venal administrator assured that New France's western posts were under the direction of administrators friendly to Marin and sympathetic to the Dakota. A new governor general, however, would find profit in other arrangements. Therefore, in 1752 when the Marquis de la Duquesne succeeded to the governor general's palace, Paul and Joseph Marin lost their official protection. The command of the critical French post of La Pointe on Lake Superior passed to Louis-Joseph La Verendrye. This very capable frontier leader was in no mood to cooperate with the Marins. While they had been profiting mightily from the Dakota trade, La Verendrye and his equally talented father and brother had been denied any position in the western trade. Far from being sympathetic to the Dakota, La Verendrye had every reason to resent them, for a Dakota war party had attacked and killed his brother and twenty-three other men at Lake-of-the-Woods in 1736. The La Verendryes were dedicated to expanding French influence west of the Great Lakes, which made them dependent upon a close relationship with the Ojibwe and the Cree. The result of La Verendrye's appointment was a quick erosion of the truce and trust the Marins had succeeded in building.[28]

The first sign of trouble was when La Verendrye notified Joseph Marin that the former's La Pointe post trading area included the entire St. Croix River Valley. La Verendrye intended to have complete control of the Upper Mississippi region. In 1753 he dispatched several of his traders to establish a wintering post on the St. Croix near the Sunrise River. La Verendrye then intercepted and turned back agents Marin had sent up the Mississippi to Leach Lake to negotiate a peace between the Dakota and the Cree. This latter action spread panic among the Dakota, who saw it as a repudiation of French friendship and the prelude to new attacks by the Cree and the Ojibwe. Bracing for war, the Dakota abandoned their hunts and suffered through the winter of 1753, bereft of game for their lodges or furs for Marin. Dakota women and children "lived on nothing but roots all winter long," they complained to Joseph Marin. "That is why today we are worthy of pity." The Dakota further complained that the Ojibwe were violating their earlier agreement to limit their hunting. As a result of these outrages the Dakota chiefs told Marin, "We cannot keep from you the fact that our young men are all beginning to mutter at seeing the Sauteux [Ojibwe] so unreasonably trying to steal territories belonging to us." The young Dakota were wise to mutter because La Verendrye clearly had the upper hand over Marin. After 1754, Joseph Marin left the Upper Mississippi country, and French commanders at La Pointe tilted their policy in favor of Ojibwe expansion.[29]

For close to a decade the French and Indian War (1754–60) and Pontiac's Rebellion (1763) interrupted the flow of trade goods into the St. Croix Valley and temporarily reduced the importance of the fur trade in the lives of the area's embattled inhabitants. The Dakota–Ojibwe conflict, however, continued. A large battle was fought near what may have been the headwaters of the St. Croix. The Ojibwe suffered heavy losses in the engagement, but the Dakota were forced to yield the field. The Dakota seem to have more than held their own throughout the 1750s and 1760s. When the first British observers entered the Upper Mississippi region in 1766, they found the Dakota disdainful of their enemies, whom they referred to as "slaves or dogs." Their lodges were well supplied with meat, and feasting continued through the winter. On the other hand the Ojibwe, still lacking access to buffalo and elk, seem to have suffered from the temporary cessation of the fur trade. Alexander Henry, the first British merchant to reach them after the fall of New France, found famine stalking the Lake Superior Ojibwe. "These people were almost naked, their trade having been interrupted," he noted in his memoir.[30]

The late 1760s and early 1770s saw an intensification of the fighting between the Dakota and the Ojibwe. Traders operating from La Pointe on Lake Superior likely encouraged the Ojibwe to increase their fur returns by expanding their hunting in Dakota Territory. Sometime around 1770 one of the greatest battles of the long war was fought at the Dalles of the St. Croix. The best account of what occurred comes from Ojibwe oral tradition recorded in 1885 by William Warren, a Métis historian of Ojibwe ancestry. According to his sources, the campaign began at the instigation of the Fox Indians, who ascended the Mississippi desirous of settling the score with the Ojibwe, their hereditary enemies. The Dakota, former enemies of the Fox, were enlisted to make a joint attack. The combined war parties made their way by canoe up the St. Croix. On the portage trail around St. Croix Falls, the Dakota and Fox encountered a large Ojibwe war party intent on raiding the Dakota villages along the lower river. The Ojibwe had been smarting from numerous successful Dakota raids on isolated hunting camps along the St. Croix and Namekagon rivers. Waubojeeg, a renowned Lake Superior leader, had gathered warriors from across northern Michigan, Wisconsin, and Minnesota. He led his large war party up a branch of the Bad River, portaging eight miles west to the sources of the Namekagon River. Waubojeeg directed the advance downstream cautiously, so when they reached St. Croix Falls his scouts informed him of the large enemy force ahead. Just before the fighting commenced, the veteran Ojibwe fighters pushed those going into battle for the first time into the river. There the novices washed off the black paint that had stigmatized them, and they were allowed to join the warriors as equals.[31]

The two forces met in combat on the portage trail. Fighting between the Fox and the Ojibwe marked the first phase of the battle. Allegedly the Fox had boasted that they would make short work of their enemies and requested that the Dakota remain aloof from the fighting. Confined by the ravines and rock outcroppings of the portage, the fighting was heated and at close quarters. About midday the Ojibwe gained the advantage and forced the Fox to flee. At the point of driving the latter into the raging river, the Ojibwe were staggered by the sudden entrance of the Dakota into the battle. After several more hours of combat, the Ojibwe, with most of their ammunition exhausted, broke away and ran from the Dakota. Victory was at hand for the Dakota when a war party of Sandy Lake Ojibwe suddenly made their appearance on the field. They had missed their rendezvous with Waubojeeg's main force and had hurried downriver, arriving at a crucial point in the contest. This reinforcement turned the tide of battle for the final time. The Dakota attack was broken, and their warriors were sent into a headlong retreat. "Many were driven over the rocks into the boiling floods below, there to find a watery grave," Warren recounted. "Others, in attempting to jump into their narrow wooden canoes, were capsized into the rapids." The Ojibwe and Dakota both suffered heavy losses, but it was the Fox who left the most dead amid the rocks and cervices of the battlefield. The victory secured for the Ojibwe the control of the Upper St. Croix Valley. An informal boundary was fixed between the Dakota and the Ojibwe around the mouth of the Snake River.[32]

English Fur Traders on the St. Croix

The first English fur trader to penetrate the Lake Superior country was Alexander Henry. Henry's access to the remote frontier was made possible by his astute alliance with Jean Baptiste Cadotte, a key figure in bridging the end of the French regime and the era of British domination. Cadotte's family had long been involved in the western fur trade. He grew up in the Lake Superior region and married the daughter of a notable Ojibwe chief. Cadotte had served as the last French governor of the fort at Sault Ste. Marie and earned the trust of the English conquerors by being one of the first French traders to embrace the new regime. When Jean Baptiste Cadotte retired from the trade, his son of the same name took his place. The latter played a lead role in reopening the fur trade of the Upper

Mississippi valley. A second son, Michel Cadotte, was active on the St. Croix and Chippewa rivers. In 1784 he operated a post on the Namekagon River, near the head of the portage to Lac Courte Oreilles. Later he established trading posts at Yellow Lake, Snake River, and Pokegama Lake. The key to the success of the Cadottes was their family relationship with notable Ojibwe leaders. They consciously identified themselves with the expansion of the Ojibwe into the St. Croix and Upper Mississippi valleys, advancing hand in hand with bands of hunters to exploit the bounty of Dakota hunting grounds.[33]

While Alexander Henry did not venture far south of Lake Superior, Jonathan Carver, another young Englishman on the make, explored the Upper Mississippi frontier. Ostensibly Carver was a mapmaker sent with explorer James Tute, a captain in the famed Roger's Rangers, to discover an inland route to the fabled Northwest Passage. Carver journeyed from Mackinac across Wisconsin and wintered among the Dakota villages of the Minnesota River valley. He was fascinated with the Dakota, whom he described as "a very merry sociable people, full of mirth and good humor."[34] He explored the Upper St. Croix Valley by portaging from Lac Courte Oreilles to the Namekagon River near present-day Hayward, Wisconsin. Carver descended the Namekagon, which he named "Tutes branch," to its junction with the main stream. James Goddard, the official secretary of the expedition, described the Namekagon as "a very pleasant country, and plenty of deer in it." Above the junction of the Namekagon and St. Croix, Carver tried his hand at sturgeon fishing. "The manner of taking them is by watching them as they lie under the banks in a clear stream, and darting at them with a fish-spear; for they will not take a bait." The region of the headwaters he laconically noted was known to the Ojibwe as "the Moschettoe [mosquito] country, and I thought it most justly named; for, it being then their season, I never saw or felt so many of those insects in my life."[35]

Although his time in the St. Croix Valley was brief, Carver's legacy lingered longer in the form of a narrative titled *Travels through the Interior Parts of North America in the Years 1766, 1767, and 1768*. This widely read and frequently translated account of the Upper Mississippi region provided generations of readers with their first exposure to the region. Like so many would-be explorers before and after, Carver mixed genuine observations with heavy doses of romantic fancy and self-promotion. Captain James Tute, who actually directed the expedition, goes all but unmentioned in the published narrative, creating the impression that Carver was the man in charge. His enthusiasm for the region, however, was effusive.

The Upper Mississippi region was "abounding with all the necessaries of life, that grow spontaneously; and with a little cultivation it might be made to produce even the luxuries of life." He did not simply note stands of maple trees but went on to predict that the "delightful groves" were present in "such amazing quantities" that they "would produce sugar sufficient for any number of inhabitants." If the principle purpose of Captain James Tute had been to evaluate the fur trade prospects of the region, the purpose of Carver's book, which he did not publish until 1778, was to promote Carver by giving the impression that he had single-handedly opened up a utopia for future settlers. He went so far as to produce a map in which he divided the Upper Midwest into a series of "plantations or subordinate colonies" so that "future adventurers may readily, by referring to the map, chose a commodious and advantageous situation." With this map Carver also established symbolic mastery over the region, inviting his numerous readers in the decades that followed to project their dreams on to what Carver portrayed as an open and free land.[36]

Carver included the bulk of the St. Croix Valley in his "plantation No. 1," which encompassed a great triangle of territory that reached as far northwest as Rainey Lake and to Sault Ste. Marie on the east. "The country within these lines," he observed, "from its situation is colder that any of the others; yet I am convinced that the air is much more temperate than those in provinces that lie in the same degree of latitude to the east of it." With this claim Carver anticipated the logic of virtually every land promoter to follow. Don't let the far northern position of the St. Croix Valley daunt settlers' dreams of an agricultural future. Not only was the climate more temperate than what people knew in the east, but "the soil is excellent, and there is a great deal of land that is free from woods in the parts adjoining the Mississippi." Of course, if the settler preferred trees, Carver allowed that the "north-eastern borders" of the region were "well wooded." In addition to abundant rice and copper, settlers were also blessed with the presence of the "River Saint Croix, which runs through a great part of the southern side of it, enters the Mississippi just below the falls, and flows with so gentle a current, that it affords convenient navigation for boats." Carver was so impressed with the prospects for the region that he passed to his heirs a fraudulent document, which claimed the cession of a large tract of land. Allegedly the Dakota granted him twelve million acres of land, including a sizable portion of the St. Croix Valley. Neither the British crown, the Eastern Sioux, nor later the U.S. Congress saw fit to recognize the Carver land grant as valid. In 1817 two of the explorer's grandsons actually journeyed back to the Upper Mississippi wilderness to have Ojibwe and Dakota elders substantiate the alleged grant. Not surprisingly, they went home with no more land than when they set out.[37]

Although the Tute-Carver expedition was supposed to be directed to discover the Northwest Passage, a large part of its real orientation was to probe the commercial opportunities of the Upper Mississippi frontier. In the 1760s, only a geographic idiot would have spent weeks ascending the Chippewa River and the Upper St. Croix in search of the fabled water route to China. The fur trade was on the mind of many of the British who went west after the French and Indian War. One of those who left a record of their efforts was a cantankerous Connecticut Yankee named Peter Pond. He described the process by which merchants of English origin began to dominate the trade in the region because of their superior ability to obtain credit and merchandise as well as their ability to act in concert with British military forces and thereby pose as power brokers between tribes. Both Carver and Pond attempted to negotiate a peace treaty between the Dakota and the Ojibwe. Each was more successful in arranging a parlay than a lasting peace. In 1775 Pond escorted a group of Dakota and Ojibwe chiefs to Mackinac, where an agreement was struck by which the Dakota promised "Not [to] Cross the Missacipey to the East Side, to Hunt on thare Nighbers Ground." That the Dakota would agree to recognize the Ojibwe as masters of the area east of the Mississippi would have been a major concession and was likely the result of Pond's failure to obtain a delegation of the eastern-most Dakota tribe, the Mdewakanton Sioux. A British military inspection of the region in 1778 led by Charles Gautier de Verville reported a large Mdewakanton village on the Upper St. Croix River. The village included a number of lodges of the Winnebago, a people who shared the Dakota's antipathy toward the Ojibwe. Clearly the Mdewakanton Sioux had not abandoned the east bank of the Mississippi to the Ojibwe.[38]

The arrival of the British regime along the St. Croix River signaled a change in the way the fur trade would be administered. The French system of leasing the right to control the trade in a large geographic area, while never able to keep all illegal *Coureurs de bois* out of business, did tend to greatly restrict the number of traders operating in any one area. Under the British and even more so under American administrations there was much less regulation of the fur trade, and an ever-growing number of participants made their way west. More participants meant more competition and less concern on the part of many traders for the long-term good of both the fur trade itself and the Indian trappers in particular. One of the early innovations of the British regime was the establishment of companies that pooled the resources and special skills of a group of fur traders. The

North West Company, founded in 1779, was an attempt by Peter Pond and other fur traders to establish greater control over competition in their business. The North West Company, based in Montreal, exerted a considerable influence over the fur trade of the St. Croix River during the years between its creation and 1816. United under its control were many of the traders who had expanded the trade since the fall of New France, including Alexander Henry and Jean Baptiste Cadotte and his brother Michel Cadotte.

A Social History of the Fur Trade in the St. Croix Valley

One of the enduring historical myths of the Upper Midwest is the heroic image of the fur trade explorer and his hardy voyageur companions. The myth rests on the very real role these men played establishing the first white businesses and settlements in the region and the romantic impulse of those who later read the traders' memoirs and journals to try and imagine just what the Midwest looked like when all was wilderness. Fur traders did act as wilderness explorers, but many aspects of their business were anything but heroic. It is vital to balance the picture of the fur trader as an explorer and pioneer with the less flattering portrait of the fur trader as a pusher of dangerous and addictive substances, a fomenter of intertribal and intratribal conflict, and a participant in environmental degradation. Nor is it historically valid to dismiss their Indian trading partners as innocent victims. The fur trade brought the Indians products vital to life in the forests of the Upper Midwest: copper kettles, steel knives, firearms, and wool blankets. The fur trade was neither a European creation nor an Indian innovation but a social and economic process forged out of all too human weaknesses and a desire for a better life. Fur traders, the Ojibwe, and the Dakota of the St. Croix were joined together in a commerce that was at once alluring, enriching, dispiriting, and destructive.[39]

Documenting the exact nature of the fur trade in the Wisconsin and Minnesota is complicated by the spotty nature of the surviving historical sources. While little can be said about the French era along the St. Croix, for example, there are other periods when the historical record opens up a window on the fur trade. One such time is the period between 1802 and 1805. George Nelson, a veteran Nor'Wester, produced a memoir of his first year as a fur trader, which he spent in the St. Croix Valley during 1802–3. Michel Curot left a journal of his year as a fur trader in the region during the winter of 1803–4. Finally, the much-studied journal of John

Sayer documents the 1804–5 trading season. All three men established posts in different parts of the St. Croix Valley: Nelson at Yellow Lake, Curot on the Yellow River, and Sayer on the Snake River near modern-day Pine City, Minnesota. The period of 1803–5 was a tense time for the fur traders because of a split in the ranks of the Nor'Westers. One of the most prominent members of the North West Company, Sir Alexander Mackenzie, along with several other partners, split with the established firm and formed the rival New North West Company. Known on the frontier as the XY Company, the upstarts went head to head in competition with the North West Company, greatly expanding the demand for the fur and food produced by Indian hunters.[40]

Easily the most striking feature that emerges from a close reading of the 1802–5 period is the importance of alcohol in the fur trade. The North West and XY companies hauled vast amounts of liquor over the portage between the Brule River and Upper Lake St. Croix. In 1803 Michel Curot reported to his superiors that the rival North West Company had brought in fifty-six kegs of high wine. This highly potent distilled beverage was akin to today's grain alcohol and like the latter had to be diluted before being consumed. Yet, even when broken down several times it remained very intoxicating. Archaeologist Douglas Birk estimated that fifty kegs of high wine could be diluted into over one thousand gallons of liquor for trade. Added to this massive amount of alcohol was the much smaller volume of rum and high wine imported by Curot and the XY Company. Operating on a much smaller scale, Curot had at least seven kegs of high wine at the start of the trading season and received an additional one in the spring. It is therefore not unlikely that between them the fur traders had close to thirteen hundred gallons of liquor for the trading year. Fur trader George Nelson estimated the Ojibwe population of the area to be "not above fifty families" with "about 60, or 65 warriors." Even if one assumes a generous estimate of a two-to-one adult male-female relationship and a two-to-one children-adult ratio, the Ojibwe in the region did not number more than four hundred people. This figure is consistent with the size of the St. Croix band in 1851 when American authorities proposed to move them west of the Mississippi River. In other words, the fur traders in 1803 were stocked with enough alcohol to provide every adult Indian with more than seven gallons of high wine.[41]

While the years 1802–5 may have been the high-water mark of the use of alcohol in the fur trade, there is little doubt that spirits occupied a critical role in the traders' inventory both before and certainly afterward. Some Ojibwe used alcohol in the same manner as most people today. The traders

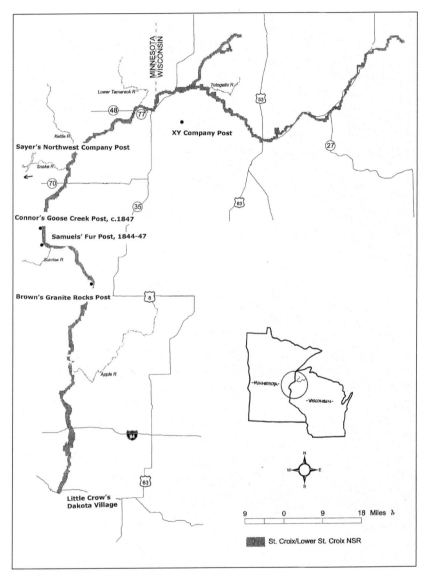

Map 4. Fur trade era historic sites, St. Croix National Scenic Riverway (Source: National Park Service).

document tastings being offered to trappers coming to trade as a social pre-lude prior to conducting business or high wine being requested by Indians upon the death of a child as a consolation. But the ugly face of what may very well have been addiction also appears in the traders' journals. When sixteen-year-old fur trader George Nelson landed at the junction of the St. Croix and Yellow rivers in 1802, he was greeted by a mob of young Indian men:

> The Indians, the moment they saw us gave the whoop. They were all drunk, the N.W. Co. had a little before given liquor. They came rushing upon us like devils, dragged our Canoe to land, threw the lading ashore, ripped up the bale cloths, cut the cords & Sprinkled the goods about at a fine rate. Such noise, yelling & chattering! "Rum, Rum, what are you come to do here without rum?"

Only a year before on that very spot the Yellow Lake band and members of the Snake River band had suffered five dead and six wounded when a drunken party led to a vicious knife fight. Nonetheless, the Ojibwe refused to let the traders proceed until they assented to "make our presents of liquor also." The result was, according to Nelson, "singing, dancing & yell-ing, & fighting too." On January 4, 1804, John Sayer noted rather casually in his journal, "Indians still Drunk & Quarrelsome amongest themselves. 2 got Stabbd but not dangerous." Just over a week later he noted, "This forenoon 2 Young Lads arrived from the Drunkards Lodge & report that the Indians were near Killing each other, at the same time requested a Small Keg of Rum which I refused them."[42]

As Nelson's and Sayer's experiences indicate, the alcohol trade brutalized the Indians, who then turned that behavior on the traders. An Ojibwe he knew by the name of Le Grand Male frequently intimidated Michel Curot. On November 1, 1803, Le Grand Male arrived at Curot's door drunk from drink he had already procured at the nearby North West post and de-manding more. Curot refused, but Le Grand Male would not take "no" for an answer. "All The night it was the same Demand and the same reply," Curot reported. "I had much trouble with this savage. I received several Blows of his fist, one especially that made my upper Lip swell up." Two weeks earlier a group of Ojibwe intent on a binge entered the North West Company post on the Snake River. They threatened to kill Joseph Reaume, the trader, tapped a barrel of pure rum, and "pillaged" the post for "ten days and ten nights." In January Curot again had problems when one of his assistants, Bazile David, refused to give rum to three Ojibwe hunters who were spending the night at the post. Curot had already given them a small keg of undiluted high wine, and the hunters were drunk. "A moment afterwards Le Jeune Razeur Like an enraged creature Struck David, saying to him, 'Dog, thou sayest that hast no Rum.'" The hunters then angrily left the post. They returned the next morning with a nine-gallon keg obtained from the North West Company, and they demanded what rum remained at Curot's post. The trader, anxious to match his competition and be rid of his troublesome guests, acceded to the request. The abuse drunken hunters inflicted upon their own families went largely unrecorded, although George Nelson noted one incident near the Brule Portage that provides an insight into what may have been all too common behavior. The traders had given rum to an Indian family with whom they had passed the night. The Indians drank through the night "very quietly & comfortably." Trouble came in the morning "when words ensued," and the son, a boy of sixteen or so years "fell upon this mother & beat her, striking with his fists & Kiicking her in the face & body!!!" Nelson's experienced companions dissuaded him from intervening, saying: "for if you do they will all three get upon you; besides it is among themselves—we dare not interfere." Shaken, the young fur trader thought, "Surely the curse of God will fall on these people." Little did he appreciate that he was that curse.[43]

How the St. Croix Ojibwe viewed the traders and the impact of the fur trade on their lives and families can only be glimpsed at through the journals of the fur traders. Traders who came to establish posts along the St. Croix River did so at the sufferance of the Indians. While the posts were a convenience to the Ojibwe, they seem to have adapted a proprietary attitude toward the goods the traders brought each fall. The Ojibwe men who tore apart George Nelson's canoes to find rum in 1802 were not humble supplicants awaiting a gift from the fur trader; rather, they were men taking what they felt was their due. A year later when Michel Curot and his men came to blows with a group of hunters determined to have a keg of rum, one of the Indians said "that it [the rum] all belonged to them, that in the Spring they would have some plus [beaver pelt]." George Nelson recorded in his reminiscences that Indians "would often . . . burst open the Shop door & take out what rum they pleased & compelled the people to mingle it to their taste." What traders regarded as begging or badgering by the Indians for something to drink was regarded by the Ojibwe as merely giving them access to those things that were meant for them to begin with.[44]

Not all Ojibwe embraced the fur trade with the same vigor, nor did all become enamored of high wines. George Nelson reported that one Ojibwe leader admonished the traders, "If you will persist to trade here, trade fairly as men & not wait till you think us too far drunk to perceive how you steal

from us & insult our females." Others blamed the fur traders for the negative impact of alcohol on their lives. "You are the cause of this blood being
shed by bringing *poisoned* rum to us," retorted one Indian after a drunken
brawl.[45]

The drinking of the Ojibwe must be viewed from the perspective of the
high level of alcohol use in general on the frontier. The fur traders, although they seldom admitted it in their journals, which might be read by
their superiors, often indulged in heavy drinking. Michel Curot noted in
his journal that his rival John Sayer had an escalating drinking problem.
"Since I have come into the fort I have noticed that Mr. Sayer is Very fond
of Drink," wrote Curot. "There has been Scarcely a night, that he has not
gone to bed Drunk." More scandalous to Curot than Sayer's habit of hid-
ing pots of alcohol for himself about the post was the latter's willingness to
drink the high wine prepared for Indian use. "I should Never have Believed
that he would be fond enough thereof To Drink the Savage's Rum." Nor
was Sayer selfish about sharing his drink with others. While his men labored to build the North West Company's Snake River post in 1804, Sayer
noted in his journal that he "gave each a Dram morning & Evening &
promised to do the same till our Buildings are Compleated provided the[y]
exert themselves." Providing men engaged in heavy labor with alcoholic
stimulants was standard practice in early nineteenth-century business and
in the armed forces. Nonetheless, Sayer was later dismissed from the North
West Company, a decision that in part reflected his heavy use of alcohol.[46]

One of the most divisive practices of the fur traders was the creation of
chiefs. Traders attempted to elevate select hunters' status by giving them
dress coats, flags, and other presents. They flattered themselves that if they
treated this hunter as special, he would be so regarded by other Indians in
the community. François V. Malhiot, a North West Company trader in the
Lac du Flambeau area, made the following speech when he created a new
chief:

> Kinsman—The coat I have put on thee is sent by the Great Trader; by such
> coats he distinguishes the most highly considered persons of a tribe. The
> Flag is a true symbol of a Chief and thou must deem thyself honored by it . . .
> love the French as thou dost, watch over their preservation and enable them
> to make up packs of furs. . . . As first chief of the place, thou must make every
> effort so that all the Savages may come and trade here in the Spring.

As Malhiot indicated, the goal was to have this chief influence others to
honor their debts at the trading post and not go to the competition.
However, far from picking the most admirable hunters (both from an

Ojibwe and a trader's perspective), men of the worst character were often selected by intimidated traders who hoped to end abusive behavior. In his memoir, George Nelson described a group of frustrated traders who decided "that by making a chief of the greatest scoundrel among them would perhaps have a good tendency." That was Malhiot's strategy in 1804, but it did not work. Similarly, Le Grand Razeur, the Ojibwe who attacked one of Michel Curot's men, had earlier been made a chief. Worst of all, chief-making caused fissures among the Ojibwe. Curot reported a stabbing among the Yellow Lake band during the winter of 1804. The "chief" refused to do anything to resolve the problem, "fearful on his own account." This caused Curot to reflect, "I believe that Band although Partly nephews and Brother in law [are] Jealous of whomever is made chief giving Preferment to any of them, Since each of them separately believes himself as Great a Man as an Other."[47]

The Ojibwe often resented the practice of making phony chiefs. When John Sayer offered the coat and flag to the hunter Pichiquequi, the latter responded angrily, even after Sayer tried to sweeten the offer with free rum. Pichiquequi "replied that he was not a chief and that Since he was thirsty he would go hunting either for a [fur] or a deer that he could trade for Rum, that he did not command any savages, that they were all Equal and [he and his people] would go where they liked to trade and that he himself would do the same."[48]

An Ojibwe hunter described in Nelson's memoir manifested this same spirit of independence. Following the formal presentation of the chief's uniform and flag, the hunter turned to the fur traders with a look of "utmost contempt":

> No doubt, you Frenchmen, you think yourselves wonderfully cunning:—no doubt you were very certain. . . . that my eyes would be blinded by the Dazzling stuff you have been Displaying here with so much ceremony before us? Undeceive yourselves. I am born free & independent. I despise those tokens of Slavery. I am not a Slave to wear oth[ers'] clothing (livery). My old clothes satisfy me; & when they are worn out I know how to procure others.

Nelson was not present at the council when the traders attempted to elevate the hunter to chief, so the exact words he recorded must be regarded as narrative license. Nonetheless, the hunter's eloquent statement of autonomy and personal independence reflected sentiments that Nelson must have seen manifested many times in his long career of trading with the Ojibwe.[49]

Another way in which the fur traders created fractures in Ojibwe society was the practice of taking Indian women as brides. These liaisons created a family connection between a trader and the Indian trappers. The trader further benefited from a woman's companionship and the help of someone skilled in contending with the North Woods wilderness and fluent in the local language. In return, her family received the assurance of the trader's material help, at least as long as he was stationed in their area. Most of these relationships were formed *a la facon du pays,* without legal contract according to the customs of the country. As a trader's wife, an Indian woman entered into a more pleasant, if more socially precarious, world. The amount of work expected of her, particularly the heavy work of portaging or moving camps, greatly decreased. On the other hand, there was the prospect that her husband might abandon her after a few years. "She will not do for me or any Indian," complained an Ojibwe who hoped to be rid of a lazy second wife, "[the] best way is to give her to the whites. With them she will have only snow-Shoes and maggacins to make, & with them she will have as much men as she desires . . . they take women, not for wives—but use them as Sluts—to satisfy the animal lust, & when they are satiated, they cast them off." This harsh statement taken from George Nelson's memoir reflected a bitter reality. Fur trader John Sayer spent the winter of 1804–5 at the Snake River post with Obemau-unoqua, his Ojibwe wife. They were married for at least ten years, and she bore him two sons, yet there is only a single reference to her in Sayer's diary: "my Squaw brot about 4 lbs [maple] Sugar." The daughter of a notable Ojibwe leader, Obemau-unoqua was abandoned by her husband in 1805 when he retired to Canada and took a white wife.[50]

Some fur traders formed loving and stable relationships with their Indian wives. Joseph Duchene, usually known by the name La Prairie, spent more than a half-century in the St. Croix Valley with Pimeegee-shigo-qua, his Ojibwe wife. Many of the employees of the North West Company and the XY Company in the region also had Ojibwe wives. Bazile David, who already had one spouse, tried to take "a Young girl 9 or 10 years old For his wife." However, his superior, Michel Curot, intervened and "sent her back." David was instructed to "take another one, who is Larger." Gardant Smith, another of Curot's *engagés,* had a very independent-minded Ojibwe wife who took the position that since a man could have two wives she chose to have two husbands. She regularly left Smith for weeks or months at a time, returning with furs or meat she traded for on her own account. In the end, two husbands were not enough for her. Two Ojibwe men contending for her favors came to blows, leading to the fatal stabbing of one.

The wives of the traders were fed from the rations of the post, and although they received no pay, they performed significant work preparing food and tending to fires. They could also significantly increase their husband's salary by snaring small animals and dressing the skins. Indian women often assisted traders contrary to the wishes of Ojibwe men. When a group of carousing Ojibwe men plotted to kill George Nelson and his three companions, it was two Indian widows, living in a tent nearby, who warned the traders of the danger.[51]

The product of fur trade marriages, the Métis, or mixed-blood offspring, significantly influenced Ojibwe society. The Métis were a significant portion of the population of the Upper Midwest. By 1820 there may have been as many as ten thousand Métis south and west of the Great Lakes. Some of the Métis were formally educated in the East, dressed and behaved like whites, and entered into fur trade society as clerks or, in the case of the women, as wives of white traders. Those denied the opportunity for education worked in the lower levels of the fur trade or simply joined their mother's people, where most were accepted as equals. John Sayer seems to have devoted little attention to promoting the prospects of his mixed-blood sons Henry and John. It is possible that Henry was attached to the Snake River post in an informal manner. At one point Sayer, in his usual delicate style, refers to "Henries Squaw." Joseph Duchene, known as La Prairie, was much more supportive of his Métis children, who played a significant role in his trading operations. One of the most successful trappers among the Snake River Band of Ojibwe was an Indian known as "Chief Marin," who may have been the son of Joseph Marin, who was active in the fur trade during the French regime. The Métis were a people capable of moving in either the white or the Indian world. They frequently participated in the councils of the Ojibwe, but their outlook was not always the same as their kinsmen, and their interests could be quite different. In later years this often played an important role in treaty negotiations with the U.S. government.[52]

Ojibwe society remained dynamic and creative throughout the period of the fur trade. The new economic conditions fostered by the European American traders within their society caused substantial material and cultural changes. Indian women, both Ojibwe and Dakota, played a new and vital role in the fur trade economy. As men focused more of their attention on trapping fur-bearing animals, the women's work of preparing pelts became more important to the family economy. The critical role this women's work played in the fur trade may have been reflected in the rise of polygamous marriages, although the evidence on this point is only suggestive.

Only very successful hunters were able to take more than one wife. Ethnologists have contended that polygamy was an example of boasting or conspicuous consumption. Yet the amount of work required to properly prepare hides was considerable, and industrious hunters could not expand their trapping without having additional assistance preparing the furs. Domestic harmony was facilitated among polygamous households by the practice of taking the first wife's sister as a second or, on rare occasions, third spouse. Close contact with the fur traders made the lives of all Ojibwe women more complex. Traditional tasks such as gathering wild rice became more important as trading posts required large amounts of rice to subsist through the winter. During the eighteenth century, the amount of rice previously gathered for simple domestic use had to be augmented by rice gathered for commercial purposes. To meet this need it is likely that Ojibwe women seeded large areas of lakeshore with rice to meet the growing demand. Indian women also took on new horticultural responsibilities. By the nineteenth century, potato patches, raised from seeds obtained from the traders, became a fixture at village sites. Together with maize from traditional cornfields, potatoes became an important means of subsistence. Even the Dakota of the Lower St. Croix, constricted by their war with the Ojibwe from ranging as far as they had in the past, began to rely more on the farming of the Indian women.[53]

As the value of furs and even deer hides increased, the use of these materials in Ojibwe and Dakota domestic life declined. Buckskin, which had been the principle material for both men's and women's outer garments, was replaced by broadcloth, augmented in the winter by capotes and leggings made of woolen blankets. Indian women adapted to the new materials artfully and by the late eighteenth century were producing warmer and more durable clothing than had been traditionally available. Long proud of their weaving skills, which produced mats made of cedar bark and swamp rushes, Ojibwe women adapted to the availability of glass trade beads to produce new bolder embroidery. While the use of beads was new, the designs followed traditional floral patterns blending old and new. Older crafts such as the original dental pictograph art, which Ojibwe women produced by biting on thin sheets of birch bark, continued in spite of the new products available and the new demands on women's time.[54]

The creative blending of old and new also marked the rise of the Midewiwin rites. The Midewiwin was a set of ceremonies performed by an organized hierarchy of priests to protect tribal traditions, cure the sick, and slay the evil. Some form of the Midewiwin evolved among most of the Indian tribes of the Upper Midwest. For the Ojibwe, the Midewiwin had both a nationalistic and religious function. Its ceremonies brought together Ojibwe from all across the Great Lakes and Upper Mississippi regions. William Warren, a part-Ojibwe historian of the 1880s, described the Midewiwin as the occasion for an annual "national gathering" when "the bonds which united one member to another were stronger." Although the evidence is by no means clear, the Midewiwin appears to have originated some time after the Ojibwe first became involved in the fur trade and may have been a cultural adaptation to the rise of individual wealth among Ojibwe hunters. The community strains brought by geographic expansion were a further factor stimulating the growth of the medicine society. The Jesuit relations that provide such a thorough look at Algonquian society between 1640 and 1700 make no mention of the Midewiwin, which supports the thesis that the society was of historic origin. Membership in the Mide society was selective and could be obtained only after long periods of instruction. After initiation a member then advanced through eight degrees or rankings, at each level learning more of tribal lore, healing remedies, and conjuring power. Midewiwin rituals were secret, and instruction was only possible after a considerable exchange of material wealth, from the novice to the priest. The Midewiwin, which continues to this day, was a creative means of redistributing the new wealth created by the fur trade and warding off the threat posed by Christianity.[55]

The Ecological Impact of the Fur Trade

The landscape of the Upper St. Croix River was changed in subtle ways by the growth of the fur trade among the Ojibwe and the Dakota. The presence of herds of elk and buffalo in the region declined dramatically as more hunters sought these large game animals with more effective weapons. There is evidence that in the seventeenth century buffalo roamed as far north as the Pine Barrens between the St. Croix and the Brule rivers. By 1820 elk and buffalo were both rare in the St. Croix Valley. Schoolcraft claimed that the last time buffalo crossed to the east bank of the Mississippi was in 1820. Twelve years later, traders reported that Dakota hunters in the Trempealeau River valley killed the last bison in Wisconsin. The numbers of elk were greatly reduced by hunting, but they survived longer. In 1854, when white settlers reported seeing several elk along the Sunrise River in Chisago County, Minnesota, it was the cause of some excitement. When the Dakota were sole masters of the St. Croix Valley, they had regularly set wildfires to enhance elk and bison habitats. The small prairie openings thus created helped to sustain the grazing animals. The Ojibwe were less aggressive in the use of fire as a tool of game management. As the fur trade grew and the Ojibwe presence in the valley increased, the herds of grazers disappeared and the prairie openings yielded to vegetation succession. Maple-basswood forests replaced many of these openings along the river.[56]

As elsewhere in the region, beaver were the most relentlessly hunted species in the St. Croix country. While beaver existed throughout the watershed, the upper portions of the valley, especially the upper Namekagon and tributaries such as the Clam, Snake, Yellow, and Totogatic rivers, were superb habitat. Nonetheless, the beaver population of the region was likely in severe decline by the beginning of the nineteenth century. A drastic reduction in the beaver population of this stretch of the St. Croix and Namekagon valleys would have had a significant impact on the landscape. The beaver, more than any other creature save humans, has the ability to consciously alter its environment. The industrious rodent does this in two ways, by impounding water to create beaver ponds and by felling trees for food.

The beaver builds dams across streams to create ponds that provide the beaver with a watery moat that keeps predators, such as the gray wolf or the wolverine, at bay. Beaver ponds render swift-flowing streams into quiet, calm impoundments of water often an acre or more in size. In a northern hardwood forest like the St. Croix, it would not be unusual to have numerous beaver dams on one very small forest stream. A study of beaver in Voyageurs National Park in Minnesota identified 2.5 beaver dams per kilometer of stream, with the result that well over half of the length of all streams was transformed into beaver ponds. The wetlands created by the beaver formed habitat for other important fur-bearing animals as well. Muskrats and otters made their homes within the ponds. Mink and raccoons hunted frogs, turtles, and snakes around the margins of the pond. The edge effect of the wetland-forest interface fostered a diverse array of other animals as well, from waterfowl to deer and moose. The hydrologic effect of thousands of beaver ponds within the valley was to slow the flow of the tributary streams and moderate flooding along the entire valley. Ponds trapped sediment carried by streams, keeping nutrients in the forest and filtering the water that was eventually discharged into the St. Croix.[57]

A beaver population of thousands within the St. Croix Valley affected the forest as well as the river. The beaver is one of the most voracious browsing animals. Although moose, deer, and elk are normally seen as the major browsing species in the forests of the Upper Midwest, the beaver has much greater impact on forest vegetation. The difference with the beaver is that unlike the other grazing animals, its impact is restricted to areas within a hundred meters of water. Beaver try to extend their range slightly by building canals, a foot or two wide, leading away from streams and into the woods. This amazing behavior reinforces the fact of the beaver's aquatic nature. Yet, in spite of this limitation, the beaver still manages to consume a vast amount of wood, leaf, and roots. A study of beaver at a single north-eastern Minnesota pond revealed that each beaver felled 1,400 kilograms (nearly 3,100 pounds) of woody biomass per year, substantially more than moose grazing in the same area. In fact, the study concluded that the beaver colony harvested twice as much biomass as a herd of Serengeti ungulates. Not only was the beaver colony an intensive grazer, but it also was very selective, favoring certain tree species such as aspen and turning up its nose at alder or conifers. After several decades of beaver activity, forests near their ponds were greatly changed, moving from aspen- or paper birch–dominated stands to a more diverse patchwork of shrubs and trees.[58]

Multiplied throughout the valley by the thousands of lodges and dams, the mini-environments created by the beaver encouraged diversity among both flora and fauna. Because of the beaver the St. Croix was clearer and less prone to flooding. The water table was higher, and springs were more abundant throughout the valley. Trappers wrought havoc on the beaver landscape. Between 1800 and 1820, the beaver was all but wiped out along the St. Croix and other streams in the region. In 1800 fur traders reported a harvest of 8,000 beaver skins for the entire Dakota trading area, of which the St. Croix was only a small part. Yet by 1820 the beaver harvest for that same area was a paltry 760 pelts.[59]

The dramatically sudden over-trapping of the beaver brought changes to the valley, but only gradually. Beaver ponds endured long after the industrious rodents had been eliminated. Not until the period after 1840 would the impact of the decline of the beaver have been fully felt, but by this time a new group of dam builders was busy on the upper river. Loggers manipulated the water levels on the St. Croix and its tributaries in ways that would have impressed *Castor canadensis,* thus obscuring from the historical record the impact of the fur trade on the flow of the river. What is clear, however, is that the loss of beaver meant an end to unique pond habitats and the elimination of the forest's most voracious herbivore. "The features of the country have undergone a change," an early settler wrote of Burnett County. "The towering pines have decayed or been leveled by the woodsman's axe. Some small lakes have receded, and tall grasses wave and willows grow where once the 'kego' [fish] sported in the clear blue waters."

Some early settlers contended that the "sun drew the waters up into the heavens," and did not see the loss of the beaver a generation earlier as the cause. All they saw was the result, dry fields ringed by the bleached shells of freshwater mussels "and by the ineffaceable mark of the water breaking upon the beach and undermining the rocky ledges."[60]

The American Fur Company Era

During the first half of the nineteenth century, competition between more and more fur traders for fewer and fewer furs threatened to make the figure of the fur trader an endangered species. This competition gave the Ojibwe and the Dakota higher prices for their furs and a greater choice of goods for which to trade. It also reduced to the lowest common denominator of behavior an exchange that had persisted for well over one hundred years as a middle ground between Indians and Europeans. Family alliances between traders and hunters became less common as both sides operated with an eye for immediate returns, the trader in the form of quick profits, the Indians in the form of alcohol. The rivalry between the North West Company and the XY Company had been ruinous to both parties. To try preventing another outbreak of that type of trade war, the Canadian-based fur traders pooled their resources in the form of a new concern, which would have a monopoly on the fur trade south of the Great Lakes. First the Michilimackinac Company and later the Southwest Company were organized to control the fur trade of the Upper Midwest. The latter company is known to have operated a post on the St. Croix. Neither company succeeded long because increasingly the main competition came not from Canada but from the United States. In 1808 the American Fur Company was charted by the state of New York. Its founder, John Jacob Astor, sought to dominate the fur trade in American territory the way the North West Company controlled the Canadian trade. After the War of 1812, when the authority of the U.S. government was firmly established in the region, Astor got his chance.[61]

As the Americans moved to assume the fur trade in the St. Croix region, they adopted the same tactic as the British a generation before. British traders like Alexander Henry formed partnerships with experienced French traders, such as Jean Baptiste Cadotte Sr., to benefit from their superior connections with the Indians. Astor's American Fur Company followed the same pattern. Their choice to head the St. Croix trading area was Joseph Duchene, known as La Prairie. He had been the North West

Company's most experienced trader in the Folle Avoine. His son of the same name, who became an interpreter, joined him in the American Fur Company. William Morrison, who had come to the region as a boy and had matured into an experienced trader, also left the North West Company and was rewarded by Astor with overall control of the Fond du Lac Department, a vast area that included the St. Croix, Chippewa, and Upper Mississippi valleys. The Cadotte family, long an aristocracy in the Lake Superior trade, was among the first to make the move toward the Americans. In 1818 Michel Cadotte employed two young Americans to act as front men for his operations in northwestern Wisconsin. Truman and Lyman Warren, the sons of a Revolutionary War soldier, eventually married Cadotte's daughters. They gradually earned the trust and support of not only Cadotte but of the family's Ojibwe kinsmen. In 1822 Cadotte and the Warrens entered the American Fur Company as traders in the Fond du Lac Department. Also entering the firm was another generation of the Cadotte clan, Michel Jr., who signed on as an interpreter, and Jean Baptiste III, who joined as a boatman assigned to the St. Croix outfit.[62]

By winning the cooperation of the most experienced traders in the region, the American Fur Company secured the bulk of the trade in the St. Croix Valley. The principle trading posts within the valley continued to be among the Ojibwe of the Snake River and in the Yellow Lake region. Under Astor's company the Ojibwe in the valley continued to be supplied and directed from Lake Superior, which mandated a continuation of the virtual alliance between fur traders and the Ojibwe bands of the Upper St. Croix. The Dakota villages along the lower river had no contact with those fur traders, and they directed their furs toward merchants operating on the Minnesota or Upper Mississippi rivers. Among the traders to winter on the Lower St. Croix and trade with the Dakota was Jean Baptiste Mayrand, a fur trader based in Prairie du Chien, Wisconsin. Mayrand operated a St. Croix post during the winter of 1819–20 and probably for a longer period. Mayrand, like Cadotte and Warren on the upper river, was attached to the American Fur Company. Based on the efforts of these men the American Fur Company was in a position to secure the bulk of the furs from the St. Croix.[63]

Dakota–Ojibwe Relations during the American Era

While the fur trade continued unaffected by political change among the European Americans, the conflict between the Indians likewise followed its bloody course. The boundary between the Dakota and the Ojibwe, which had gradually settled on the areas between the Snake River and St. Croix Falls, divided the valley into a northern zone oriented to Lake Superior and a southern zone looking to the Mississippi. This division, which first occurred during the fur trade, would long mark the history of the St. Croix and would effect the development of transportation, agriculture, and the tourist industry along the river. The division endured in part because the warring parties' territories were separated along an environmental fault line, a vegetative transitional zone between the rich soils and prairie openings to the south and the mixed coniferous forests to the north. Within the transition zone deciduous forests dominated, although the landscape was a mosaic of marshes, savannahs, and forests, all in all a fine range in which to stalk deer or gather wild rice, berries, and maple sugar. Indeed, it was the attractiveness of the region from a subsistence point of view that kept both the Dakota and Ojibwe in abrasive contact within the zone.[64]

While the bountiful landscape of the St. Croix tended to draw the Dakota and Ojibwe into conflict, Indian political structure did little to moderate conflict. The highly individualistic Ojibwe lacked formal mechanisms to broker and enforce adherence to a boundary line. Bands acted in the manner they saw fit. Dakota leaders, while exercising somewhat more centralized authority, also had a problem achieving individual compliance. Warfare was an established feature of each society. It was a vital theater of action in which individual young men could establish status in their community. Recognition as a successful hunter or a brave warrior was all the more important because for both the Ojibwe and the Dakota, unlike European American society, it was not accumulated wealth or inherited position that conferred status but individual accomplishment. Young men looked forward to war and were always difficult for elders to control. The ominous warning of a Dakota chief to Joseph Marin in 1754 that "we cannot keep from you the fact that our young men are all beginning to mutter" was a frequent prelude to war. George Nelson reported a fellow fur trader's complaint to the Ojibwe: "It is the young men who are too ardent . . . they are afraid of being looked at as cowards if they have not a Scalp to shew & contrary to the advice of the old & experienced, & to the great injury of all, they make a descent upon their enemies & plunge both nations into war!" Peace for either community was often at the mercy of individual ambition or family obligation. One of the defining features of both Ojibwe and Dakota life, tremendous individual autonomy balanced by community responsibility, encouraged the continuation of the conflict. Family members, after a period of mourning, had the right, some would say obligation, to avenge the dead. This was not something that was subject to interference by political leaders. Revenge was the most persistent reason for war parties to embark on the river each spring. Every fallen family member who was avenged called forth a retaliatory raid by the enemy, a dreary, deadly cycle.[65]

For every epic battle, like that at St. Croix Falls, where warriors fought warriors in desperate battles, long remembered around winter campfires, there were hundreds of wretched ambushes leaving a child or elder murdered in the brush. Brief periods of peace, brokered by hunters anxious to utilize the rich borderland region, sometimes resulted in Dakota–Ojibwe intermarriages. More so than at any other point of contact between the Dakota and the Ojibwe, a considerable exchange of kinsmen occurred along the St. Croix. To live in the lodges of the enemy was to occupy a precarious position, yet custom dictated that men live with their wives' family. William Warren related the fate of one St. Croix Ojibwe dwelling with the Dakota. At a war dance an over-excited Dakota warrior fired an arrow into the Ojibwe, who previously had been accepted as a member of the tribe, so as to "let out the hated Ojibway blood which flowed in his veins." This reckless act led the wounded Ojibwe to later seek vengeance by leading a war party against the village in which he had formerly lived. Over time the number of people in the St. Croix Valley who were of mixed Dakota and Ojibwe ancestry became quite large. This sometimes led to poignant encounters. One of the leaders of the Rice Lake Ojibwe during the early nineteenth century was Omigaundib, whose father had lived for a time among the Dakota. When he later returned to the Ojibwe and became chief of his band, he left behind a Dakota family. His Dakota sons later became leaders of their village. For his lifetime there was peace between the Rice Lake Ojibwe and the chief's Dakota kinsmen. Even after the peace eroded, the sons of a common father avoided participating in raids against each other. Omigaundib, nonetheless, eventually was drawn into the war. A Dakota war party proceeded to Rice Lake and killed three children playing on the shore. One of the dead was Omigaundib's daughter. Rather than call for a war party and vengeance, Omigaundib placed his slain child in a canoe, covered her body in the black paint of mourning, and proceeded down the St. Croix to the Dakota villages at Point Douglas. His appearance there cut short the celebration of a successful war party. His arrival, quiet and dignified, made clear he had come not as an enemy chief but as a kinsman. The scalp of the little girl, proudly being paraded among the lodges, was suddenly transformed from a trophy to a cause of lamentation.

With tears in their eyes the Dakota pressed Omigaundib to accept gifts to cover his tragic loss. "I have not come amongst you, my relatives to be treated with so much honor and deference," he said. "I have come that you may treat me as you have treated my child, that I may follow him [her] to the land of the spirits." In the end, a young Dakota girl was presented to Omigaundib to return with him to Rice Lake.[66]

While those of mixed Dakota and Ojibwe heritage were the most vulnerable when fighting broke out, their kinship ties allowed them to function, as Omigaundib did, as conciliators. "The occasional short terms of peace which have occurred between the two tribes," William Warren noted, "have generally been first brought about by the mixed bloods of either tribe who could approach one another with greater confidence than those entirely unconnected by blood." Because of these ties the St. Croix Ojibwe were much less active in organizing war parties against the Dakota than their tribesmen who lived along the Chippewa River. In 1818 the U.S. Indian agent in the region reported that eight Ojibwe from the upper St. Croix actually went so far as to warn the Dakota downstream of the approach of a large Ojibwe war party. Armed with this information, the Dakota "were preparing to give them a warm reception," the agent concluded. Such incidents were rare. On most occasions, the St. Croix Ojibwe were powerless to stop war parties directed downstream, even though such attacks opened them up to retaliation. The Dakota, particularly the Mdewakanton chief Little Crow, also were open to peace overtures. During the winter of 1801–2, a Dakota war party captured the North West Company trader La Prairie. They treated him well and presented him with a "Pipe of Peace" and a tomahawk to give to the Yellow Lake Ojibwe. "Let them chuse, & decide whether they will accept the Pipe & Smoke with us as friends, or take the tomahawk," the Dakota leader told La Prairie. "We are ready for either, but we would rather have them be our friends." Unfortunately, La Prairie, who clearly should have known better, repeated the Dakota message verbatim but kept the peace pipe for himself. With only the tomahawk before them, the Ojibwe decided the Dakota message was intended as an insult and answered it with a war party.[67]

The Dakota made frequent forays into the Ojibwe lands along the St. Croix. In addition to the hunting, the Mdewakanton often entered the valley in the fall to harvest the region's abundant wild rice. Perhaps in appreciation of how precarious an undertaking this was, Chief Little Crow's band often demonstrated restraint against enemies who fell into their hands. While death or capture was the usual punishment for a warrior caught alone in the forest by his enemies, Little Crow's people sometimes contented themselves with merely breaking the guns of the Ojibwe. The Mdewakanton, however, also were determined to maintain access to their traditional lands and often made demonstrations of their ability to project war parties throughout the St. Croix Valley. These ventures did not always end in violence. Often when they discovered Ojibwe trap lines, they merely broke the traps, thereby providing a warning that the hunter risked the wrath of the Dakota.[68]

Like all wars difficult to bring to a conclusion, the conflict continued not merely because of blood feud and misunderstanding, but because the Ojibwe and the Dakota were locked in a territorial struggle closely linked to each side's survival as a people. Little Crow, the Dakota chief who had participated in numerous peace conferences and who was the author of numerous personal attempts to conciliate, nonetheless understood that war made rational sense. "He observed that a peace could easily be made," American Indian agent Thomas Forsyth reported in 1819, "but said it is better for us [Dakota] to carry on the war in the way we do than to make peace, because, he added, we lose a man or two in the course of a year, and we kill as many of the enemy during the same time; and if we were to make peace, the Chippewas would over-run all the country lying between the Mississippi and Lake Superior." Little Crow and the Dakota basically faced the question "should we give up such an extensive country to another nation to save the lives of a man or two annually?" The Dakota response was similar to that taken by the United States throughout its history: land is worth blood. Thomas Forsyth ended his report by noting, "I found the Indian's reason so good that I said no more on the subject to him."[69]

While the Ojibwe and the Dakota were locked in "a war for land," the U.S. government gradually established its political hegemony over the Upper Midwest region. The American flag first flew over St. Croix waters in 1805 when Lieutenant Zebulon M. Pike led an expedition of twenty soldiers into the Upper Mississippi region. His purpose was to make it clear to British fur traders that the region was under American control. The Dakota drew a different conclusion from Pike's visit. At a time when the number of Dakota available to continue the war against the Ojibwe was becoming lower due to western migration, it is probable that Little Crow viewed the Americans as potential allies. Anxious to secure regular access to American trade goods, something the North West Company provided to the Ojibwe, Little Crow agreed to the cession of two tracts of land for future American forts. One tract, at the mouth of the Minnesota River, became Fort Snelling, the principle U.S. military base in the region. The second tract Pike deemed strategic was the mouth of the St. Croix. In exchange for this territory the Dakota received a mere $200 worth of trade goods and a small amount of liquor. It is likely that Little Crow viewed this transaction as down payment on future military help from the Americans. He scarcely anticipated that the negotiation with Pike had set in motion a chain of events that would lead to the defeat of his grandson and namesake in 1862 by the very soldiers Little Crow the elder viewed as allies.[70]

Pike also attempted to mediate the conflict between the Dakota and the Ojibwe. Although he reported boastfully to President Thomas Jefferson that he had brought about peace, the best he was able to do was halt the progress of a single Dakota war party and receive Little Crow's promise to try and restrain his young men. When the Americans strengthened their hold over the region after the War of 1812, they intruded themselves more aggressively into the long simmering war. In 1820 Lewis Cass, the governor of the Michigan Territory, which then included Wisconsin and part of Minnesota, attempted to broker a peace between the Dakota and the Ojibwe. Unfortunately, Cass managed to bring with him only 150 Ojibwe, mostly from Sandy Lake. The equally small number of Dakota present manifested "indifference" to the prospect of a treaty. Cass succeeded in having the few chiefs present assent to a peace as "lasting as the sun," but Henry Rowe Schoolcraft, the future Indian agent who was a member of the Cass expedition, remained justly skeptical. He recognized that the conflict was based on "a dispute respecting the limits of their territories, and favorite hunting grounds, but if so, nothing was agreed upon in the present instance to obviate the original causes of enmity." Schoolcraft concluded, "Whether the peace will prove a permanent one, may be doubted."[71]

The Cass expedition set in motion a series of virtually annual convocations between the Ojibwe and the Dakota organized by the Americans at Fort Snelling. Established in 1820, Fort Snelling was the northernmost military establishment in the United States. The fort served as neutral ground where the Ojibwe and Dakota could usually meet in security under the supervision of U.S. Indian agent Lawrence Taliaferro. A proud, intelligent Virginian, Taliaferro worked tirelessly to reduce the intertribal warfare. He also established strong ties with the Dakota by taking as his wife the daughter of the war chief Mahiyawicasta. An 1821 council held by Taliaferro brought together more than eight hundred Dakota and Ojibwe, but it was followed by a year of severe fighting that resulted in nearly a hundred casualties. Taliaferro followed this up with a formal peace treaty in 1823. The following year he sought to impress upon the Dakota the extent of American power by taking a delegation to Washington, D.C. In 1825 he helped to arrange a major meeting of Mississippi valley Indians at Prairie du Chien. Unlike earlier efforts to bring peace that had been based on engendering goodwill, the Americans finally tried to solve the root of the problem—the territorial conflict between the Dakota and the Ojibwe. The Dakota delegate protested bitterly when the Ojibwe presented their claims to all lands east of the Mississippi. Little Crow had no intention of granting to the Ojibwe the lower St. Croix homeland of his people. Finally, after badgering by the Americans, the Ojibwe recognized the Dakota claims to the lower river. The St. Croix boundary between the two peoples was ruled to be at "a place called the standing cedar, about a day's paddle in a canoe, above the Lake at the mouth of that river; thence passing between two lakes called by the Chippewas 'Green Lakes,' and by the Sioux 'the lakes they bury the Eagles in.'" In modern terms the line ran from a point on the river known as Cedar Bend, near the Chisago–Washington county boundary, northwest past Lindstrom, Minnesota, to the upper reaches of the Rum River. Little Crow signed for his people, while Peeseeker, known as Buffalo and Naudin, the Wind, signed for the St. Croix Ojibwe.[72]

Within a year violence again flared, and in 1826 the peace was shattered when several Dakota warriors shot and killed two Ojibwe trading within the shadow of Fort Snelling. By 1830 even Little Crow's Dakota were sending war parties across the boundary against the Ojibwe. Some American leaders took the pragmatic, if somewhat cynical, view that while the Ojibwe and the Dakota were determined to fight each other, "they will not feel a disposition to disturb the peace and tranquility of our exposed frontier settlements on the St. Croix and Chippewa rivers." Those officials who worked to stop the violence learned to adopt more modest goals, and after 1826 they focused on simply trying to keep the warring parties apart. The Sioux Agency remained at the mouth of the Minnesota while the Ojibwe of the Upper Mississippi were removed from Taliaferro's responsibility. Instead of being required to go to Fort Snelling, they were directed northward to Lake Superior and the Indian superintendency of Henry Rowe Schoolcraft. Government agents were so anxious to reduce the chance of conflict that in 1832 Schoolcraft burned the temporarily abandoned trading post of Joseph R. Brown because it was located "at a point where the Sioux and Chippewas" were "improperly brought into contact."[73]

In spite of Schoolcraft's punitive action, Joseph Renshaw Brown was destined to have a long and important involvement with the St. Croix River Valley. He had lived on the Minnesota frontier since he was fourteen years old, when he came west as a drummer boy in the army. In 1825 he put away his uniform and entered the fur trade. There was an unsavory taint to

Brown's fur trade career. This may simply be because he was imprudent enough to have gotten on the wrong side of Indian agents Taliaferro and Schoolcraft, whose voluminous writings greatly influence the historical record. But even with that bias taken into account, Brown's callous treatment of Indian women is disturbing. During the late 1820s, when Brown was engaged in trade with the Dakota, he took as his bride a Métis woman of Dakota ancestry. She may not have been the most faithful wife, but Brown nonetheless ended the marriage after only five years. While trading on the St. Croix with the Ojibwe, Brown took, first as his mistress and later as his wife, Margaret McCoy—an Ojibwe–French Canadian Métis. After a little more than a year he abandoned Margaret, even though she was pregnant, when he decided to recross the Indian boundary and trade again with the Dakota. While trading at Lake Traverse on the Minnesota River, he enjoyed the favors of Winona Renville, the "second" wife of a fellow fur trader, Joseph Renville. At the same time, Brown courted Winona's seventeen-year-old daughter by a previous marriage, the Dakota Métis Susan Freniere. Winona, Susan, and Brown all resided together in a small cabin at the trading post, which must have made for some interesting domestic arrangements. Although Brown was by now known among the Americans as, in the words of one traveler, "a gay deceiver amongst the Indian fair," Susan Freniere agreed to become his wife. As he had never bothered to divorce Margaret McCoy, this left Brown with two wives, a circumstance he did not legally fix for five years.[74]

Brown's initial post on the St. Croix was located about four miles upstream from St. Croix Falls, on the Minnesota side of the river. The spot was then known as Granite Rocks in reference to the boulders in the stream there that would in the future cause great log jams in the river. Brown was in competition with the American Fur Company's St. Croix traders, Lyman Warren and Thomas Connor. He had a distinct advantage over his rivals. By the early 1830s U.S. Indian agents such as Schoolcraft and Taliaferro had forced the American Fur Company to reduce the amount of alcohol used in the trade. After 1832 Congress supported this policy by making it illegal to use alcohol in the Indian trade. Some sprits were still smuggled into Indian country, but the volume necessarily declined. Brown was one of those smugglers. His partner in his venture, Joseph Bailly, had purchased twenty-seven kegs of alcohol for the trading season. St. Croix Ojibwe abandoned the American Fur Company post at the south end of Pokegama Lake and flocked to Brown at Granite Rocks and his branch trading post on the Snake River. The American Fur Company formally complained to territorial officials that Brown had "large quantities of whiskey and the consequence is a heavy loss to our people who had none."[75]

Brown was also an irritation to the Dakota. Although his Granite Rocks post was well within the Ojibwe side of the border, it was considerably closer to Dakota country than any previously established Ojibwe trading post. The Dakota leader Little Crow was concerned that the post would encourage the Ojibwe to hunt and trap on Sioux lands. The fact that Brown encouraged the Ojibwe to settle around his post sites, helped them to plant large fields of corn, and encouraged them to reside there during the summer made Little Crow's fears seem all the more real. Eventually Brown had more than one hundred acres of corn planted near his post. When Schoolcraft encountered Brown on the St. Croix River in 1832, he ordered a careful search of his canoes for alcohol. Frustrated in this search, Schoolcraft nonetheless revoked Brown's license to trade at Granite Rocks. Neither this action nor Schoolcraft's burning of his buildings at Granite Rocks much perturbed Brown. He had already resolved to close that post and confine his Ojibwe trading activities to the Snake River outpost and perhaps a small outpost on the St. Croix opposite the mouth of Wolf Creek. By 1833 Brown opened a new post dedicated to the Dakota trade near the mouth of the St. Croix at a place called Oliver's Grove, near the present site of Hastings, Minnesota.[76]

Brown's new trading post caused dissention among the Mdewakanton Dakota. The aging and increasingly less energetic chief Little Crow had his village located on the Mississippi River not far from Fort Snelling. Brown's new post offered a convenient trading site closer to the band's traditional St. Croix hunting grounds, and two of the rising young leaders of the village defected from Little Crow and moved to the area near the mouth of the St. Croix. This greatly nettled Little Crow. First Brown had encouraged the Ojibwe to live and hunt on the very edge of the boundary, now he was drawing the Dakota into closer proximity to the area of contention. Little Crow protested bitterly to Schoolcraft. Although the danger was real, Little Crow may have complained about Brown in order to enlist the Indian agent in bolstering his sagging prestige among his people. The fur trader, who by now had had both Dakota and Ojibwe wives, clearly knew the risks of his actions, but in quest of short-term profits, he was heedless of the consequences. Brown continued to trade on the St. Croix in 1833 and 1834, after which time Little Crow and Indian agent Lawrence Taliaferro were able to force him to temporarily remove himself from the seat of conflict to the Minnesota River valley.[77]

The Treaties of 1837

Land cession treaties made it impossible to maintain even modest controls on fur traders. The Dakota and the Ojibwe reluctantly accepted the treaties because of the growing environmental degradation of their embattled homelands. The elimination of beaver during the 1820s had been one warning sign of the change. The high-value beaver allowed Indian hunters to obtain their annual wants with a very limited effort. Much more hunting time was required and less food was obtained when the smaller, less valuable muskrat was trapped as a replacement. The same was true of the increased difficulty encountered and energy required when subsistence hunters were forced to rely more on white-tailed deer than a herding animal like elk. The Dakota were particularly affected by these changes because the Ojibwe had steadily encroached upon their territories along the St. Croix and Mississippi. The Dakota no longer had the broad sweep of territory over which they had traditionally ranged during their annual subsistence cycle. As early as 1812, the Dakota were forced to adapt to these changes. Agriculture, which was unimportant to the Dakota at the height of their power in the 1600s, became increasingly significant as a way to bridge the nutritional gaps in their seasonal subsistence cycle. When the Cass expedition in 1820 visited the Dakota villages on the St. Croix, one of its members, Charles Trowbridge, observed that corn was "almost their only food." While this was certainly an exaggeration, based on a misunderstanding of the Dakota's seasonal movements, it nonetheless reflected a profound change. Trowbridge further noted that Dakota hunters did not even try to hunt in the vicinity of Little Crow's village near the mouth of the St. Croix, a clear sign that game in the region had been seriously depleted. In 1820, buffalo, which were once abundant along the Upper Mississippi, could not be found short of a two days' journey beyond the river. Over time buffalo were found only after increasingly farther journeys to the west. Lawrence Taliaferro, a careful observer of Dakota life, frequently noted that they suffered from starvation, not merely in the winter but even during the summer. In August 1835 he noted that they were completely without wild rice and that "to go out to hunt is for them to go off to starve." A year later he reported that the St. Croix Dakota had completed their summer hunt without killing a single deer. Requests for government supplies from the Indian agent became more frequent, as did complaints of Ojibwe intrusions on Dakota lands.[78]

The Ojibwe, who were accustomed to operating within a smaller territorial range than the Dakota, suffered less than their rivals. Even so, adjustments were forced upon them. In 1832 hunters along the Upper St. Croix still found deer near the river, but moose, which had been abundant, were eliminated from the area and could only be found along the remote headwaters of the Brule River. The Ojibwe, who in the early 1800s had provided fur traders with most of their provisions from hunting deer and gathering wild rice, also became more dependent on agriculture. Joseph Brown, making no mention of his large supply of illegal whiskey, claimed that the way he won the Snake River Ojibwe to his posts was by showing the Indians how to plant fields of corn. "The failure of the [wild] rice crop in the fall made the corn, potatoes, pumpkins and turnips very valuable to the Indians and probably saved many from starvation during the winter," he claimed in a letter to a missionary. When Schoolcraft visited an Ojibwe village near Big Fish Trap Rapids in 1832, he found "corn and potato fields." Not only were such fields new to that generation of St. Croix Ojibwe, but Schoolcraft also noted that the fur trade had also brought "a considerable change of habits, and of the mode of subsistence; and may be considered as having paved the way for further changes in the mode of living and dress." By 1832 as much as half of the trade goods brought from Mackinac for the Ojibwe trade were foodstuffs and clothing. A gradual but nonetheless decisive change in the fortunes of the St. Croix's Indian peoples was the transition, made apparent by the 1830s, from being providers of subsistence to the whites to becoming dependent upon fur traders for their subsistence.[79]

The dependency of both the Dakota and the Ojibwe on the European Americans for alcohol, trade goods, and, increasingly, food was the knife's point used to push them into treaty negotiations that led to the loss of the territory over which they had fought for so long. In July 1837 Henry Dodge, governor of the Wisconsin Territory, negotiated the cession of Ojibwe lands along the Upper Mississippi and St. Croix rivers. In return for these hard-won lands, the Ojibwe received annual access for the next twenty years to a paltry $19,000 worth of trade goods, a mere $9,500 in cash payments, and government-supported blacksmith shops and model farms. They retained the use of their old lands for hunting, fishing, and gathering, but only at "the pleasure of the President of the United States," which in effect meant as long as they did not get in the way of European American settlement and industry. Of greatest immediate importance, however, was the provision that paid $70,000 to fur traders who had claims against the Ojibwe. The decline in stocks of fur-bearing animals on their lands and their growing need for European American manufactured goods had gradually put the Ojibwe in the position of annually accepting more goods in advance than they could pay for at the end of the trapping season. The accumulated debt was substantial, although the payment of $25,000 to Lyman M. Warren, the principle trader on the St. Croix, seems excessively high. Settlement of this debt was an important matter, in part because of the presence of traders like Warren at the negotiation. But the Ojibwe themselves were well aware that they needed the treaty to keep open their source of credit. Chief Shagobi, of the Snake River band, bluntly admitted that he and the other leaders were "afraid to return home if their traders are not paid. They fear they should not survive the winter without their aid."[80]

Even considering their dire necessity, the St. Croix Ojibwe made a poor bargain when they agreed to the 1837 treaty. The payment received for their very extensive and valuable lands was too small to make a significant difference in the band's quality of life. Indian agent Henry Rowe Schoolcraft, who did not participate in the treaty negotiation, complained, with some exaggeration, that the payment "would not exceed a breech cloth and a pair of leggings apiece." Fur trader Lyman Warren blamed the bad deal on bribes paid by Lawrence Taliaferro to the leaders of the Pillager Ojibwe, who lived at Leech Lake. The Pillagers then used their influence to encourage the St. Croix and Chippewa river bands to accept a hasty settlement. Schoolcraft, who was strongly biased against Taliaferro, found this tale convincing. "The pillagers certainly do not," Schoolcraft wrote in his memoirs, "as a band own or occupy a foot of the soil east of the Mississippi . . . but their warlike character has a sensible influence on those tribes quite down to the St. Croix and Chippewa Rivers." Even the location of the treaty payment sites favored the Pillagers, not the eastern bands. The St. Croix people were forced to travel several hundred miles to the Crow Wing River to receive their meager payments.[81]

The Dakota were also forced to part with title to their share of the bloody St. Croix in 1837. The American Fur Company clamored for the treaty, claiming in excess of $50,000 in Dakota debts. The Dakota were loath to sell their St. Croix lands, but there also was a strong feeling that they could not go on as before. Little Crow, the son, namesake, and successor to the leadership of the St. Croix Dakota, found it harder and harder to locate game within their old hunting grounds. In the fall of 1835, after nearly dying of exposure in a snowstorm, he staggered into Taliaferro's agency in desperate need of food. Wearily Little Crow agreed to the necessity of taking up the plough. A smallpox epidemic in 1836 further demoralized the Dakota, leading one chief to say to Taliaferro, "the land is bad . . . and your advice is this day asked for my people." Taliaferro's advice was to go to Washington, D.C., and negotiate the secession of some of their lands. By advocating the journey, it may have been the Indian agent's intention to over-awe the Dakota with American power. If so, the strategy worked, for after a week of negotiations the Sioux representatives were forced to accept the government's price of one million dollars for all tribal lands east of the Mississippi, more than $600,000 less than the Dakota asked for. Unlike the Ojibwe treaty, the 1837 Treaty of Washington did not grant the Dakota a limited right to hunt and fish on the ceded lands. From the government's point of view, the Dakota's tenure on the banks of the St. Croix was at an end.[82]

In agreeing to the 1837 land cessions, the Ojibwe and Dakota people embraced an unavoidable contradiction. Their lifestyle and culture were based upon their occupation of the land. Yet, to maintain that lifestyle they had to give up their ownership of the land. Far from being dupes of the government negotiators, the Indians knew the Americans were anxious to exploit the pine forests of the St. Croix and Rum rivers. Ojibwe chiefs very sagely had proposed ceding the territory for sixty years, the amount of time they estimated for its forest resources to be exhausted. Flat Mouth, chief of the Pillager Ojibwe, observed, "It is hard to give up the lands. They will remain and cannot be destroyed but you may cut down the trees and others will grow up." Governor Dodge, however, rejected that proposal and insisted on a traditional cession. What the Dakota and the Ojibwe received in return were treaty annuities that allowed the Indians for twenty years the opportunity to continue the pattern of living they had adopted since the beginning of the fur trade. The Ojibwe retained a limited right to use the forests and the waters of the St. Croix. A Leech Lake Ojibwe captured this sentiment when he told the treaty council, "We wish to hold on to a tree where we get our living and to reserve the streams where we drink the waters that give us life." In that sense the treaty was a conservative solution to a looming crisis. Life would go on as before. But for how long? The president of the United States could order them off the lands at any time, and the annuity payments that kept the flow of trade goods coming were due to last only twenty years. The treaty was a blend of trying to protect a fading old way of life and tentative embrace of a new, scarcely imagined way of living. Annual payments for farm implements, seed, and the assistance of experienced farmers and the less specific provision for the establishment of schools clearly demonstrated that Dakota and Ojibwe leaders understood that they were fast approaching a threshold of change, and that they would need help to cross to the other side.[83]

Strangers on the Land: The St. Croix Indians in the Settlement Era

In the wake of the treaties several new kinds of European Americans came into the St. Croix country. Lumbermen were the largest group, followed by farmers and merchants. Of most direct interest to the Ojibwe were the missionaries. Showing much less scruple for the division between church and state than modern public officials, the U.S. commissioner of Indian affairs relied upon missionaries to carry out the transformation of the St. Croix bands from hunters and gatherers to sedentary agriculturists. The work had actually begun four years before the treaty, in 1833, when Reverend Frederick Ayer established a mission school at Yellow Lake, about a mile from the trading post. Ayer was a Presbyterian sent west by the American Board of Foreign Missions. After two years of difficult work trying to win the support of the Yellow Lake band, Ayer moved the mission to Pokegama Lake. The soil there was much more conducive to agricultural experiments, and the supplies of wild rice and fish were reputed to be more reliable. These factors made the Snake River band more sedentary than the Yellow Lake Ojibwe. Best of all, Ayer received an invitation from the Snake River people to bring his school to their band. In time Pokegama became the most successful mission in the region. In 1838 the Presbyterian missionaries working among the Ojibwe agreed to consolidate their efforts at that site. Ayer was joined at various times by William Boutwell, Edmund Ely, and Sherman Hall. The government lent support to their effort by locating one of the official Indian model farms at the south end of Pokegama Lake. Jeremiah Russell, of the Indian Bureau, sought to carve a farm out of the wilderness. He hoped that in time it could be a nursery for Ojibwe schooled in European American agriculture.[84]

As agents of change, the missionaries caused tension and division among the ranks of the Ojibwe. No two Ojibwe responded to the presence of these new strangers in the same way. The leaders of the Snake River band saw the mission school as a positive development that would give their children the means to learn the white man's letters. Others may have accepted the missionaries out of regard for their farming efforts, which after all provided a backup source of support during times of famine. The Yellow Lake band was deeply divided by Frederic Ayer's initial mission. At a council soon after his arrival Ayer was told in no uncertain terms he was not wanted there. "The Indians are troubled in mind about your staying here," said one speaker, "and you must go—you shall go." But a second faction in the tribe felt the contrary and the next day told Ayer that they were grateful for what he had done: "You have clothed and provided for us. Why should we send you away?" Ayer was invited to stay, but in the months that followed he was constantly unsure of his position: "Things were not as they should be." The band chief remained constantly, in Ayer's words, "on the fence," as he tried to maintain a consensus among his badly divided people. When the missionary left Yellow Lake, the chief must have been greatly relieved. Reverend Boutwell had an even more difficult time with the Leech Lake Ojibwe. After receiving several warnings, the Ojibwe poisoned the missionary's daughter. Fortunately the girl recovered, and Boutwell quickly left for the friendly clime of Pokegama Lake.[85]

The modest success enjoyed by Ayer and Boutwell was partially based on the care each took to cultivate the fur trade elite who had long influenced life along the St. Croix. Ayer became a friend of Lyman Warren. The veteran fur trader was a devout Presbyterian who used his money and influence to help Ayer build his base among the Snake River band. Boutwell earned entry into any trading post in the region by marrying the daughter of Ramsay Crooks, the managing partner of the American Fur Company. This Ojibwe Métis woman was described by one contemporary as "a commanding figure" who did much to win her husband a hearing among her mother's people. Even so, the missionaries often skirmished with their Indian neighbors across a cultural divide. Frederic Ayer, at great trouble and expense, brought farm animals to the lake mission. His efforts to have a proper American farm were sometimes frustrated by Indian hunters who, when hungry, did not differentiate between wild game and domesticated animals. "At Fond du Lac and Pokegama," wrote the Reverend Sherman Hall, "they have been much tried this summer with the Indians. They have killed several cattle at the latter place for the mission, and one at Fond du Lac. Some have appeared otherwise hostile." Nonetheless, the missionary was convinced he and his colleagues would "persevere in efforts to save these wretched heathen." On another occasion Ayer lost considerable face when he accused an Indian woman of stealing several shirts left out in the sun. He went so far as searching, and none to gently, her lodge, only to find out that Mrs. Ayer had simply misplaced the items. The Indian woman felt disgraced by the affair, although she never took action against the missionary. "Some of the Indians laughed heartily," at the crestfallen man of God, while "others made remarks rather sarcastic."[86]

It was not, however, the cultural barriers separating the Ojibwe from the evangelical Christians that led to the demise of the mission in the St. Croix Valley. In the end it was the rekindling of the ugly war between the Dakota and the Ojibwe that broke up the mission and its agricultural experiment. With the withdrawal of the Dakota to the west side of the Mississippi with the 1837 treaty, there was hope that commerce with European Americans could expand in the region and that the chronic wars might be brought to an end. This hope was shattered in 1839 when four Leech Lake Ojibwe killed a Mdewakanton leader at Lake Harriet, the site of a successful Protestant mission to the Dakota. The attack was the action of a few rogue warriors. The bulk of the Ojibwe wanted to maintain peaceful relations. Two large delegations of Ojibwe, one from Mille Lacs and the other from the St. Croix, had just met with Dakota leaders at Fort Snelling, where they smoked tobacco and pledged amity. When news of the murder reached the Dakota, they vowed to reward treachery with treachery. Dakota war parties fell on the Ojibwe returning unsuspectingly from the Fort Snelling conference. The St. Croix people were surprised at the present site of Stillwater, Minnesota, and twenty-three Ojibwe, mostly women and children, were killed. "I was on the battle-field of Lake St. Croix soon after the conflict," recalled a missionary, "and saw the remains of the slaughtered Ojibweys scattered in all directions. The marks of bullets were upon the trees, and the shrubbery was all trodden down. Some of the dead were suspended upon the branches of the trees." A new round of vengeance raids followed. One ambush led to the deaths of two of Little Crow's sons in the forest between the Snake River and St. Croix Falls. The scalping knife fell on the Pokegama Lake settlement in 1841.[87]

The mission was located on the east side of Pokegama Lake, although the majority of the Snake River band lived on an island in the lake. The island village gave the Ojibwe extra protection from Dakota raiding parties. A few of the Snake band, however, trusted the protection of the mission and had settled in cabins on the mainland. The evening before the attack a large Dakota war party secreted themselves in the brush adjacent to the mission. Their plan was to wait for the Ojibwe to commence work in their fields and then fall upon them. This ambush, like so many others, was spoiled by several overly anxious warriors. That morning the Ojibwe were late in canoeing from the island to the mission, and those on the mainland did not go to the fields. When a solitary canoe of two men and two young girls approached the shore, it was fired upon. The Ojibwe were thus alerted to the danger. Those on the mainland barricaded themselves in several cabins while those on the island took up arms. The Dakota laid siege to the cabins for several hours before giving up in frustration. At least one Dakota was killed in the fighting as well as two young Ojibwe girls. The missionary E. F. Ely found the little corpses on the shore. "The heads cut off and scalped, with a tomahawk buried in the brains of each, were set up on the sand near the bodies," he later recalled. "The bodies were pierced in the breast, and the right arm of one was taken away."[88]

Although the Snake River band had successfully defended their village, they feared a return by the Dakota. The band broke up into family groups and retreated into the wilderness. The mission was abandoned by its acolytes. "The Indians were scattered," recalled Elizabeth Ayer, "and dared not return." For a time Reverend Ayer tried to visit the scattered members of the band in their isolated camps, but when it became clear they did not intend to return to Pokegama, the Presbyterians had no choice but to abandon their mission. In 1842 the mission was removed to La Pointe. Not until the spring of 1843 did the Ojibwe return in force to Pokegama Lake. The mission was briefly reestablished. But the rapid increase in the number of European American lumbermen and a handful of settlers in the region made the missionaries lose faith in the location as an effective base from which to convert the Ojibwe to the white man's God and a farming lifestyle. The Reverend William Boutwell, who also served as a field agent for the commissioner of Indian affairs, encouraged the Snake River Ojibwe to abandon Pokegama Lake and locate at Mille Lacs, where wild rice and fish were abundant and contact with whites less frequent. The mission in the St. Croix Valley was abandoned in 1845.[89]

The missionaries also soured on their prospects along the St. Croix, Rum River, and other areas ceded in the 1837 treaty because of the pervasive presence of whiskey traders. While the St. Croix had been Indian territory, the agents of the Office of Indian Affairs had the power to regulate who traded there, where they traded, and with what wares. After 1837 the valley was simply another part of the Wisconsin Territory, a vast region with large opportunities and little in the way of civil administration. Alcohol, which in times of competition between fur traders had always greased the wheels of commerce, now became the principle article of trade for men intent on separating the Ojibwe from their annuity payments. By 1844 William Boutwell complained to a fellow missionary that the ceded lands were "inundated with whiskey."[90]

Among the unsavory traders who entered the St. Croix at this time was Joe Covillion. He was a Métis who took over the former mission school at Yellow Lake and used it for his post. Located on the Yellow River just where it leaves Little Yellow Lake, the trading house was the scene of many drunken reveries and a key location in the first murder mystery in the St. Croix Valley. In 1845 Albert McEwen hired Covillion to guide him to timberlands in the Yellow Lake region. McEwen had a large amount of gold coin he hoped to use to secure title to lands upon which a profitable speculation might be made. McEwen never returned from the trip. Covillion explained that he had not been with McEwen and cast suspicion on an Ojibwe who was alleged to have actually served as guide. Not long afterward McEwen's body was found stuffed in a hollow tree about ten miles from Covillion's post. Preliminary investigations revealed that Covillion had in his possession a large amount of gold coins, McEwen's watch, and a fistful of land warrants. Calmly the trader explained that he obtained these from the Ojibwe in trade. Later that winter the Indian whom Covillion had claimed guided McEwen was found dead in his camp. Covillion, the owner of "considerable property," retired to Taylors Falls, where he died in 1877.[91]

Another less than worthy trader of this period was Maurice Mordecai Samuels. In 1846 he had a trading post at the mouth of the Sunrise River. In time-honored fashion Samuels established himself with the Ojibwe by taking one of their women as his wife. Later he relocated to St. Croix Falls, where he operated a "ball alley" and trading post. Samuels was described by fellow pioneer William H. C. Folsom as "a shrewd man and an inveterate dealer in Indian whisky." No friend of the fur trader, Folsom accused Samuels of being "unprincipled" and "repellant" to the "moral sense of the community." There can be little doubt about how repellant was the type of whiskey sold by Samuels. He did not trouble himself to import the product from the Ohio Valley where whiskey was abundant and cheap. It was far less expensive to use grain alcohol and then attempt to impart the right flavor and color by artificial means. Samuel's recipe included boiled roots and tobacco, which according to Folsom poisoned many whites and Indians. One consumer of the concoction reputedly went insane and leapt from a high point of the Dalles to the falls below. Samuels profited handsomely from his trade with the Ojibwe and in time became a leader of the community of St. Croix Falls.[92]

Whiskey was an important commodity at all trading posts, but the whiskey shops of men like Samuels and Covillion in particular were the scenes of many degrading and deadly spectacles. Bad liquor sold with no restraint led to trouble at Alexander Livingston's grog shop on the St. Croix at the mouth of Wolf Creek. Livingston, who may have operated in cooperation with the veteran fur trader and whiskey dealer Joseph R. Brown, was gunned down in 1849 after a "drunken melee in his own store." Livingston died of his wound, while his killer, a Métis named Robido, escaped prosecution. Another whiskey dealer to die as a result of his own greed was Miles Tornell, a Norwegian operating near Balsam Lake. Tornell refused to back down in the face of competition from a German American whiskey dealer, a man identified only as Miller, who operated a post on the lake.

The German resolved the competition by hiring an Ojibwe to murder Tornell. When the crime was detected, the Ojibwe was executed, while Miller was merely flogged. In 1847 one of Samuels's subordinates, Henry Rust, was killed in a brawl with a drunken Ojibwe, Notin. Unlike most such cases, this one came to trial. The verdict reflected the outrage many early settlers felt toward the whiskey traders. Notin was found not guilty, and a criminal complaint was issued against Jake Drake, the Samuels employee who sold Rust his stock of booze. Drake himself fell victim to foul play shortly thereafter, when an inebriated Métis slayed him near his Wood Lake post.[93]

The presence of the whiskey dealers and the availability of treaty money accelerated the abuse of alcohol among the Indians of the valley. James Hayes, Indian agent to the Ojibwe, complained of the "cupidity and heartlessness of the whiskey dealer," which he blamed for the "accounts of outrages and crime" that washed over the St. Croix frontier in the wake of the treaties. Among the Dakota, who had formerly lorded over the St. Croix, the impact was even more pathetic. "They *would* have whisky," wrote missionary Gideon Pond. "They would give guns, blankets, pork, lard, flour, corn, coffee, sugar, horses, furs, traps, any thing for whisky." As a result, "They killed one another . . . they fell into the fire and water and were burned to death, and drowned; they froze to death, and committed suicide so frequently, that for a time, the death of an Indian in some of the ways mentioned was but little thought of by themselves or others."[94]

Between the rapacity of the whiskey dealers and the incompetence of federal authorities, the St. Croix Ojibwe benefited little from the financial terms of the 1837 land cession. In 1838 the Office of Indian Affairs bungled the first payment due them. The Ojibwe had been told to gather on Lake St. Croix near the future site of Stillwater, Minnesota, to receive their payment in goods and supplies. The Ojibwe began to gather there in July. Every steamboat ascending the river was besieged by anxious Indians who sought their due from white immigrants, not appreciating that they "had nothing to do with payments." All summer and most of the fall the Ojibwe waited, faithful and famished. The large congregation of Indians stripped the surrounding area of both firewood and game. Only in November, with the Indians starving and freezing, did the promised goods finally arrive. One hundred barrels of flour, twenty-five of pork, bales of blankets, boxes of guns and ammunition, even casks of gold dollars were all unloaded while thick snow flakes settled on the ground. Desperately hungry, the Ojibwe tore into the food. Many ate too much too fast and suffered agonizing cramps for their trouble. According to one witness, "many of the old as well as the young died from overeating." In the meantime, ice formed on the St.

Croix, rendering useless more than a thousand canoes the Ojibwe had brought to transport their goods. They were forced to destroy the craft rather than let them fall into the hands of the Dakota. Only that which they could carry on their backs could be taken north to their winter camps. Much of the food, money, and goods had to be left behind. During the long, agonizing march up river and during the harsh winter that followed, many Ojibwe perished. As pioneer chronicler Folsom noted, "Their first payment became a curse rather than a blessing to them."[95]

In this manner the thousands of dollars of federal assistance to the Ojibwe that the chiefs had seen as the means to maintain their fur trade lifestyle only further impoverished the Indians. J. F. Schafer, who distributed supplies to the Ojibwe in 1851, complained of "the introduction of liquor among the Indians immediately after issuing provisions." When Schafer saw the Ojibwe trading "their Blankets &c. for liquor," he tried to suspend the distribution of goods until the whiskey dealers left the payment site at the mouth of the Snake River. Indian agents frequently referred to the St. Croix Ojibwe as "exceedingly poor, and naked and needy." William Warren, who had spent his life living among the Ojibwe, advised the governor of Minnesota, "There is not under the sun a more wretched people than they are & will continue to be so as long as they remain in close proximity to a bad white population." Governor Alexander Ramsay described the St. Croix band as "the most miserable and degenerate of their tribe."[96]

The condition of the Indians excited more fear than pity among the European American settlers and lumbermen who were quickly moving into the ceded lands along the St. Croix. There was little attempt on the settlers' part to understand the customs and traditions of the Indians they found living in the valley. Typical of these cultural clashes were the numerous stories of Indian men barging into the cabins of white settlers and demanding food. Ojibwe etiquette required visitors, however uninvited, to be fed. That kindness, of course, required some reciprocation, but not immediately. Whites regarded these visits as intimidation and complained to Wisconsin and Minnesota officials of "marauding Indians." Whenever something went missing, Indians were the first suspects. When early settlers in St. Croix Falls were missing a pig of lead, they accused the Ojibwe of the theft. The Indians denied the crime, although the whites claimed to notice "that all their war clubs, pipes and gun stocks had been lately and elaborately ornamented with molten lead." These types of actions, in addition to encounters with lumbermen, inclined federal officials to revoke the provision of the 1837 treaty that allowed the Ojibwe to remain on the ceded lands.[97]

On February 6, 1850, President Zachary Taylor issued an executive order ending the Ojibwe's right to hunt and fish on the ceded lands. Local Indian agents were given the responsibility of determining which Ojibwe were to be removed and where they would be relocated. The news caused considerable consternation among the Ojibwe of Lake Superior, but among the St. Croix bands there was some interest in removing to another area. Only a month before the president's order the Snake River Ojibwe had petitioned their agent for removal to the Crow Wing River in the Minnesota Territory. Portions of the band had already left the valley and crossed over the divide to Mille Lacs. Plans were made to remove all of the Ojibwe from the valley, but typical of the slipshod manner in which Indian removals were managed, federal authorities were unable to gather together the majority of the Indians in the region. After working all summer to make the move work, Indian Agent John Watrous only effected the removal of 288 St. Croix residents to the Crow Wing River. Few of these Ojibwe remained there long; nearly half were gone in a month. No concerted efforts were again made to remove the St. Croix bands, nor were they awarded reservation lands in subsequent federal treaties. In the wake of President Taylor's order and the botched removal program, the St. Croix Ojibwe were left in a legal limbo. They were not recognized as having rights in the St. Croix Valley, yet there they resided for the next eighty years on lands unused or abused and abandoned by European American settlers.[98]

With no more remaining legal claim to the St. Croix than the Ojibwe, the Dakota were barely tolerated by the inrushing horde of white farmers. In 1855 a large band of Dakota established a winter camp in the valley near Marine Mills. At first residents regarded the Dakota as interesting exotics. "They were really a curiosity to many of our citizens," reported one townsman. In seeing the Dakota "dressed in pure Indian winter style," the people of Marine shared with each other "not a few half supressed, half frightened remarks [of] ridicule." The merchant in charge of the local general store brought out a large barrel of crackers that the Indians "devoured" with the noise of "a flock of hungry geese." It was not long, however, before the Dakota ceased to be interesting and were regarded by most people in the area as a nuisance. One farmer complained the Dakota were "frightening our wives and children, plundering our premises, laying vicious hands on every thing their savage eyes crave, and not leaving unmolested the domestic sanctity of our potato hoes." Without the least irony, the settlers complained, "And what is worse they are killing all our deer,—this last offense amounts to an unpardonable crime."[99]

The opportunity to hunt in the under-utilized forests of the St. Croix is what lured the Dakota back across the Mississippi River. What recently arrived farmers regarded as "our deer" were, of course, resources the Dakota had relied upon for generations as part of their seasonal subsistence cycle. Changes in the population and ecology of the Upper Mississippi country made their old hunting grounds on the St. Croix more attractive than ever. Game populations along the Mississippi were taxed by the growth of settlements such as Red Wing, Hastings, and St. Paul and their adjacent agricultural districts, where by 1850 more than five thousand European Americans resided. Development along the St. Croix was focused more on logging, with Swedish immigrants only just beginning to establish farmsteads north of Stillwater. The presence of these whites was not yet enough to deplete the game resources of the long-contested region. The Dakota may also have felt somewhat shielded from Ojibwe attack by the population of newcomers. Every January or February in the 1850s, the Dakota undertook hunts in the valley. These were male-dominated hunting parties, with only a handful of women and children in the company. In addition to helping to prepare the deer hides, the women made moccasins that they sold for bread in Stillwater. The Apple River was a particularly rich hunting preserve. "They were heavily laden with skins, game, &c., and seemed to be well pleased," recorded the *St. Croix Union* in January 1857, at the conclusion of that year's hunt. The amount of game brought down by these hunting parties was indeed prodigious. "How many deer did you kill?" asked a reporter who visited a Dakota camp in 1855. In answer one of the hunters "held up both hands, and motioned with them quite deliberately, ten times—indicating, as we interpreted it, One Hundred." A year later when the Dakota left their hunting camps near Marine, the local populace estimated, with perhaps some exaggeration, that between eight and twelve hundred deer were taken.[100]

The hunting success of the Dakota perturbed the European American settlers because they counted on game as a source of food and barter during the first years of farming. "It is hard for the industrious and poor white settler to have his wood and stacks of hay burnt up," the *St. Croix Union* editorialized, "his traps and their booty stolen, and his game shot down, and much of it wasted." The settlers formed committees, signed petitions, and lobbied the territorial governor, but to no avail. The new white residents of the St. Croix complained that the Dakota had not become sedentary and blamed the government that "allowed a set of scheming rouges with a pittance of whiskey to cheat them out of their annuities." Nothing, however, was done to stop the Dakota visits, which continued until the

1860s, when their villages were pushed far up the Minnesota River valley and the St. Croix ceased to be a lucrative hunting ground.[101]

Tragically, throughout the painful twilight of Indian tenure, while English and Swedish voices replaced those of the Ojibwe and Dakota along the St. Croix, the vicious intertribal war continued. The conflict was no longer really about territory, as treaties with the United States had awarded the valley to others. Vengeance, however, continued to exert a powerful spell. Remembering the wrongs of the past helped to obscure the problems of the present. Just as important was the need of young men to find a way to assert their manhood in a traditional way. Economic decline narrowed their range of opportunities to win distinction, so the feud continued.

In March 1850 a war party from the village of Little Crow, the son of the Dakota leader who had first negotiated with the Americans, surprised an Ojibwe camp on the Apple River. The ambush was a complete success. Eleven Ojibwe women and children and three men were killed as they made maple sugar. One boy was captured. The next day the jubilant Dakota passed through Stillwater on their way west. They "went through the scalp dance, in celebration of their victory—forming a circle round the Ojibwe boy—their prisoner—and occasionally striking him on the face with their reeking trophies," recorded the *Minnesota Chronicle*. The encounter was no different from hundreds that had come before and others that would follow. The times and the river, however, were different. With hope, boldness, and perfidy, new people and new ways were dominating the valley. What once was seen as the way of wilderness was now, with the passage of the frontier and the disinheritance of a people, deplored as simple, tragic, murder.[102]

The wretched attack at Apple River was one of the concluding scenes in the long history of Dakota dominance of the St. Croix Valley. After the tragic Sioux Uprising of 1862, the Dakota were removed far from the border river. Indian voices continued to be heard along the waterway, but after 1862 those people were the Ojibwe. They outlasted their ancient enemies by sheer persistence, and they endured in the valley after the 1837 cession of their lands to the United States by practicing that same virtue.

The majority of the old Snake River band of Ojibwe abandoned the valley during the 1850s, relocating to Mille Lacs. The bands at Yellow Lake and along the headwaters of the St. Croix, however, remained where they had always lived. Lacking land tenure, they lived as squatters on government or lumber company lands. Wild rice and cranberry harvests remained vital to their subsistence and were supplemented with the yields of hunting and fishing. Furs continued to be traded, although the exchange now took place with small-town merchants at a general store and not with red-sashed voyageurs at a trading post. During the late 1860s the U.S. government began to move the Ojibwe onto designated reservations. Most of the St. Croix Ojibwe were related to tribal members living at the Lac Courte Oreilles Reservation. A smaller number had family connections to the Ojibwe of the Bad River Reservation. In time, the reservations were subdivided into individual family allotments. The St. Croix Ojibwe were not based on any reservation, and most received no allotments and little in the way of educational or health services. While Bad River and Lac Courte Oreilles were recognized Indian communities, the St. Croix Ojibwe pursued an independent existence largely unknown to the government. People in northern Wisconsin began to refer to the St. Croix band as "the lost tribe."[103]

Of course, the Ojibwe were the last people in the valley to be "lost." They adapted to the rise of the logging industry by utilizing it as a source of wage labor. Ojibwe frequently worked as lumberjacks and river drivers. In the latter task they excelled. In 1902 the loggers Gear & Stinson employed an entire crew of Ojibwe to bring their drive down the Clam River.[104] A resident of Shell Lake later recalled that "the young men, many of them, are our best drivers on the river; quick, sprightly, active."[105] Métis logger Edward St. John employed a large number of his Ojibwe kinsmen in his forest operations. During the last years of the nineteenth century, between 150 and 175 Ojibwe continued to reside along the St. Croix.[106] Trouble for them came when the pine forest was cut, and there no longer were log drives on the river. This period coincided with the rise of fish and game regulations that made it difficult for Indians to live off the land on a full-time basis. Private ownership of land was also at its peak during the final years of the nineteenth century, restricting the ability of the Ojibwe to gather wild foods. Cranberry marshes that had been utilized for generations, for example, were increasingly drained to grow hay for dairy cows. White farmers, often from foreign lands, sometimes nursed fears about the native people who lived around them. In 1878 several Swedish settlers started a panic that spread like wildfire through Burnett County, Wisconsin. A large gathering of Ojibwe was exaggerated into the beginnings of a concerted attack by both the Ojibwe and the Dakota on all settlers. Scores of farms were abandoned in anticipation of an attack. The governor called upon General Philip Sheridan to dispatch federal troops to restore order. The army exposed the entire affair to be a misunderstanding, although it did recommend to Wisconsin that the Ojibwe not be allowed to "roam about in bands."[107]

As squatting on private lands became problematic, some of the Ojibwe bought parcels of land where families erected wooden shanties and invited friends and kinsmen to settle as well. One such collection of wigwams and houses was located about a mile up the Namekagon River from the St. Croix. Called Dogtown or Ducktown, it was home to as many as fourteen families and was occupied as late as 1938. John Medoweosh, a band leader, owned a tract of land at the junction of the Yellow and St. Croix rivers. He lived there for many years with an extended, multi-generational family. Augustus Lagrew, a Métis with a full-blood Ojibwe wife, owned land a few miles from Shell Lake, Wisconsin, that also served as a place of congregation for the Ojibwe. Gifts of food or small loans by white neighbors helped the Ojibwe get through hard winters, although rarely did the Indians beg for handouts or apply for formal aid through the county poor fund.[108]

The U.S. government rediscovered the St. Croix Ojibwe in 1910, when Senator Robert M. La Follette held a Senate hearing on the condition of Indians in Wisconsin. The fact that the St. Croix Ojibwe had in the past received little in the way of annuities prompted several congressional efforts to provide them with federal relief. The St. Croix Ojibwe, however, were not given what they needed most: a guaranteed land base within their homeland. Not until 1934, with the passage of the Wheeler–Howard Act (Indian Reorganization Act), did the St. Croix Ojibwe receive a federally recognized reservation. After eighty landless years the St. Croix people could not be brought together at a single location. Instead, the new three-thousand-acre reservation was spread out over eleven separate Burnett County locations.

In the years that followed, the Ojibwe grew more and more like their neighbors whose ancestors hailed from Europe. Most of the St. Croix band became practicing Christians. One of their numbers, Philip B. Gordon, became one of the first Indian priests in the United States. He served not only his own people but for many years was the beloved pastor to a largely white parish in the St. Croix Valley.[109] Ojibwe children participated in the same rural schools as the sons and daughters of farmers. Yet, in spite of these marks of assimilation, the Ojibwe remained anchored in their Indian identity. This identity became more important in the 1970s when the Red Power movement sparked greater political assertiveness. One result of this was the so-called Walleye War that was triggered in 1983 when the federal court established the rights of the Ojibwe to fish outside of state regulations. The decline of agriculture and forest products in the region had forced both whites and Indians to rely more on jobs in the tourism and recreation fields. Whites feared that the Ojibwe's exercise of treaty rights would degrade stocks of fish that were critical to maintaining tourism. These tensions, which became violent in some parts of the North Country, were largely restrained in the St. Croix Valley.[110]

A more important assertion of Native American status came with the establishment of casino gambling. In 1974 President Richard Nixon approved changes in federal Indian policy that sparked a general move toward greater independent control of reservation lands by the tribal community. Although unanticipated at the time, this led to a gradual expansion of restricted enterprises, from garbage dumps to gambling, on Indian reservations. The St. Croix band of Lake Superior Ojibwe took advantage of this change to establish two casinos at Turtle Lake and Danbury. In a stunning turnaround, the St. Croix Tribal Enterprises became the largest private employer in Burnett County. Hundreds of white as well as Indian people found jobs in the gaming rooms and hotel complex. Profits from gambling led to an expansion of family, housing, and health services for the tribe.[111]

At the 1837 council that resulted in the cession of their St. Croix lands, the Ojibwe chief Maghegabo tried to explain to Governor Henry Dodge that his people would endure in the valley. "Of all the country that we grant you we wish to hold on to a tree where we get our living & to reserve the stream where we drink the waters that give us life." The chief then placed an oak sprig, the germ of new life, on the council table. "Every time the leaves fall from it, we will count it as one winter past." After the leaves have fallen from that symbolic tree more than 150 times, the St. Croix Ojibwe are more numerous and more economically successful than at any point in their history. Through exercise of the same patience and persistence that had served them so well in the long twilight struggle with the Dakota, the Ojibwe survived the wave of white emigration that broke over the valley in the mid-nineteenth century. For them the St. Croix and the Namekagon rivers remain "the waters that give us life."[112]

2 | River of Pine

For nearly 150 years European and American merchants had passed through the St. Croix Valley, attentive to the number and location of Indians within the valley, mindful of the presence of wild game, sometimes observing its agricultural prospects, but largely unconcerned about the timber resources of the region. The bright and articulate George Nelson, who first entered the river valley in the fall of 1802, was an exception. The "beautifully wooded" islands and hills of the valley impressed him, and with a merchant's eye he predicted they could be as commercially important as the timberlands of the St. Lawrence valley. The St. Croix's "splendid groves of pine," he wrote, "could as easily be floated down the Mississippi as from Chambly to Sorel." More typical, however, was U.S. Army Lieutenant James Allen's terse dismissal of the Upper St. Croix landscape as "poor, and pine; none of it fit for cultivation."[1]

Interest in the region's forest resources dramatically increased during the 1830s. A rush of entrepreneurs anxious to exploit St. Croix forests pushed the federal government to clear Indian title to the valley. Forests hardly worth noting a generation before had been rendered into promising assets by the growth of towns and farms in the valley of the Mississippi. The St. Croix, Chippewa, Red Cedar, and Rum rivers, all tributaries of the Mississippi, boasted vast forests of pine that helped to build downriver towns such as Winona, Rock Island, Davenport, and St. Louis. In the post–Civil War era, the demand became even more insistent and the market more lucrative as the treeless plains were surveyed into 160-acre homesteads. The exchange of a sod house for a frame home built of Wisconsin or Minnesota pine was a badge of success for the homesteader and the basis of fortune for many a lumber baron.[2]

The fur trade had divided the St. Croix Valley between an upper river, dominated by the Ojibwe and economically tied to Lake Superior, and a lower river, home to the Dakota and linked to St. Louis–based traders. In terms of transportation geography, the logging frontier would restore the unity of the valley. The entire river system would be harnessed to bring the

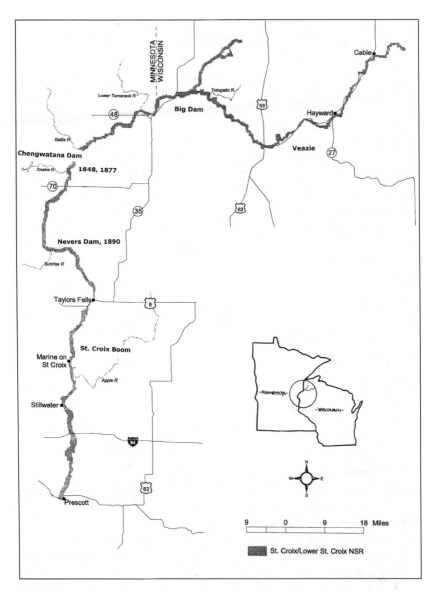

Map 5. Logging era historic sites, St. Croix National Scenic Riverway (Source: National Park Service).

winter's harvest of logs to the collecting booms along the lower river. Like a funnel, the St. Croix River was used to concentrate the wealth of the entire valley at its mouth. The forests of the upper river played a large role in building towns and industry along Lake St. Croix, as well as the nearby cities of St. Paul and Minneapolis. The logging frontier first made manifest the dichotomy of a thinly inhabited upper river resource frontier and the prosperous urbanized lower river. In the course of doing so, it wrought a massive transformation of the valley's landscape and severely, in some cases irrevocably, altered its ecosystem. Through its involvement in the lumber industry the St. Croix played its most important role in American history, but at a cost still being exacted today.

Lumbermen attempted to transform the free-flowing wild river into a disciplined industrial waterway. Never before and never again would the river be used so intensely. Mill operators began each day by studiously noting its fluctuations in level. Around blazing campfires, log drivers endlessly debated the ebb and flow of its current. By building dams as assiduously as the all-but-eliminated beaver, by blasting boulders and constructing booms, the lumbermen made each mile of the St. Croix's 165-mile length serve the purpose of delivering logs to mill and market. Like the tentacles of some great industrial monster, the lumber industry probed, damned, and controlled even the remotest of the river's tributaries, bending their wild reaches to its commercial purpose. The early lumbermen more than doubled the natural transportation capacity of the St. Croix watershed to 330 miles of water capable of carrying logs to market. When the industry expanded further in the wake of the Civil War, more splash dams and stream improvements brought the size of the St. Croix system to a staggering 820 miles of useable waterway. The St. Croix was more than a logging river, for better than a half century; when the ice went out each spring, from its headwaters to Stillwater, it became a river of pine.[3]

From Fur Trade to Fir Trade

Fur traders, as businessmen familiar with the region and its resources, seemed to be in an excellent position to profit from the rising market for lumber in the 1830s and 1840s. Yet few were able to make the transition. Many were too old and set in their ways to adapt. The Warren and Cadotte families that had so long controlled the trade of the valley from Lake Superior struggled to make the transition to logging. Lyman Warren saw the handwriting on the wall in 1838 when he left the American

Fur Company. Leaving the Lake Superior country, he settled on the Chippewa River near the falls and established a sawmill. He was, however, struck with illness in 1847 and died before becoming deeply involved with logging.[4]

Those fur traders who were in a position to profit from logging were men with trade contacts with the downriver towns that comprised the market for St. Croix pine. Logs, unlike furs, could not be carried over the Brule portage to Lake Superior. The bulky commodity had to follow the dictates of gravity and go south with the river's flow. Fur traders tied to Mackinac like the Cadottes lacked the market and supply contacts to make the transition from furs to logs. Joseph Renshaw Brown, a less than scrupulous trader who had lived among both the Dakota and the Ojibwe of the St. Croix, had the necessary downriver contacts. He was a classic frontiersman on the make, anxious to win his fortune, be it by furs, land, or timber. As early as 1833, Brown had been cutting pine on the Upper St. Croix. Three years later Brown positioned a logging crew at the current site of Taylors Falls, Minnesota. He likely had additional logs cut upstream and floated down to the falls. Upwards of two hundred thousand feet of pine were cut by his men, but before the bulk of it could be floated downriver the logs were burned in a forest fire. The remainder of his cut was simply abandoned on the riverbank when Brown, with a characteristic sudden change of direction, decided to quit the St. Croix and resume fur trading on the Minnesota River.[5]

Brown's early logging on the St. Croix had been an illegal intrusion on Ojibwe land. Indian agent Lawrence Taliaferro tried to ward off European Americans seeking to supply lumber to the growing Mississippi valley settlements. In 1836 former fur trader Joseph Bailly complained to Congress, "A few years back the labor of a few Lumbering parties operating with whip saws was sufficient to supply the wants of that market, but now that the country is settling with a rapidity unexampled in the history of our country it requires greater supplies." Was it the government's intention, Bailly asked, to let the whole population of the Mississippi valley "suffer for want of Lumber because a few miserable Indians hold the country"? Even without government sanction, fur traders and others attempted to make their own agreements with the Ojibwe to secure access to the pinelands of the Upper St. Croix. In March 1837 three of the American Fur Company's former lions in the region, William Aitkin, Henry Hastings Sibley, and Lyman Warren, brought together a conference of St. Croix and Snake River Ojibwe for the purpose of securing a ten-year lease on the forests of the upper river, only to be foiled by Indian agent Taliaferro. Men with more modest expectations simply discretely made their way upriver with a handful of laborers and, after offering gifts to the local Ojibwe band, began to cut pine.[6]

Failure to come to terms with the Indians could be quite costly, as John Boyce of St. Louis discovered in 1837. He led eleven men past the falls in the autumn of that year. With logging equipment, six oxen, and a mackinaw boat, he pushed up the St. Croix to the vicinity of the Snake River, where he established a logging camp. The Snake River band, which well understood the great value the white man placed on pine lumber, protested Boyce's activities. "Go back where you came from," ordered Little Six, a bandleader backed by more than one hundred of his people. Boyce sought the mediation of the Presbyterian missionaries at Pokegama Lake. They advised him to leave, and the Ojibwe threatened to prevent Boyce from removing any of the pine. "We have no money for logs; we have no money for land. Logs cannot go" was their firm policy. Boyce persisted in spite of all threats and harassments. In May, after a winter of logging, Boyce tried to raft his harvest downstream. The Ojibwe, in a manner Boyce regarded as menacing, followed the drive. High water and hungry, unpaid, thoroughly dispirited men led to the loss of most of the logs. With the Ojibwe looking on, Boyce also lost most of his logging equipment when the mackinac boat upset while being lined down the falls. The boat itself was saved when the shrill whistle of the steamboat *Palmyra* "broke the silence of the Dalles." Aboard were other lumbermen who rendered Boyce assistance, saving his boat and recovering some of the logs. Perhaps most importantly, the steamboat bore the news that the U.S. Senate had ratified the 1837 treaty of cession, appeasing the Snake River Ojibwe. It was too late, however, for Boyce. The few logs that were recovered and sold did not come close to meeting the expenses of the venture. Pioneer chronicler Folsom recorded that "Boyce was disgusted and left the country."[7]

The Treaty of 1837 opened the St. Croix Valley to European American occupation, and loggers surged into the valley to exploit the new frontier. John Boyce had paid a high penalty for beginning operations prematurely, as did a number of the other lumbermen who were on the river immediately in the wake of the treaty. In September 1837 Franklin Steele and several partners ascended the river in a bark canoe and a scow loaded with supplies and men. They built several cabins at the falls, filed land claims on the best mill sites, and scouted good timberlands upriver. Four other groups of lumbermen arrived that fall. Steele's group organized themselves as the St. Croix Falls Lumber Company, and construction was begun on a $20,000 sawmill. They controlled an important waterpower site but retained inexperienced millwrights and lumbermen. Their mill was not completed until 1842, in part because the site posed considerable construction challenges. Franklin Steele soon sold his share of the company, and the firm fell under the control of absentee owners and men more interested in land speculation than logging. The most important of the former was Caleb Cushing, a prominent Democratic politician in Massachusetts and a veteran U.S. diplomat. As logging expanded on the St. Croix, it became clear that the head of the Dalles was not the best place to locate a mill, and subsequent lumbermen elected to drive their logs farther downstream to lower Lake St. Croix. This did not deter Cushing from continuing to make sizable investments in the site and in timberlands upriver, as well as wrangling with his partners in costly lawsuits. The latter worked their way up to the U.S. Supreme Court. Cushing was not able to establish his control over the company until 1857. By that time the bloom had long since left the bright promise of the falls site. Other towns had been platted and emerged as logging centers. The *Polk County Press* mocked the hopes that St. Croix Falls would become the industrial center of the river. "The ruthless hand of time has made sad ravages, and though the industrious relic hunters might find there a dam by a mill site, they would not find a mill by a dam site."[8]

The ill-fated St. Croix Falls Lumber Company did succeed in making history in its career of failure and misfortune. Quite unintentionally the company caused the first rafts of lumber to be sent downriver to St. Louis. In 1843 the company's boom below the falls gave way before high water. The entire corporate stock of logs was borne away in the flood. While this meant the newly completed mill could not cut the lumber, the company could still salvage something if the logs could be caught downstream. John McKusick, a young logger just arrived from New England, collected about two million feet of logs. These were assembled into rafts and floated down the Mississippi to St. Louis. In the years that followed, thousands of rafts of logs would follow in their wake. John McKusick used the proceeds from his share of that log sale to purchase the machinery for a water-powered mill. That mill was established at Stillwater, where it played a major role in making that site, as opposed to St. Croix Falls, the lumber center of the river.[9]

The surest way to making money in this early stage of the logging boom was to keep things simple—get a crew of men to the Upper St. Croix, cut several hundred thousand board feet of trees, drive them over the falls, assemble them into a raft, and float them to growing river towns, preferably one not too far south. In May 1838 Lawrence Taliaferro reported that two hundred men were at work in the pineries of the St. Croix and the Chippewa; by October the number had grown to five hundred.

Figure 3. A lumber raft. Rafts assembled on the St. Croix for transport down the Mississippi were often much larger than the one pictured here. *Harper's Magazine,* March 1860.

These small-scale operators benefited from what amounted to a free re-source. The patchy pine lands of the lower river were not surveyed by the General Land Office until 1847, while the tall timber of the upper river was not mapped until the 1850s, so even if an individual wanted to pur-chase the land he was logging, it would have been impossible. Even pre-emption claims were not possible on unsurveyed lands in Minnesota until Congress extended that privilege in 1854. With federal land policy making legal purchase impossible and the market clamoring for more lumber, the pioneers of the pineries responded in the best, unscrupulous tradition of the frontier—they took what they needed and damned the consequences. The era of free timber on the Lower St. Croix lasted at least a decade, from 1838 to 1848, and on the upper river private fortunes were made off public lands well into the 1850s.[10]

Frontier Logging: Life in the Forest

Logging on the St. Croix during the 1840s and 1850s was a primitive, small-scale, frontier enterprise. The size of a logging crew was small—between ten and fifteen men. Typical was the eleven-man team de-ployed by the ill-fated John Boyce in 1837. That year there were at least five crews operating on the St. Croix. By 1854 this number had swelled to eighty-two crews, twenty-two of whom were established along the Snake River, where some of the finest white pine was to be found. Each of the crews included several oxen. In 1837 Boyce had six of the beasts of burden. They were used to drag felled pine from where the logs were cut to the river landing where they were stacked for transportation in the spring. The early loggers had the advantage of being able to work stands of pine adjacent to the river. Since the logs were only dragged a short distance, a simple wooden travois, called a *go-devil,* attached to a single oxen and was the only equipment needed to transport logs. The crews were divided into a few specialized tasks. Most valuable were the choppers, men skilled with using an axe and in making a tree fall where they wanted. They were usually the best-paid men in the crew. Once a pine was felled, the swamper came in and cut off any branches. Another man called the barker peeled off the bark on the underside of the tree, to create a smooth surface for dragging on the snow. Next the chainer connected the oxen to the log and guided it through the drifts to the landing. During this early period of logging, axes and handspikes were the only tools used. Brute strength and teamwork made up for the lack of technology. After a long day of working in the woods in sub-freezing temperatures, the men were exhausted and cold. One contemporary remarked that the "boys" had been "transformed into men like unto Abraham of old"—the black whiskers of youth having been made "white as the driven snow" by the frost.[11]

Logging camps from this primary, primitive period of the logging frontier rarely consisted of more than one or two buildings, erected near the banks of the river. The men lived and ate in a single shanty constructed of rough logs cut on the site. A journalist who visited a logging shanty on Wood Lake in 1855 found "it to be about 25 feet square with a roof running almost to the ground. The gables were built up with logs, and no windows. A door opened into the domicile and was secured with a wooden latch." The bunks were aligned under the low hanging eves of the shanty, with a deacon's seat at the foot of each bed. "This is a seat running on both sides of the fire from one end of the camp to the other." In the center of the camp was a great open hearth, while "on the roof was a chimney, and the smoke receded from the center to the cavity above without the aid of a back wall." Some of these camps were built without the aid of a single nail, from the materials available on the site. There was no illusion of building for the future; after a winter, or at most two, the buildings were abandoned.[12]

The biggest challenge to the early lumbermen was supplying their camps. An advance party would be sent upriver in late autumn, usually via a bateau or some other river craft. They would build the shanty and bring in a store of preliminary supplies. Later, after the snow fell, overland transportation to the camps became possible. Lumbermen maintained a large warehouse in Taylors Falls that would be stocked by steamboat deliveries during the fall and would serve as the starting place for winter supply sleds. Platted in 1851, the village of Taylors Falls was an important supply center for the logging camps because it stood at the head of steam navigation on the St. Croix. The town's merchants had to carve their town out of the trap rock at the foot of the falls of the St. Croix River. Taylors Falls was also a useful supply center because of its location on the Point Douglas–Superior Military Road. Completed in 1858, this road served loggers by providing partial access to the pineries along the Snake River with tote roads blazed by the lumberjacks branching out from it and leading deep into the forest to the camps. Crude shelters for man and beast were set up at several spots along the road to provide a safe overnight site for the supply sleds.[13]

Because of the rough track over which the supply sleds had to traverse, the goods they brought were only the bare necessities. These were purchased at a very dear price. The lack of agricultural development along the river in the 1840s and 1850s meant that most of the food had to come from downriver, often from as far away as Illinois or St. Louis. During his first year on the St. Croix, Franklin Steele had to pay $4 for a barrel of beans, $2 for a gallon of molasses, $11 for a barrel of flour, and a whopping $40 for a single barrel of pork. Those lumbermen who did not send buyers downstream for supplies found it difficult to secure edibles at any price. In 1846 Stephen B. Hanks, a cousin of Abraham Lincoln, purchased supplies for a Stillwater lumberman. In St. Louis he bought several tons of beans, hominy, eggs, and dried apples. On his way upriver he stopped in Bellevue, Illinois, and secured fifty barrels of flour and several more of whiskey. This still left him short of a critical item—pork.[14]

Pork and beans were fuel that powered the pioneer logger. "Pork and beans are all the go," recorded the visitor to an 1850s camp in the valley. The cook would prepare these in a Dutch oven over an open fire in the shanty. "He baked his beans thus wise: a hole made in the earth floor near the fire, was partly filled with live coals and the oven set upon them. In time the lid would be removed and beans would be nicely baked." Stick-to-your-ribs staples such as pork and beans and potatoes, washed down with black tea, dominated the menu at breakfast, lunch, and supper. A crew of fourteen men had no trouble devouring ten bushels of beans and six barrels of pork over a season. One lumberjack later joked that when the men awoke they would "tremble and start from the land of dreams to the land of pork and beans."[15]

The demand for local produce by the logging camps on the Upper St. Croix provided an opportunity for the Ojibwe. With the decline of the fur trade, the St. Croix bands had expanded their involvement in agriculture and looked to hunting and gathering to provide new products with which to procure European American goods. Camp cooks sometimes traded maple sugar, wild rice, cranberries, and venison for salted beef, pork, and flour. Not all encounters in the forest, however, were this positive. In 1864 two Ojibwe shot and killed Oliver Grove and Harry Knight near Pipe Lake, in Polk County, Wisconsin. The incident was a crime of opportunity motivated by a desire to rob the two lumbermen, who had been cruising for timberlands. The bodies of the murdered men were cut into pieces, weighted with rocks, and sunk to the bottom of the lake. After they went missing, as many as three hundred loggers participated in a search of the forest. After several months a rumor circulated among the Ojibwe that soldiers were on their way to investigate and, if necessary, punish the perpetrators. This led to an unofficial confirmation of the identity of the principle perpetrator. Lumberman James Bracklin, supported by his loggers, took it upon himself to seize the suspect. A tense standoff followed in which Bracklin tried to prevent several hundred Ojibwe from retaking the man. Fortunately, there was no further violence. The incident ended when the accused Ojibwe shot himself, "fearing the vengeance of the white man." Several hundred dollars and the personal effects of the victims were recovered.[16]

While whites emphasized the murder and robbery aspects of this case, there were other reasons for tension between the lumbermen and the Ojibwe. The dams built by loggers to ensure the transport of their winter's cut played havoc with the Indians' use of the river. Large amounts of logs sent downstream on a head of water damaged canoes, swept away fish weirs, and made river travel hazardous. In 1851 Indian agent John S. Watrous experienced this firsthand when a dam at the foot of Cross Lake was opened; his canoe was wrecked, and all of his supplies were lost. The 1864 standoff concerning the deaths of Grove and Knight occurred at a time when the Ojibwe were protesting a dam built on Rice Lake that had raised the water level and drowned the all-important wild rice crop. Not only did lumberman James Bracklin refuse to do anything to modify the Rice Lake dam, but he set to work that summer on a second dam at Chetek Lake, knowing full well this would destroy more Ojibwe rice beds. It required considerable restraint on the Ojibwe's part for the incident to conclude with only the death of the one suspect. The potential for conflict with loggers inclined the U.S. government to remove the St. Croix bands from their

Figure 4. Oxen haul a big pine log to the river landing. *Harper's Magazine*, March 1860.

homeland. A half-hearted attempt was made in 1851, but it met with little success and removal was abandoned. Lumbermen and the Ojibwe continued to share the river. Unlike in the Mille Lacs region, where lumbermen provided financial compensation to the Ojibwe for the flooding of their rice marshes, the St. Croix and its tributaries were manipulated to suit the lumbermen with little concern for the interest of the Ojibwe.[17]

Most of the loggers operating in the valley during the mid-nineteenth century had little contact with the Ojibwe. They did not journey into the upper river valley until late November or early December, a time when most of the Ojibwe had returned to their family hunting grounds. Men from Maine, with a sprinkling of Germans, Canadians, and Swedes, dominated the crews. They were mostly young men, in their twenties or early thirties. "Boys they were," recalled one pioneer from that period, "willing to toil at the most strenuous labor if it brought but a reasonable promise of return." During this pioneer era in logging only a fine line separated the crew from the boss. "Most of the early lumbermen," another early settler recalled, "were young men of limited means who came to better their condition. A man with capital enough to buy a couple of yoke of oxen could get credit for supplies and hire a crew of men and cut a million feet of logs or more in a winter." The result of such opportunity was hundreds of small camps operating independent of one another. The small size of the crews often made them close-knit and very efficient. In 1855 the foreman of a

Figure 5. During the heyday of logging in the valley in the 1880s, *Harper's Magazine* saluted the lumberjacks and rivermen with this illustration titled "Wisconsin—On the Lumber Drive in the St. Croix."

crew on the Groundhouse River, a tributary of the Snake River, wrote back to his hometown newspaper in Maine boasting that his men had put up an estimated two to three million board feet in 117 days, a record he challenged any of the eastern crews to match.[18]

Card playing, pipe smoking, and occasionally signing were recreations common in the shanty house after meals. "Some sang songs," recorded a journalist visiting a camp in 1855, "and it is but justice to them to say that we were agreeably disappointed in finding some fine singers there. The songs are principally of a love nature and usually to the better feelings of mankind and sympathy." Nils Haugen, a Norwegian immigrant, recalled the camp in which he worked during the winter of 1866–67 as "primitive," but the crew was "clean, and not a cootie or other bug was discovered all winter." He spent his winter evenings reading from the boss's collection of Sir Walter Scott novels. The linkage of lumberjacks and literature was not entirely exceptional. James Johnston, a Canadian immigrant who spent the winter of 1856–57 logging on a branch of the Snake River, was delighted to find in camp a copy of *Ivanhoe* and a collection of Captain Maryatt novels. Later he "made it a custom to have some book in camp and sometimes at the request of the boys would read aloud while the crew would listen." On one improbable occasion he was reading *Jane Eyre* to the men who sat in rapt attention as the young orphan Jane was humiliated by one of her teachers. One of the men, who regarded Jane Eyre as "one of God's little lambs," shouted out a curse "from the very bottom of his soul" at the insult to the heroine. The rest of the crew then "broke out in cheers and laughter."[19]

Frontier Logging: The Importance of Waterpower

The scale of logging on the St. Croix increased steadily through the 1840s and 1850s. The 8 million feet produced by the valley in 1843 was typical of output during the 1840s. By 1855 production had greatly increased to 160 million feet. Less than ten years later, the amount of logs floated down the St. Croix topped 200 million feet. Logging operations became both larger and more complex. To increase the harvest, double camps, with crews of twenty-five to thirty men, became the rule. To move an ever-increasing amount of logs, more men were required as river drivers and more and better dams were needed to increase the flow. By 1864 more than $600,000 was invested in forest operations along the St. Croix. The number of men employed in the woods swelled to fourteen hundred loggers, more people than had ever before lived in the valley. Little wonder the Ojibwe were forced to yield before the advance of this axe-wielding army.[20]

While hundreds of European Americans flocked to the St. Croix to participate in the logging boom, the bulk of the land in the valley was falling into the hands of a small number of men with access to capital. Typical of these was the partnership of veteran Maine lumbermen Isaac Staples and Samuel F. Hersey. Staples was the resident partner who oversaw operations from their Stillwater, Minnesota, mill site, while Hersey was the out-of-state investor who used his profits from the Maine woods and capital connections in Massachusetts to purchase extensive pine lands along the St. Croix. Between 1853 and 1864, Hersey, Staples and Company purchased forty thousand acres of timberland. They were too experienced in the ways of the industry to make all of these purchases for the government minimum of $1.25 per acre. Rather, three-quarters of their empire was secured much more cheaply through the use of land warrants. These were notes redeemable in public domain land. The federal government offered these to veterans of the Mexican–American War. Of course, most veterans did not want to begin life anew on the frontier. Historians have estimated that only one in five hundred veterans cashed in their warrants for land. More commonly, the warrants were sold at discount, on an average of 75 percent of the value, to real estate speculators. During the early 1850s land sales via warrants outpaced cash sales. By a combination of warrants and cash, Hersey, Staples and Company was able to secure vast tracts of contiguous land. When Knife Lake Township, in Kanabec County, was offered for sale in 1859, the firm was able to secure twenty-one of its thirty-six sections. These block purchases were important to economical logging. Access roads and dams, and sometimes even camps, could be reused season after season because of the firm's long-term involvement in the area.[21]

While Hersey, Staples and Company became the largest single owner of timberland in the valley, there were numerous other eastern-born men who established themselves in the industry. The first sawmill at Stillwater was erected in 1844 by a partnership made up of John McKusick from Maine, Elam Greeley of New Hampshire, and Elias McKean of Pennsylvania. Socrates Nelson, another early mill operator in Stillwater, came to the town from Massachusetts as a merchant but soon joined the lumber rush. Daniel Mears, another Bay State native, followed the same progression from merchant to lumberman, first in St. Croix Falls and later at Hudson, Wisconsin. In 1839 Illinoisan George B. Judd partnered with Walker Orange of Vermont to establish the first mill in the valley at Marine-on-the-St. Croix

Figure 6. The upstream face of a small rafter dam. This style of dam, including the lift gate, was very common within the St. Croix and Namekagon valleys. Ralph Clement Bryant, *Logging* (1914).

(see map 5). William H. C. Folsom, who came from Maine to the St. Croix Valley in 1845, was also a typical frontier lumberman. In less than a year he went from being a hired hand to part owner of a small mill. Typical of the opportunity that existed in the valley, Folsom was able to establish himself in business simply by filing a preemption claim on a waterpower site on the west bank of the river a few miles above Stillwater. Three other partners provided the capital, while Folsom contributed the site and his labor. After a year working to establish the mill, Folsom sold out to his partners for a cash profit.[22]

In the minds of these first lumbermen, waterpower sites were of paramount importance in determining where to locate their mills. Waterpower had historically been the principal force driving America's early industry. U.S. surveyors carrying out the job of locating section lines in the American wilderness were under orders to note all potential waterpower sites. Before the Civil War, sawmills on the St. Croix were largely dependent upon a steady, fast flow of water to transform logs into lumber. For this reason St. Croix Falls was considered the prime location for industry in the entire valley, and it became a bitter bone of legal contention. Another obviously good mill site was Marine, which also became the site of conflicting claims. Unlike St. Croix Falls, however, the partners who established the first mill at Marine quickly dispatched with their rivals. When Orange Walker and his Illinois partners arrived at the site with their milling and logging equipment, they found two men camped on the site, ready to contest that they had staked first claim to the site. Rather than squabble over the squatters' assertion, the Illinois partners paid $300 to establish their clear title to the waterpower site. It was a smart investment, and within a few months they had built dwellings for themselves and their workers and erected the first sawmill on the river. An overshot mill with buckets attached to the wheel was built beside a small stream entering the St. Croix. The water wheel powered a heavy, slow-moving muley saw. It produced no more than five thousand feet of lumber per day, but it was the beginning of a revolution on the river.[23]

The lumber produced by the mills still was a bulky product and, therefore, expensive to move to markets located anywhere but downstream. St. Croix mill owners had their cut assembled into rafts that would then be floated to market towns along the Mississippi River. The rafts were carefully sectioned together through the use of large wooden stakes driven into holes augured into the boards. The holes damaged the wood and lessened its market value, but they securely kept the raft together. Large oars, forty to fifty feet long, at the bow and stern of the raft provided means to steer the makeshift craft. The completed raft might consist of a series of sections, totaling hundreds, sometimes thousands, of feet in length. A steady river current was critical to successfully rafting boards to market. Lake St. Croix on the Lower St. Croix River and Lake Pepin, a twenty-seven-mile section of the Mississippi just downstream from the St. Croix, were the bane of the raftsmen. Broad slack water was prone to heavy winds. When the breeze was in the raft's favor, sails could be put up and the craft could be easily advanced. Head winds could delay a raft for days, with the men helplessly hung up or struggling desperately with lines along the muddy bank, trying to pull the raft to a point where the current resumed. The Mississippi's normal steady flow of a mile or two an hour was ideal for rafting, although fast places where a narrowing of the channel or obstructions in the river bed caused rapids to form could be as detrimental to rafting as slack water. The Upper Rapids on the Mississippi consisted of fourteen miles of fast, rocky water ending at Rock Island, Illinois. The smaller Lower Rapids near Keokuk, Iowa, were less of a challenge but still consisted of twelve miles of dangerous water, very capable of drowning a careless crew and busting up a raft worth thousands of dollars and scattering its boards over hundreds of miles of banks and sloughs. While on smooth water, the rafts were kept moving twenty-four hours a day. A trip from the St. Croix to St. Louis, the largest of the downriver markets, would take about three weeks.[24]

Rafts of logs were much more difficult to control than lumber. The logs were larger, irregular in shape, and harder to secure into a manageable craft. Both rapids and slack water were more difficult to manage with log rafts, yet skilled pilots could bring the logs down to St. Louis. Log rafting expanded the possibilities for milling St. Croix lumber from sites within the valley to virtually any likely location downstream from the pineries. St. Croix logs were regularly rafted to sawmills in Winona, La Crosse, Rock Island, Keokuk, Quincy, as well as St. Louis. The St. Croix Valley's proximity to the unparalleled transportation opportunities offered by the Mississippi River, a virtue shared by the Chippewa River, made these areas extremely attractive to lumbermen during the pioneer phase of logging in the region. Later, in the 1870s, as railroads began to expand in the area and offer an alternative transportation system, access to the Mississippi became somewhat less important. During the era before the Civil War, however, when logging was dominated by the use of waterpower, rafting was the sole means for moving logs and lumber to market.[25]

The reliance of lumbermen on rafting logs and lumber created a strong seasonal labor market for men willing to work on the river. In the early days of the industry, an unlikely relationship grew up between the little Illinois town of Albany and the lumbermen of the St. Croix. Located on the Mississippi River across from Clinton, Iowa, the town of Albany produced many of the best pilots on the upper river. River men from Albany took charge of many of the early raft flotillas sent from the valley. Stephen Hanks, who piloted the very first raft of logs from the St. Croix to St. Louis, was from Albany, as were all the river men in that flotilla. That summer of 1846 Hanks piloted three rafts down to St. Louis, each round trip taking close to thirty days. While the pilots had to be men who knew the river, the crews who manned the sweeps merely needed to be strong and willing to work long hours under the open sky. Scandinavian and Canadian immigrants often took to the rafts when the spring rafting season began. A crew of as many as ten men would be necessary to take a raft south. In rapids at least two men were needed to handle the long oars through the powerful current. Between manning the St. Croix boom and downriver rafts, the lumber traffic at Stillwater alone gave employment to more than twenty-five hundred men in 1860.[26]

The most vital use of waterpower was not in sawing the logs or shipping the lumber to market, but the transportation of logs from the forests of the upper river to the mills and boom on the lower river. The pine forests of the Upper St. Croix would have remained wilderness had the river not been harnessed to drive the winter's cut downstream. Nonetheless, log driving was the most expensive, the most difficult, and the most vexing aspect of logging in the St. Croix Valley. The main river was blessed with a strong, steady current but also with numerous rocky passages that proved to be troublesome chokepoints. Save for the Namekagon, the St. Croix's numerous tributaries were small, winding forest streams with limited flow. Success at moving a winter's cut from the pineries to the mill required a mix of appropriate weather conditions, skillful planning, and exhausting, cold, wet work.

Throughout the winter logging season the wool-clad lumberjacks stacked the pine logs in large piles at a streamside landing. When the ice went out in April, that tributary stream would be used to carry the logs to the main river. Some of these streams were so small that a logger could nearly straddle them with a foot on each bank. The ideal size for a logging stream was just slightly wider than the longest log at the landing. For streams of such size to move thousands of feet of logs, and even more so for those that were smaller, "improvements" were needed. This meant straightening several ox-bow bends and sometimes removing a few boulders. It was expensive, time-consuming work, and the lumbermen always tried to get away with undertaking the most minimal improvements. Their goal was to remove logs from a tract of land perhaps on a single occasion, at most for only a few years. They were not interested in investing in long-term commercial improvements. One expense that could seldom be avoided was the construction of dams to raise the water level of the stream in its narrow banks and increase the rate of flow enough to move the bulky logs. Ideally the dam could be a crude, hastily constructed splash dam that could quickly back up a head of water and then be chopped open to release its flow. Frequently, however, a formal dam with a lift gate that could be opened and closed would be required. The cost of a formal dam could be substantial—from hundreds of dollars during the 1850s to thousands of dollars by the turn of the century. The outlet of a pond or small lake was the ideal site for such a dam, as the lake could be used as a reservoir for the backed-up water. A couple of days of high water would usually be enough to clear a landing of its harvest of logs and send the mass down to the St. Croix or one of its major tributaries such as the Snake or the Kettle River. Where small watercourses had to be driven long distances, it was necessary to build an additional dam halfway downstream. When all the logs reached the second impoundment, that dam would be opened, and the logs would surge on with the crest of the flood.[27]

During the early years of logging in the St. Croix Valley, the value of even the best pineland was greatly influenced by the location and character of the area's watercourses. Hersey, Staples and Company, the Stillwater logging giant, made large purchases in Kanabec County, Minnesota, with the intention of using the Groundhouse River to carry the logs down to the Snake River. Some of the firm's partners were dubious of this plan. "I trust Genl Hersey before he consents to have any more land entered on the G House [Groundhouse River]," wrote Dudley C. Hall, "will be satisfied himself, as to the capacity of that river for driving logs." During the winter of 1855–56, the company set two teams of oxen and about fifteen men to work logging about halfway up the Groundhouse. A dam was built near the camp, and when spring came, the company tried to drive the winter's cut to the Snake and from there down the St. Croix to Stillwater. But things did not go as planned. The head of water from the dam dissipated before the log drivers could get the bulk of the logs down the torturous stream. Precious weeks went by as the drivers struggled to refloat logs left stranded by the drop in water level. Partners like Dudley Hall peppered the company's managers with requests for updates on the disastrous drive on the Groundhouse. The delayed drive, according to Hall, was "a thousand times more important than the mill . . . I trust you will . . . get them in if money can do it." Disgusted with the problems on the Groundhouse, Hall plainly stated, "I for one will never give my consent to cut any more logs on that river."[28]

The only thing that prevented the Groundhouse problems from ruining the entire season for Hersey, Staples and Company was the fact that they had operated camps on other more manageable streams and that harvest gave the mill a modest supply of logs. That year they also operated a camp on the Beaver Brook and another on the Namekagon River. These camps successfully sent their logs down to the St. Croix. Once they reached the main river, however, their logs became mixed with the winter's cut of scores of other lumbermen operating camps on the Sunrise, Kettle, Clam, Tamarack, and Upper St. Croix rivers. This was a problem that lumbermen in the eastern states had faced before, and they transferred their solution to western waters. Every log put into the river was impressed with a distinctive mark hammered into the butt end. To sort out the logs, lumbermen working along the river pooled their resources to fund a common retrieval system. Initially this was a simple association in which each lumberman reported how many logs he put into the river. When the mass of timber reached the lower river, it was assembled into rafts and counted. If a lumberman rafted more logs than he put into the river, as often happened, then he owed the others a debit to be paid in cash or logs. The system relied upon honesty and trust and could not survive the expansion of logging during the 1850s.[29]

The St. Croix Boom Company

The formation of the St. Croix Boom Company, chartered by the Minnesota Territory in January 1851, marked the beginning of a new, more sophisticated approach to the management of a common waterway as a conduit for thousands of individually owned logs. The boom company was given the right to capture all logs passing over the falls of the St. Croix, sort them according to the owner's mark, and then give them back to the rightful owners in return for a fee of forty cents per thousand board feet delivered. Initially men from Marine, Osceola, and Taylors Falls dominated the boom company, so they located the collecting pens near those towns. This site retarded the development of the boom company because it was too far upriver to effectively serve loggers on the Apple River. This stream, which enters the St. Croix south of Marine, drains a large area, reaching deep into the lake country of Polk County, Wisconsin. Loggers were operating along seventy-two miles of improved river, and its output in the late 1840s and 1850s was second among St. Croix tributaries only to the Snake River. An even bigger problem with the original site of the boom was that it was inconvenient to Stillwater, Minnesota, the town that emerged during the 1850s as the valley's lumber center. Stillwater mill owners had to pay twice to receive their logs—once to the boom company for collecting and sorting their logs and then again to the rivermen who organized and floated their logs twenty-one miles downstream to the Stillwater mills. Isaac Staples, a partner in Stillwater's largest mill, was anxious to manage the river to his advantage. His opportunity came in 1856 when the original St. Croix Boom Company went bankrupt. Staples and a group of Stillwater-based partners took over the boom for fifty cents on the dollar and relocated its main operations to a site just outside the limits of their town, at the head of Lake St. Croix. Until its demise in 1914, the boom company controlled the upper river, taking charge of every log, making every lumberman pay its fees, bending the St. Croix to its will.[30]

The inspiration for the St. Croix boom had been the efficient organization of log transportation by the citizens of Oldtown, Maine. Isaac Staples, who had lived in Oldtown, had seen its boom in operation. With an experienced eye he selected a superb location for the new St. Croix boom, a narrow, high-banked stretch of river where the stream was divided into several channels by small islands. The boom itself was made up largely of logs chained end to end, anchored to piles driven into the streambed to form a floating fence. There were a series of these fences that acted as a conduit, leading logs to holding pens. Into these pens went the logs of a particular company. Collected there would be the logs splashed several weeks before into some remote tributary stream in the upper valley, minus those logs lost in back channels or sunk to the bottom of the river. Catwalks built along the boom allowed loggers to easily move from one part of the boom to the next.[31]

The St. Croix boom was the most profitable in the Midwest region. This was partially because the state of Minnesota had written a generous fee into its charter. Just as important, however, was the unique construction of the boom that allowed for the bulk of it to be closed off when the number of logs in the river was low. By opening or closing channels, the boom could be expanded or contracted. This meant that during slack periods the boom could operate with only a skeleton crew, holding down labor costs, but maintaining a continuous service for lumbermen. The true measure of the boom's effectiveness, however, was its ability to handle a high volume of logs. In 1853 the river at the head of the boom constituted a solid packed mass for three or four miles. This was a common site during the 1850s, and one year the owner of a particularly nimble horse offered "to cross the St. Croix River . . . on horseback, driving his horse over upon the floating saw logs that in some places absolutely covered the face of the stream." By the 1870s the mass of logs waiting sorting during mid-summer stretched for fifteen miles. Hundreds of men worked long hours to sort through the mass and send the logs downstream to waiting mills. But with two to three million feet of lumber to sort for some 150 to 200 different lumber companies, the backlogs were inevitable.[32]

The highly profitable St. Croix Boom Company in time became a hated, if powerful, influence on the St. Croix. Lumbermen anxious to start milling their winter's cut fumed over delays at the boom and resented that they had to dig deep into their pockets to pay the boom for sorting their logs. More irate still were the steamboat men who often found the channel above Stillwater completely blocked with logs. Towns like Taylors Falls,

Marine, and Franconia suffered economically as they were shut off from downriver trade. Farmers between Stillwater and Taylors Falls were upset to have a low-cost means of shipping their crops to market endangered by the powerful boom company. Those located directly on the river suffered a further indignity when the mass of logs so blocked the river as to cause the stream to overflow its banks and flood their homes and fields. During the 1860s and 1870s the boom company tried to moderate these problems by constructing a shipping canal on the Wisconsin side of the river that would bypass the bulk of the boom works. At times the company would furnish teams and wagons so that cargoes could be portaged around the logs. It also made available to travelers its small steamboat positioned above the jam. This willingness to work with people and communities impacted by the amount of logs in the river went far to holding down the volume of discontent. In the end, the townspeople and farmers inconvenienced by the boom were forced by the boom's economic importance and Stillwater's political muscle to accept that logs and lumber were crucial to the region's growth.[33]

In 1865 the editor of the *Taylors Falls Reporter* captured the dependence upon the lumber industry that was gradually settling over the towns, both below and above the boom.

> Merchants furnish men who go into the woods to cut the timber, with supplies, and wait the arrival of the logs in market for their pay. Laborers work in the pineries, and eagerly watch the coming of the logs to secure their wages, while their better halves wait until the logs come in, for the minor luxuries, which succeed such occasion. If the water is low business is dull, money is scarce, consequently, Lawyers, Doctors, Editors, Ministers, Office-Holders, &c. have to 'live on the interest of what they owe,' until better times are here.

Equally as interested in the success or failure of the lumber industry were the farmers of the valley. Providing food and fodder for the lumber camps was the critical local market that made pioneer agricultural activities viable within the valley. As long as the boom company expressed a willingness to try to moderate their interference with river commerce, the majority of people within the valley supported transforming the St. Croix into a river of logs.[34]

In later years the St. Croix Boom Company would be referred to as the "Octopus" because of its power over the river. Yet the St. Croix boom had much less power over the river than the boom companies organized by lumbermen in Michigan and Wisconsin. The St. Croix boom only handled logs that came over the falls and had no authority to operate on the upper river. In contrast, the Menominee River Boom Company in Michigan not only sorted all logs to reach the boom but also took charge of driving all logs put into that river from its headwaters to the boom near Lake Michigan. The same was true of the famed Tittabawassee Boom Company and the Muskegon Boom Company. Eventually all log driving on the Chippewa River in northwestern Wisconsin was put under the control of a single company. In contrast, the St. Croix lumbermen remained determined to control the fate of their logs for as long as possible. An attempt in 1872 to form a company to drive all logs on the St. Croix came to naught when the loggers working in the upper valley could not agree on a fair price to pay. Special log-driving companies did successfully operate on the Apple River and the Snake River, but on the upper St. Croix scores of independent loggers resisted the control of a single authority in charge of the river. Frequently, lumber companies operating in proximity to one another might band together on a temporary basis to drive their logs to the boom, but these were just short-term alliances. Log driving remained in the hands of rugged individualists.[35]

Industrial River

The Civil War marked a significant benchmark in the development of the lumber industry in the St. Croix Valley. From 1837 to 1865 a pioneer industry gradually took root in the valley and flourished. During this time the role of the various towns in the valley was determined. Marine and St. Croix Falls, which had been so promising during the 1840s, had been forced to take a secondary position as production centers to Stillwater and other towns on Lake St. Croix. Land that had belonged to the Ojibwe and the Dakota had been acquired by the United States and then hastily transferred to private hands, most of it for the minimum price. Under the Indians the valley had been shared, sometimes quite grudgingly, in common by whole communities; now it had been privatized, with the will of a few industrialists shaping the future of the land and the river. The demand for St. Croix lumber grew during the Civil War, in spite of the massive disturbance of military operations on the life and economy of the lower Mississippi valley. Three major developments, each enhanced by the Union victory in the war, helped to drive the St. Croix lumber industry in the years after 1865: (1) the settlement of the sparsely treed Great Plains; (2) the expansion of the national rail network, which created the conditions for a genuine national lumber market; and (3) the industrialization of American life, which created both the demand and the means to realize greater lumber production.

The expanded reach and inflated ambition of St. Croix lumbermen had a direct and immediate impact on the character of the river and its tributaries. Between 1849 and 1869, for example, the lumbermen greatly increased the amount of water they needed for log transportation. On the Snake River, the river driving company charged with managing the flow of logs expanded the drivable length of the river from 50 miles to 80 miles. The Wood River was expanded from 16 miles of useable stream to 50 miles. The main branch of the St. Croix itself was expanded from a mere 80 miles to well over 100. Just as important were the new tributaries that were damned and channelized to fulfill the needs of loggers. Within a few years of the close of the Civil War, lumbermen were driving logs on 75 to 80 miles of the numerous side streams, lakes, and branches of the Kettle, Yellow, and Namekagon rivers. Simple forest streams such as the Tamarack and the Totogatic were made navigable for logging, the latter utilized for better than 50 miles of twisting streambed reaching through what is today Wisconsin's Burnett, Douglas, Washburn, Sawyer, and Bayfield counties. Dams and stream-clearing teams ensured that no sooner did loggers open a small tributary of the St. Croix for their use than they would begin to employ the tributary's tributaries for the same purpose. The main branch of the Kettle River, for example, was used for more than 85 miles, deep into the Minnesota wilderness, to within less than 25 miles of Lake Superior. Its principal tributaries, the Pine, Willow, and Moose rivers, hardly capable of floating a canoe today, were used to reach even further into the interior.[36]

The experience of the lumberman Elam Greeley on the Clam River in 1875 is illustrative of the manner in which logging was expanded on the St. Croix's numerous tributaries. Greeley's lumberjacks had made a large cut that winter, but by June, when most of the region's harvest had been passed through the boom at Stillwater, his logs were hung up on the Clam River. Greeley ordered his foreman, Andrew McGraw, to put the driving crew to work cutting out a canal eighteen feet wide, twenty-five feet deep, and two hundred yards long between Beaver Lake and the river. An additional eighty-foot-long canal connected Greeley Lake with the river. Controlling dams were put in where the canals reached the lakes. When the dams were opened and the canals were connected to the lakes, the flow of the river was powerfully augmented. On this head of water the lumberjacks were able to drive all of the logs down to the St. Croix River. While the *Minneapolis*

Tribune toasted Greeley as "a most enterprising lumberman," no one recorded what the Ojibwe, who had harvested wild rice from the lakeshores for generations, thought of the manipulation of the water levels.[37]

What made this expansion of the log transportation in the valley possible was the increased number and sophistication of the dams constructed by loggers. By 1889 there were between sixty and seventy logging dams within the St. Croix watershed. Small headwaters dams, such as five located on the upper Snake River costing only between $500 and $2,000, were typical of the majority of the river improvements. Dams located on the St. Croix or its principal tributaries, however, required considerable engineering skill and a formidable capital investment. In 1871 Isaac Staples invested $10,000 to have a dam built on the St. Croix River just downstream from Upper Lake St. Croix. The dam facilitated the transportation of logs from the Moose River, an area highly prized for the superiority of its pine. Logs sluiced through the dam were assessed a fee to allow Staples to recoup his sizeable investment. Sometimes lumbermen would pool their resources to undertake such construction activities. The Namekagon Improvement Company, for example, was capitalized at $25,000 to operate a logging dam on the main branch of the Namekagon River. Working under a charter from the Wisconsin Legislature, the company went on to build seventeen dams on the river.[38]

River improvements represented a formidable portion of the price of doing business. Between 1879 and 1884, lumberman Edwin St. John logged on the Lower Tamarack River, a tributary of the Upper St. Croix in Pine County, Minnesota. In order to bring out a total harvest of close to forty million feet of logs, St. John had invested $3,000 in camp buildings; $3,000 in road building; $16,000 in horses, oxen, and logging equipment; and $5,000 in getting the Lower Tamarack and its tributaries in shape to drive logs. In 1880 the *Burnett County Sentinel* estimated that to build the thirteen biggest dams in the St. Croix watershed, loggers invested a total of $385,614. The most expensive was Big Dam on the Upper St. Croix, a twenty-four-foot-high barrier that cost $94,319. Moreover, dam building was not a one-time expense. These works required annual maintenance and usually needed to be rebuilt every ten years. Therefore, a figure of more than one million dollars would be a conservative estimate of how much money lumbermen invested in St. Croix dams between the Civil War and the end of river driving.[39]

Of necessity, dam building in the St. Croix watershed became more sophisticated because of environmental changes wrought by the first generation of loggers. Smaller dams begot larger dams in part because the volume of logging also accelerated greatly during the 1860s and 1870s. Another factor was the effect of repeated log driving on rivers never intended by nature to carry large volumes of logs and a rapid flow of water. The surge of water flowing downstream from logging dams had the effect of eroding natural riverbanks. The Snake River and its tributaries such as the Anna and Knife rivers, as well as the Upper St. Croix itself, became wider streams after a decade or so of log driving. Yet, while the streams became wider, they also became shallower during the bulk of the year when log driving was not taking place. Logging also accelerated siltation. In 1875, for example, the drive on the Snake River was disrupted near Pokegama by sand blocking the channel. More problematic was the disruption of the natural flow of water downstream by dams closed for long periods to build a head for log driving. The broader shallow rivers, deprived of the protective shade of large pine forests, lost more of their volume to evaporation. The need for an increased investment in dam building was part of the legacy bequeathed by the pioneers to those businessmen who followed them into the pineries.[40]

The Log Drives

Dams made possible the most colorful, dangerous, and difficult phase of logging in the St. Croix Valley—the annual spring log drive. In April of every year the best of the lumberjacks were engaged to escort the winter's cut down the small winding headwaters streams to the main branch of the St. Croix and from there to the head of the boom at Stillwater. It was a job where time was of the essence. The long drive had to be completed before the water level, swelled by melted snow, splash dams, and spring rains, fell, leaving valuable logs stranded in water too shallow to float the fallen monarchs of the forest. It was cold, wet work performed by rugged men clad in two or three red woolen shirts and fitted with caulked boots. The most experienced of the river men were outfitted with long pikes, and they rode the slippery logs in the van of the drive. They were known as *river pigs,* a title in which they took perverse pride, and their job was to keep the logs from snagging on sand bars, sharp river bends, or shoals. At obviously difficult spots on the river several men would be stationed throughout the drive to prevent logjams. These men, together with those who floated majestically on their logs, were known as the *jam crew.* The least experienced men were given the coldest and meanest work on the drive, the *sacking crew.* This entailed following in the wake of the drive and

Figure 7. Wanigans are tied to the bank of the St. Croix. These floating supply wagons provided river drivers with the food and shelter needed to sustain them during their cold work on the river. Wisconsin Historical Society, WHi-9931.

wading into the shallows to wrestle stranded logs back into the current. Several wooded boats, know as *bateaux,* sharply pointed at the bow and stern to ward off floating logs, were part of the drive and could be used to transport men to trouble spots as they developed. Even more important was the *wanigan,* a covered flat-bottomed boat that served as a mobile cook shack. The wanigan provided hot food each morning and evening, although many of the men in the jam crew took their midday meals with them in little backpacks they nicknamed *nose bags.*[41]

The river men had to be exceptionally hardy fellows. In what sounds today like the perfect conditions for triggering hypothermia, they labored in air temperatures of thirty to forty degrees while regularly plunging into snowmelt waters that were even colder. In 1867 Nils Haugen, a young Norwegian immigrant, won a place in the jam crew. He prided himself on his ability to ride a log, but on the second day of the drive received a "good wetting." He remembered the "water was icy cold," although his first thought was about saving face:

Fortunately no one saw it, so I was saved from being guyed. It was always a matter of merriment to see one fall in. I had on three woolen shirts at the time; I took them off and wrung them out, put them on again, and wore them for the next three weeks, never suffered a cold or other inconvenience from the mishap.

How a man coped with the sudden chill of a spill in the river was more important than finding river men who did not fall from their logs. A rookie river driver who had fallen into the Willow River came out of the water cold, badly frightened, and bedraggled. "I had lost my hat and hand spike and must have been a pitiable looking object." He was sent to warm up by a fire, but after his dunking, "I was so scared that I was not much good on the drive." Some men felt that "whiskey helped them to stand the cold water, ice and snow of the early spring," but few foreman allowed their men regular access to strong drink for fear of the consequences to workforce discipline.[42]

At night the drive crews would establish a camp on the riverbank. The evening meals generally featured better fare than camp dinners; fried fresh pork was a favorite. As with camp meals, the men received as much as they wanted. Nils Haugen recalled:

> We slept in tents. The blankets were sewed together so that we were practically under one blanket, the entire crew, the wet and the dry. Steam would rise when the blanket was thrown off.

The workday would begin for the river men about three in the morning. This allowed the lumbermen to take full advantage of the water conditions but exposed the crews to considerable danger working among the rolling, grinding logs in pitch-blackness.[43]

River men were paid substantially more than other forest workers because of the hardships and dangers they endured. Young James Johnston recalled his first day trying to ride logs on the Willow River. A branch hanging low over the stream swept him from his precarious perch, and he fell "head first into the river." Before he could rise to the surface "some half dozen logs ran over me." Gasping for air, he swam for a break in the mass of logs. "When I came up I grabbed the side of a log and, of course, my weight rolled the log toward me and I went down again and a few more logs rolled over me." Only the fact that the current took him to a shallow place in the river saved Johnston's life. An accident recounted in the *Stillwater Lumberman* in 1875 underscores the danger faced by the men working the drives:

Last night Ed Hurley was brought down from the drive on Clam River in a badly mangled condition. A log rolled on him and broken [*sic*] his right leg in several places. Dr. Hoyt, of Hudson, was examining him this morning in consultation with city physicians, and found it necessary to amputate his right leg close to the body, which was done this forenoon. It is thought he cannot survive this day. He is a married man and his folks reside here. He is a first class lumberman and will be sadly missed by the river men.

The prospect of earning as much as $2.50 per day ensured that there was a steady stream of men willing to take their chances with the churning logs and to replace Ed Hurley and the other men who went down on the drive. Even the most skilled log riders fell at some point; most trusted their luck that it would not be where the logs could crush or drown them.[44]

The danger and difficulty of trying to harness the natural power of the St. Croix to move millions of feet of bulky, heavy logs encouraged lumbermen to work cooperatively on the drive. Lumbermen who sought to pursue their success at the expense of others were a menace to the industry. On headwaters streams a logger with a heavy cut could ensure the success of his drive by getting all of his logs into the river ahead of his rivals. The result, however, might be that those rival's logs would be blocked from heading downstream and were in danger of missing the high water and being stranded in the forest, far from the lumber market. To avoid this unpleasant prospect, foremen were tempted to begin their drive at the first sign of the break of the ice. Premature drives forced the men to work harder in lower, colder water conditions, with misery and risk as the reward. Cooperation was much more desirable for the men who worked on the river and for the lumbermen anxious to get their harvest safely to the Stillwater boom. One method of cooperation was for all of the lumbermen working that winter on a certain stream to agree to pay one of the firms to take charge, for a certain fee per log, of all of the cut. In 1877, for example, the lumbermen working on the Knife River all contracted with Charles F. Bean to drive all logs on the river. Bean had operated his own camp that winter and in addition to all of the other logs had 1.7 million feet of pine stacked on his own rollaway. When loggers did operate their drives independently they would establish informal, ad hoc alliances with rival crews they encountered on their downstream journey. In 1875 P. Fox and A. P. Chisley cooperated with each other and jointly drove their logs down the Snake River. Once on the St. Croix they encountered Charles Bean and his crew of rivermen driving 7 million feet of pine from the upper river. It was agreed to combine their crews and proceed down river with fifty drivers managing 12 million feet of pine.[45]

The goal of the rivermen was to bring most of the logs down to the boom by the end of May. Much more skill and cooperation were necessary to bring a large amount of logs downriver in June and July. In 1877 some twenty million logs were brought down the Snake River in late June. It was necessary to boom the logs at the junction of that river and the St. Croix due to low water on the main river. There the logs waited several weeks until heavy rain brought a rise in the water level. As more and more dams were built throughout the river system, lumbermen learned how to extend the log-driving season by coordinating the opening and closing of dams throughout the valley. In July 1875 a drive of eighteen million feet of logs from the Upper St. Croix was hung up near Rush Creek in Chisago County, Minnesota. There simply was not enough water in the river to carry the logs farther. The foreman in charge of the drive arranged to have dams on the Snake, Yellow, Clam, Namekagon, and Upper St. Croix opened in sequence, providing a head of water sufficient to bring in the valuable drive.[46]

The numerous large lakes connected to the tributaries of the St. Croix were critical to logging in the valley. The lakes were natural reservoirs for storing large amounts of water for logging purposes. The logging dam built in 1848 by Elam Greeley at the outlet of Cross Lake was one of the first important dams in the valley. Its ten-foot head of water was critical to logging on the Snake River. Of course, the dam was also a bottleneck through which all logs cut in the Snake valley had to be laboriously sluiced. This operation greatly slowed the process of bringing Snake River logs down to Stillwater. In 1875 thirty-eight million feet of pine were hung up in the lake. The week required to sluice through these logs occupied a full one-third of the time required to complete the entire drive.[47]

A River Jammed with Logs

Every spring the newspapers of towns along the lower St. Croix River and in the various trade journals of the lumber industry all fixated on a single question, "whether or not the logs will come out." Most lumbermen entered the months of May and June heavily mortgaged to pay for the costs of a winter of cutting and hauling logs and preparing their mills for sawing. All chance for profit was hostage to the delivery of the logs by the drive. Any news concerning the status of the drives was eagerly grasped at and widely repeated. Lumbermen fretted when a low snowfall over the winter inhibited hauling to the river bank because it directly impacted

Figure 8. The chaos of the almost annual log jam at the Dalles of the St. Croix. *Outing Magazine*, March 1890.

the amount of logs harvested and indirectly made the task of driving all the more difficult due to less snowmelt and lower water levels. Speculation concerning the status of the drive on a major river like the St. Croix or the Chippewa had a direct effect on the price of lumber in St. Louis and Chicago. Among the most discouraging and dismaying reports that could be received in Stillwater, Hudson, and other mill towns was the news that there was a logjam on the St. Croix.[48]

There were many locations throughout the St. Croix Valley that were notorious among the river drivers as likely spots for a logjam. The Big Falls on the Apple River and the Kettle River Sloughs were regularly anticipated to hang up logs, so men were stationed there to try to keep the pine moving. The most troublesome location in the valley, however, was on the St. Croix itself at its dramatic rock-walled Dalles. Some of the most spectacular and intractable jams in logging history occurred in the St. Croix Dalles located below St. Croix Falls. The Dalles is a very narrow stretch of river where fast water confined by the sheer walls of trap rock creates the conditions for a tremendous bottleneck. Frequently jams resulted at a spot known as Angle Rock—a large promontory that juts out from the Minnesota shore into the channel and forces the river to make a sharp right-angle turn. It was during high-water years, when the river pigs were able to quickly and with relative ease bring down the harvest from the upper river, that jams were most likely to occur. Under such conditions, the river pushed the mass of logs with greater than usual power. Small jams could occur more quickly, and each had the potential to become a massive pile-up as millions of board feet of pine flowed relentlessly toward the blocked channel.[49]

The first great jam on the river occurred at Angle Rock in 1865. It was a high-water year, and the river below the promontory became clogged. The current unceasingly forced sixteen-foot-long pine logs forward. Logs were sent shooting upward, piling high atop the jam. Others were driven below the surface of the churning water, abutting the bottom of the streambed and all but damming the river. Behind the tumult of Angle Rock millions of feet of logs accumulated. The Clam River drive crashed into the Kettle River logs, while the Snake River drive and all other logs on the river were borne inexorably toward the jam, which with each day became larger and larger. The 1865 jam extended from Angle Rock to the falls of the St. Croix, a distance of one and a quarter miles.[50]

The 1865 jam attracted attention throughout the region. Excursion boats were run up the river from Hudson, Prescott, and St. Paul to allow town folk to gawk at the mighty river of logs frozen in suspended animation. Photographers set up their cumbersome box cameras on the shore, and the more adventurous of the tourists clambered out to the middle of the river to have their pictures taken amid the expanse of timber piled pell-mell. Unfortunately for the lumbermen of the Upper Mississippi, the 1865 jam was not a freak occurrence. It was rather the warning of what would become an annual danger as more and more logs were forced into the narrow river and more and more dams, all opened at the same time, forced a greater flow down the stream. During the 1870s the St. Croix Lumbermen's Board of Trade pooled their resources and dispatched crews of lumberjacks to choke points like Angle Rock. These men usually broke up jams before they could blockade the entire river, although in 1877 and 1883 two great jams occurred, closing the St. Croix for several weeks and backing the river up to the falls.[51]

The worst jam occurred in 1886, after the lumbermen had nearly a half-century of experience driving the river. Angle Rock was again the culprit, although in 1886 more than 150 million feet of pine were confined behind the jam. The masses of logs extended all the way to the falls and two miles beyond. The great clog in the lumbermen's main artery sent a panic through the valley. Apoplectic mill owners hurried to the scene and shouted themselves hoarse. A babble of instructions rained down on the rivermen desperately trying to break the jam and echoed through the St. Croix country, repeated by every small-town newspaper. A journalist for the *Stillwater Gazette* described the mess as "the jammedest jam [he] had ever saw." From the humblest mill worker laid off by the dearth of pine to the banker holding lumbermen's past-due notes, the entire valley fixated on the mammoth jam, their lives, like the river, held motionless by the impasse. This threat to the livelihood of most, however, was a boon for the village of Taylors Falls. It had never been much of a mill town, and during the 1880s its commercial significance as a local farm service center was rapidly eclipsed as new villages sprung up in the interior, along railroad lines. Jams like the one in 1886 gave the village a foretaste of what life could be like as a tourist destination. Special railroad excursion trains brought as many as a thousand people per day to view the site. "Give us each year our yearly jam" became an innkeeper's irreverent prayer. The boarding houses of the town were overflowing with overnight guests, and their dining rooms were crowded for three luncheon seatings.[52]

Sharing center stage with the miles of log-clogged river was the sight of hundreds of men laboring to break the jam. Six weeks of work went into the effort. No technology was spared to open the river. The head of the jam was repeatedly dynamited to no avail. Steamboats were brought upriver to try to pull the logs free. Overhead wires were installed to lift logs up out of the chaos. Teams of horses as well as two steam donkey engines supplied the necessary power while large electric lights, the first many country people had ever seen, shone down from the overhanging rock ledges, allowing the men in the gorge to work in twenty-four-hour shifts. All the while the weekend excursionists opined loftily on how the men were missing the mythical "key log," the single strategically located straw in the haystack that would release the jam. Few river pigs believed in the key log myth, and it was doubtful that there were any believers among the weary crews who sought to break the 1886 jam. They cleared the river through the consistent application of ingenuity and determined labor, gradually pulling logs off of the pile at an accelerating rate until the mass once more began to move.[53]

Industrial Logging

Jams such as the 1886 disaster were a reminder that using the St. Croix to transport logs left lumbermen at the mercy of natural forces. The acceptance of natural cycles and adaptation to their fluctuations had been an important part of life in the valley from the days of the Dakota. First fur traders and later farmers bowed to the seasons. The first generation of lumbermen also learned to adjust to the fluctuations of snowfall, water level, and river flow. By the 1880s, however, the patience and resignation that was necessary to work with nature had been replaced by a restless determination to bring an industrial efficiency to the business of logging on the St. Croix. The building of sixty or more dams in the valley was only one example of an application of technology to expand the logging industry's rate of output and margin of profit.

In forest operations, the increased productivity was realized by the adaptation of the crosscut saw. During the mid-1870s the axe, long the symbol of the skilled logger, was replaced as the principal tool for felling trees by a new type of saw. Loggers had used crosscut saws to "buck" or saw in half the felled trees, but such saws were not used to topple great pine trees because the sawdust would be packed by the weight of the trees against the teeth of the saw. During the 1870s a new type of saw blade, six to seven feet long, with "cleaning teeth," was introduced. The teeth of the blade were made of alternating size and shape, some designed to cut, others to rake away the sawdust. The result was a major increase in productivity. J. C. Ryan, a veteran Minnesota logger, contended that two men with a crosscut saw could bring down one hundred white pine per day. Just as important was the fact that a team of sawyers could do so with less waste to the butt end of the log, ensuring that a small but appreciable portion of the log previously lost could now be made into board lumber.[54]

Loggers were able to move a greater number of trees cut from the forest to the stream by the use of another improvement in forest operations, the ice road. Once the pine adjacent to rivers and streams had been cut, roads became essential to expanding into the remoter sections of the forest. Turning the harsh north woods winters to their advantage, loggers covered over the crudely grubbed and leveled rights-of-way with a coat of ice. Sleighs pulled by a four-horse team were capable of handing between ten and fifteen thousand feet of logs. A full-time road crew operating a water tank sleigh, for laying down a new coat of ice, and a rut-cutting sleigh, for grooming the roads, worked constantly to keep the ice roads in top condition. During the 1870s heavy draft horses, most of them Percherons weighing more than a thousand pounds, replaced oxen as the principal draft animals in the camps. After the fall harvest, farmers from throughout the valley and from as far away as Illinois would ship their horses by barge and rail to the St. Croix to be leased by the logging camps.[55]

Logging camps became larger and the structures more specialized. In addition to the men actually engaged in cutting logs and crews maintaining the ice roads, there were full-time filers, who kept the crosscut saws sharp, teamsters, and blacksmiths. In February 1877 the *St. Croix Lumberman* reported several logging camps containing as many as three hundred men. Camps such as these were a far cry from the simple one-shanty operations of the 1850s. St. Croix logging camps became small settlements with numerous special-purpose structures. One or two bunkhouses, depending on the size of the crew, were the center of the camp. A separate cookhouse was the domain of the cook and his chore-boys. All meals were prepared and consumed there. In keeping with the emphasis on efficiency, many of the larger camps observed a rule of no talking at the dinner table. This lessened the opportunity for brawls and got the first shift of diners out of the cook shack more quickly. An office became a feature of the big camps. It was the headquarters for the foreman and the log scaler. This executive staff was sometimes expanded to include a camp clerk to take charge of all record keeping. Sometimes the clerk also operated a small store in the office, selling tobacco, stamps, and clothing. One or two barns were necessary for the horses. These would be equipped with stalls and a storage area for hay and oats. Occasionally the barn might have a small bunkhouse attached to it for the special use of the teamsters or road crew. A blacksmith shop, maybe a filers shack, and the latrines would round out the complement of camp buildings.[56]

As the scale of logging increased, so too did the emphasis upon efficiency. As efficiency became a virtue of management, logging was increasingly mechanized. During the late 1880s and into the early twentieth century, some logging operations augmented their use of horses on the ice roads with specially designed steam haulers. Essentially they were traction engines equipped with treads and used to haul a virtual train of sleighs over ice roads, from the cuttings to the river landing. Many of the early steam haulers were manufactured in Eau Claire, Wisconsin, although it has been speculated that the first ones were merely converted threshing machines. Around the same time, steam power was also harnessed to operate jammers whose job was to load or unload logs from sleighs to riverbank or railway car.[57]

Lumberman William Hanson employed one of the first steam haulers used in the valley. In 1888 he brought one to his camp near Lake Namekagon. A lumberjack named "Wild Bill" Metcalf was one of the men trained to operate the hauler. Metcalf was more than willing to take on the new technology because the responsibility carried with it a salary of twenty-five cents an hour, a big improvement over the dollar a day he had been making while driving horse teams on the ice roads. The hauler would handle an average of eight sleighs per run, and because the iron horse never became tired it was on the go during the peak period twenty-four hours a day. Metcalf counted his pay and tried to keep himself awake during the long days, taking naps while the hauler was being loaded or unloaded. As the season wore on, however, he found it harder to keep up with the machine's relentless schedule, particularly in the spring when the afternoon sun made him warm and drowsy. "I'd go to sleep steering the hauler," recalled Metcalf. "The engineer'd toot his whistle and I'd jump a foot."[58]

In numerous and often inventive ways, lumbermen experimented with the "high tech" possibilities of mechanized transportation. Even river transportation was improved by the application of steam power. The large lakes of the valley had always been a challenge to river drivers. Small steamboats became a common sight on headwater lakes. The *Katie R*'s shrill whistle was first heard on Cross Lake during the 1880s, and the boat became a fixture of the Snake River drive through the end of the century. During the 1890s lumbermen used a small steamboat on the Upper St. Croix, speeding the movement of logs as far upriver as Upper Lake St. Croix, more than 125 miles above the previous head of navigation on the river at St. Croix Falls. As early as 1876, Martin Mower, the iron-fisted power behind the St. Croix Boom Company, tried to develop a steam-powered ice boat that he hoped to operate on the frozen river. Mower eventually built and operated a prototype, but its lack of success ensured no further experiments along those lines.[59]

The railroad was the most obvious application of steam technology to impact the St. Croix. In many parts of Michigan and Wisconsin, railroads replaced rivers as the principle means of transporting logs to mill and market. This did not happen within the St. Croix Valley. A few logging railroads were established in the valley. The Wisconsin Lumber & Manufacturing Company operated a network out of Cable, Wisconsin, while the Drummond & Southwestern operated an extensive network of standard-gauge track along the headwaters of the Namekagon River.[60] Still, for all the problems with dams and jams, the river remained the cheapest means of moving the great sixteen-foot white pine logs that were stacked at forest landings throughout the logging season. Yet, the iron horse was integrated into the forest operations of the St. Croix logging district in a variety of ways that fulfilled the lumbermen's growing passion for efficiency.

The first railroads completed in the St. Croix Valley were the Lake Superior & Mississippi Railroad (completed 1871), which linked Stillwater with

Duluth, and the North Wisconsin Railway (completed 1883), which ran from Hudson to the Lake Superior town of Bayfield. The right-of-way of each line was laid out well away from the river, creating rival transportation corridors and a string of new towns. The villages of Pine City and Hinckley soon boasted new steam-powered sawmills, although their production did not come close to rivaling the volume produced by the Stillwater mills. As a class, nineteenth-century lumbermen were not particularly distinguished for their foresight, so it is not surprising to find that the majority of them opposed the expansion of railroads into the remote region of the upper valley. This was particularly true for men who owned extensive tracts of Wisconsin forestland and who feared the cost of county bond initiatives to encourage railroad construction. On the other hand, lumbermen were pragmatists and began to make use of the North Wisconsin Railway long before it was ever completed. During the long construction phase of the line, lumbermen provided cordwood, poles, lock-downs, and thousands of railroad ties. Even with the line only partially completed, it gave the lumbermen an inexpensive and timely means to supply their logging camps operating on remote headwaters and the upper Namekagon River. The Northern Pacific Railroad, whose main terminus was at Duluth, ran a spur line to Grantsburg, Wisconsin.[61]

While railroads did not supplant the river for log transportation, the train did augment the network of natural and improved waterways. The railroad could be pressed into service when there were problems with driving logs by water. In 1877 low water made driving on the Willow River problematic for large logs. The solution was to pull 350,000 board feet of the best logs out of the river where it passed under the tracks of the North Wisconsin Railway and ship them by rail to the mill. In a similar fashion Snake River loggers occasionally made use of the railroad during difficult driving conditions. Lumbermen also used the railroad to provide a steel extension from the farthest reach of the waters to the untapped reserves of white pine in the interior. During the 1890s the Empire Lumber Company operated in Douglas County, Wisconsin, beyond the headwaters of the St. Croix. They built a short-haul railroad between their camps and the river. Logs would be loaded onto this railroad and hauled to the river, where they would be dumped into the stream and driven down to Stillwater. At the boom, the logs would be formed into rafts and then floated downriver to Winona, Minnesota, where the company built a new steam-powered mill. This movement from rails to river by the Empire Lumber Company reflected the economic advantage of moving logs by water. The cost of shipping logs by rail was 60 percent greater than conducting a river drive.[62]

Yet, if railroads could be used to bring logs to the river, they could also be used to take them away from the St. Croix Valley. As the value of St. Croix pine increased and mills in other parts of Wisconsin and Minnesota began to suffer from a shortage of logs, the economics of the lumber industry began to dictate new applications for the rail network. The opening of the Lake Superior & Mississippi Railroad greatly hastened the destruction of the vast forest of Pine County, Minnesota. Only two years after the line was completed, twenty-one million board feet were shipped out of the valley to St. Paul. By the late 1880s several railroads bisected the St. Croix, including the Minneapolis, St. Paul & Sault Ste. Marie and the Chicago, St. Paul, Minneapolis & Omaha railroads. These and other lines created numerous small sawmill centers that could tap the flow of pine that previously had flowed uninterrupted to Stillwater. Forest product firms were established at Turtle Lake, Comstock, Shell Lake, and Hayward. Railroads spread the impact of the St. Croix pinery outward, like ripples on calm water. By 1895 those ripples had reached as far east as the northern Wisconsin town of Rhinelander, where St. Croix logs kept the local mill humming through the use of the Minneapolis, St. Paul & Sault Ste. Marie Railroad. Diversions such as this became increasingly common and eroded the link between the logging camps of the upper river and the mill towns downstream.[63]

By the end of the nineteenth century, the towns along the lower St. Croix, from Prescott upriver to Taylors Falls, had made a substantial investment in large, modern, steam-powered sawmills. The old reliance upon waterpower ended in the 1860s, a casuality of the need for a more powerful and consistent energy source. With this change came a substantial increase in the scale of investment needed to compete in the lumber industry. In 1825 James Purinton built a sawmill and waterpower dam at Hudson, Wisconsin, for $25,000. By the 1890s the cost of a new mill might exceed $300,000. For that price tag the lumberman bought a sophisticated industrial complex including mammoth steam engines and boilers; a network of specialized mill buildings, including facilities for making dimension lumber, shingles, and posts; kilns for drying lumber; and extensive yards for storing the finished product. At the heart of such a complex was a giant band saw—the last word in fast and efficient lumber production after 1880. Scores of mills such as these, from Stillwater, Winona, and Red Wing all the way down the Mississippi to Muscatine and St. Louis, were dependent on St. Croix logs. Giant mills also arose on the banks of the St. Croix. Between 1860 and 1900, sixteen steam-powered mills were built on the St. Croix River. In addition to these large mills were scores of smaller ones built in interior sections of the county, some merely to provide lumber for settlers, others positioned to use the railroad to ship their product. Between the Civil War and the beginning of the twentieth century, at least fifty-five such mills were established.[64]

Corporate Control of the St. Croix

Of all the lumber barons of the St. Croix, the most powerful was the barrel-chested, bald-headed Isaac Staples. A contemporary described him as "restless, alert, far-seeking, systematic, and persistent." No single man controlled more of the vast pine forest than he did. No single man logged as much timber each year. No single man milled as much lumber. He had created the powerful St. Croix Boom Company and controlled its vital operations through his ownership of its stock. He was the valley's biggest and most successful farmer, and as president of the Lumberman's National Bank, he was one of its most important bankers. Minnesota politicians anxious for advancement courted Staples's favor while throughout the valley thousands of breadwinners traced their paychecks to his enterprises. He arrived as a stranger in 1853 and remained to flourish as the presiding patriarch of St. Croix logging for the next generation.[65]

Isaac Staples came to the St. Croix as the representative of eastern investors. It was their financial resources that allowed him to build a firm foundation for his logging venture. His share of the profits, however, was substantial enough that over time he was able to purchase a controlling interest in Hersey, Staples and Company, in partnership with Maine businessman Samuel F. Hersey. With Hersey in Maine, Staples had a free hand in directing the firm's business on the St. Croix. Nonetheless, he eventually cut his ties with the Hersey family and by the 1880s operated his extensive business interests with complete independence. Of all the young Maine men who came to the valley in the wake of the 1837 Indian treaty, Isaac Staples had scrambled to the top of the sawdust mountain. This success and his overarching influence made him as resented as he was respected.

Yet, for all of his success, by 1885 Isaac Staples, perched in his Victorian mansion high on a bluff with the valley spread out before his feet, represented the past, not the future, of the St. Croix lumber industry. The sixty-nine-year-old magnet perceived issues from the older, frontier-era perspective of individual ownership and influence. The patriarch did not appreciate that since steam power had welded the entire nation into a single market, the real opportunities in business would in the future go to those who could combine small local producers into efficient and rational combinations. Most of his rivals made the same mistake. During the 1880s the parlors, restaurants, and exchanges where Stillwater lumbermen met were abuzz with speculation that Isaac Staples was vulnerable. In looking so intently for the patriarch's weak spots, the Stillwater men did not see their own vulnerability. When Staples allowed the irascible and eccentric Martin Mower to purchase a controlling interest in the St. Croix Boom Company, it seemed to confirm that the patriarch was losing his grip. In 1885 the Stillwater lumbermen fought a prolonged duel over who would hold the lucrative post of surveyor general in the Stillwater District. The surveyor scaled all logs entering the St. Croix boom. Although it was a state position, the surveyor was paid by the lumber companies a hefty price for all logs scaled. The Lumberman's Board of Trade favored Judson McKusick, a member of a clan of mill operators and politicians involved in the city since its founding. Isaac Staples sought to demonstrate his suzerainty over his rivals by forcing the appointment of Adolphus Hospes, his son-in-law. When both men tried to carry out the office, the shipment of lumber downriver was disrupted by sheriff's actions, court orders, and lawsuits. The dispute ended with Staples imposing his will upon his rivals and forcing them to capitulate.[66]

While the Stillwater lumbermen wrangled among themselves, a new force began to assert itself on the St. Croix—the Mississippi River Logging Company. Mill owners along the Mississippi between St. Louis and Winona, Minnesota, formed the company in 1870 to guarantee a steady flow of logs from the forests of Wisconsin to their large steam-powered mills on the great river. During the 1870s they had fought a bitter duel with lumbermen based in Eau Claire, Wisconsin, for control over the Chippewa River. The Eau Claire men resented that the Mississippi River Logging Company intended to take the lion's share of Chippewa River pine to out-of-state mills. Wisconsin timber, they reasoned, should be used to provide Wisconsin jobs and build Wisconsin communities—and enrich Wisconsin businessmen. But after a decade of cold war on the river and all-out opposition in the courts, the Eau Claire lumbermen finally agreed to accept the Mississippi River Logging Company. The man behind this successful campaign was a German immigrant of true business genius, Frederick Weyerhaeuser. Not only did he harmonize the diverse interests of the many Mississippi mill owners, but Weyerhaeuser was also able to convince the Eau Claire lumber companies of the advantages of bringing the entire Chippewa River valley under the control of one rational, systematic management. Under Weyerhaeuser's guidance, Chippewa River logging was done in the usual independent manner. However, once logs were put into the river, they became the charge of a jointly owned company that managed all work on the river. Mill owners were delivered a guaranteed percentage of the drive. Conflict was minimized, and everybody made a healthy profit. As the Chippewa valley was quickly transformed into a cutover wasteland, however, the partners of Weyerhaeuser's emerging trust began to look to the St. Croix as their principal source of supply.[67]

One of the first steps taken by the Weyerhaeuser partners in the St. Croix Valley involved the founding of the town of Hayward, Wisconsin. It was merely a railroad siding on the recently completed North Wisconsin Railroad until the flamboyant lumberman Anthony J. Hayward convinced the Weyerhaeuser associates Laird, Norton and Company to help him form the North Wisconsin Lumber Company. Hayward's grand style and constant interest in new ventures led to his being dubbed "the Grand Duke" by his associates. He was supposed to provide the practical experience necessary to get the operation up and going, but he was too often absent from Hayward or inattentive to important details. For example, in 1883 he was away in Madison, Wisconsin, for several weeks. During that time he successfully lobbied the legislature to create Sawyer County out of Ashland County, but he failed to secure such necessities as an adequate cook and oxen for his logging camps. Hayward also secured the legislature's permission to build a dam across the Namekagon River, but then after selecting an appropriate site for the works, he allowed it to be carelessly constructed with brush and earth. The dam, which would later fail spectacularly in 1907, created Lake Hayward to serve as a millpond for a water-powered sawmill constructed at the site in 1882. It took years of constant effort to make the North Wisconsin venture successful. In 1885 Frederick Weyerhaeuser bought out "Grand Duke" Hayward's share of the venture, although it was not until 1891, ten years after they started, that the partners began to see a return on their investment.[68]

Weyerhaeuser and his associates became involved in St. Croix logging slowly and discreetly. In November 1883 Weyerhaeuser purchased from Isaac Staples ten thousand acres of forestlands on the Moose River, a branch of the Kettle River. The sale netted Staples the princely sum of $262,819. This transaction did not arouse great curiosity in part because it was made not by Weyerhaeuser's timber trust but by an ambitious young Stillwater logger, William Sauntry. Staples did not understand that Sauntry was Weyerhaeuser's agent in Stillwater.[69]

The Moose River deal proved to be very profitable for all of the parties concerned. Sauntry eventually delivered 9.8 million board feet of logs to the downriver trust. In 1886 Weyerhaeuser took another large step into the St. Croix Valley when he engineered the formation of the Musser-Sauntry Land, Logging and Manufacturing Company. Capitalized at one million dollars, the company began to make pineland purchases throughout the St. Croix Valley. Once again, the Mississippi mill men supplied the bulk of the money, and Sauntry took charge of forest operations. Between 1888 and 1907, the company produced nearly eight million board feet of logs. Through these deals and numerous smaller ones he participated in with Weyerhaeuser, Sauntry had advanced to the front ranks of lumbermen in Stillwater. The opportunity for the biggest coup of all, however, came in the fall of 1887. Sauntry informed Weyerhaeuser that Isaac Staples was prepared to exit the field. If the right offer were made, he would liquidate his 50,400 acres of pineland, his river improvements, and his shares in the St. Croix Boom Company. Here was an opportunity to control the future of logging on the river, and the downriver mill owners did not hesitate. With Laird, Norton and Company, one of Weyerhaeuser's staunchest associates, taking the lead, Staples was bought out for the sum of $650,000.[70]

The Weyerhaeuser syndicate made its move into the St. Croix Valley at a time of growing anxiety within the lumber industry concerning the future availability of timber. The old myth that the Lake States forest was inexhaustible had been belied by the speed with which trees fell during the decades after the Civil War. During the 1880s a scramble ensued within the industry. Prudent lumbermen realized that only by securing large tracts of forestland could they guarantee a future supply of logs. The speed and scale with which Frederick Weyerhaeuser and his associates moved into the St. Croix Valley, securing a half-dozen major tracts, surprised and dismayed their rivals in Minneapolis and Stillwater. Minnesota newspapers compared their arrival to a sudden "scourge of locusts."

Yet when the Weyerhaeuser men inspected their St. Croix lands, they found that Isaac Staples had sold them tracts "poorly located, inferior, and picked over." Instead of finding tracts of fine, straight white pine, much of the land contained tamarack, balsam, and hardwood. The best of the pine was not white pine but its inferior cousin, red pine, or as it was known in Wisconsin, Norway pine. White pine was present on the lands but not in dense stands. With the lower value of the timber on their lands, the mill owners needed to be very efficient in their logging operations to turn a profit. Great jams such as the one in 1886 were an anathema to men with thousands of dollars invested in giant steam-powered mills. They watched the spectacular jam of that year with a disbelief that turned to disgust when they heard the Stillwater lumbermen ruefully predict that a similar jam would likely occur again. Having secured control over a vast portion of the valley's forest resources, the Weyerhaeuser syndicate turned their attention to seizing control over the river.[71]

Although it was a bigger and more complex river system, Weyerhaeuser wanted the St. Croix to be run with the efficiency and regularity of the Chippewa River. On that river Weyerhaeuser controlled the driving company that manipulated all 149 dams in the valley and took responsibility for the delivery of all logs put into the water. To prevent any disruption of the flow of logs during dry seasons, Weyerhaeuser had a massive dam, the biggest logging dam in Wisconsin, built on the Chippewa River at Little Falls. The structure could send a fifteen-foot head of water into the river, enough water to refloat logs one hundred miles downstream. The dam cost the lumbermen $147,457, but Weyerhaeuser believed it was better to make the upfront investments necessary to ensure smooth operations than suffer uncertainty and delay later. The problem with trying to bring a new emphasis on efficiency to the St. Croix was that the industry had been too long under the control of the same men. Weyerhaeuser had already shown Isaac Staples the door, but to improve log driving on the St. Croix he had to get around Martin Mower.[72]

Even his friends and associates had to admit that Martin Mower was "somewhat eccentric." Irascible, grouchy, and litigious were words that most outsiders applied to the veteran lumberman. Mower had come to the St. Croix with his brother back in 1842, and within five years he was the proprietor of a string of logging camps. For more than thirty years he had been intimately involved in the management of the St. Croix Boom Company, and after 1879 he was the controlling owner. In personal appearance he was a disheveled bachelor, "sot" in his ways. But woe to the rival who failed to appreciate that he was a "capable and shrewd" businessman. Mower ran the St. Croix Boom Company, not to facilitate logging or milling endeavors but with an eye to his personal prosperity. Since he controlled the funnel through which all logs cut in the valley had to pass, Mower sat atop a money-making machine. Under his management the St. Croix boom, unlike most boom companies in Minnesota and Wisconsin during the 1870s, did not increase its investments in river improvements. While great jams regularly delayed for weeks the shipment of logs, Mower awarded generous dividends, often over 30 percent, to the boom's stockholders, of which he was the largest. For several years Frederick Weyerhaeuser had tried to prod the gray-bearded old man of the river to improve and expand the boom company's operations. Throughout the 1880s the Weyerhaeuser syndicate tried to purchase Mower's stock, but he would not part with his guaranteed source of wealth. There is every reason to believe that he would never have sold to the downriver mill owners. Yet once more Weyerhaeuser's stealthy approach won the day. In 1889 Mower was approached by a group of young lumbermen, headed by William Sauntry, to "lease," not sell, control of the boom company to them for twenty years. In return for a generous annual fee, Mower yielded to the younger generation, probably not knowing that in doing so he was giving way to Weyerhaeuser.[73]

Immediately upon securing Mower's exit, the Weyerhaeuser syndicate met and authorized extensive river improvements on the St. Croix to prevent future jams. The plan was to build a large dam that could control the rate of flow in the river and regulate the passage of logs over the falls and through the narrow confines of the Dalles. Weyerhaeuser, however, was not the only one to make such a plan. Isaac Staples emerged from the shadows with a challenge. Although it had appeared to all that Staples had retired from business when he sold off his forestlands to the syndicate, the patriarch of the St. Croix had one last bolt to shoot. In 1887 Staples sank $50,000 into the purchase of property and water rights at St. Croix Falls. It was the site of high expectations and grandiose dreams since the days of Caleb Cushing's failed investments. The hardheaded old lumberman had fallen under the spell of those dreams, but typically he dreamed with his eyes open. Staples planned to build a toll dam just above the falls. He would prevent logjams, but more importantly he would be in a position to collect substantial fees from every logger on the river. Even in his rocking chair, Isaac Staples sought recognition that he was the "boss logger" on the river. It was a likely plan save for the one element in the equation on which Staples had failed to reckon: Weyerhaeuser. In the German immigrant turned "Robber Baron," Staples finally met someone who could outmuscle and out-money him on the St. Croix. Weyerhaeuser insisted that if there was going to be a dam on the river, he would control it, and he used his established ties with the Wisconsin State Legislature to secure a charter for his dam; he used that charter as leverage to prevail upon the Minnesota Legislature for a similar charter. Then, with construction crews already at work on his dam, Weyerhaeuser beat back Staples's last-ditch court challenges. Had Weyerhaeuser sought public notoriety he could have used the struggle to prove that he was now the "boss logger" on the river; instead he let William Sauntry take the credit while he merely drew contentment for being able to impose a more rational, harmonious, and efficient order on the river.[74]

The new dam was a milestone in the lumber industry's campaign to transform the St. Croix. It was built eleven miles upriver from St. Croix Falls near Wolf Creek at the farm of homesteader Charlie Nevers. Joseph Renshaw Brown, who operated a trading post near the site sixty years before, would have truly been shocked to see the massive works of the dam rise up twenty feet from the bed of the river. It stretched 614 feet across the St. Croix and backed up the river for ten to fifteen miles. Although it was much larger than any other dam in the valley, it was constructed in a manner similar to other logging dams. It was secured by wood pilings and was controlled by a bear-trap gate.[75] It was widely regarded as the largest wood-piling dam in the world, and the eighty-foot gate was the largest in the world. The Nevers Dam cost close to $250,000 after the construction fees, legal expenses, and riparian rights were totaled, making it the most expensive logging dam ever built in the North Woods.[76] More importantly, the broad waters impounded by the dam became a vast holding pen for all logs cut in the valley. When the men at the St. Croix boom were ready for more logs, word would be sent up to Nevers Dam, and the lumberjacks there would open up several of the fifteen sluicing gates and pass the appropriate number of logs into the river. Similarly, if the men on the boom experienced low water, the great bear-trap gate could be lowered, and within hours the level would rise on the lower river. In May 1890 the *St. Croix Valley Standard* boasted about the impact of the new dam:

> The benefits of the Nevers Dam are already being felt. . . . Only one day has the Boom (here at Stillwater) shut down this season so far, and today all the gaps will be at work again. . . . The success of the mills this year is phenomenal, not a break to detain them . . . the cut will be large, several of the mills showing much better record at this date than for several years past.

Upon command, the river drivers stationed at Nevers could send as many as four million feet of logs down to the boom each hour. Electric lights installed at the dam provided the option of operating the sluiceway twenty-four hours a day. With the completion of Nevers Dam, the lumbermen took a large step away from being bound by the irregular patterns of nature and toward the industrial management of the St. Croix.[77]

When the great bear-trap gate of Nevers Dam was closed, it was possible for lumbermen to reduce the flow of water in the St. Croix River to a mere trickle. Although this was rarely done, the lumbermen's new power over the river reignited their lingering feud with the steamboat men on the river. The feeling that the lumbermen had become overweening in their

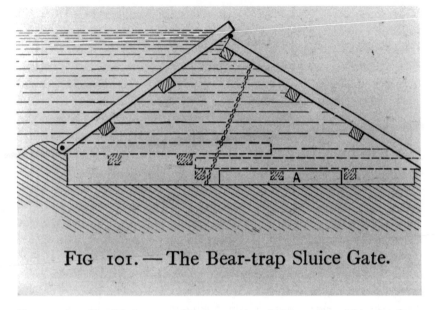

FIG 101. — The Bear-trap Sluice Gate.

Figure 9. A profile of the bear-trap sluice gate used on the Nevers Dam. This type of gate was developed in Pennsylvania, although Nevers Dam was the largest dam structure for which it was ever used. Ralph Clement Bryant, *Logging* (1914).

power had grown during the 1880s among residents of the villages upstream from the boom. Martin Mower's crabbed management of the boom company had made a bad situation worse. Unlike his predecessors he did nothing to try to ameliorate the inconvenience caused to river traffic by the mass of logs at the mouth of the boom. In April 1891 the steamboat men won an important victory when the Wisconsin Legislature amended the Nevers Dam charter to provide some protection for steamboats on the river. Twenty-four-hour notice was required before the dam could completely shut its gates. Steamboat men tried and failed to force the lumbermen to ensure a three-foot navigation channel in the river as far as St. Croix Falls. An attempt at more aggressive action to curb the loggers' control of the St. Croix began in 1900 when steamboat men and their supporters in the villages above Stillwater tried to involve the U.S. Army Corps of Engineers. Major Frederic Abbott of the corps proposed restricting the times when either the boom or the dam might be able to limit other river users. The lumbermen, however, were once again a match for their opponents. They were able to document that thousands of mill and boom workers in Stillwater and numerous other towns throughout the Mississippi valley would lose their jobs if the flow of timber were interrupted. The attempted revolt against the lumbermen eventually led to the creation, in 1909, of the St. Croix River Improvement Association, an organization that would foster public concern for the well-being of the river. But at least until 1912 the needs of lumber continued to dominate the St. Croix.[78]

The clash over navigation on the St. Croix indicated that while logging remained very important to the St. Croix Valley, it was by no means the only economic interest vying for control over resources. By the beginning of the twentieth century, farmers, steamboat men, a growing tourist industry, and sportsmen all began to challenge the lumber industry's hegemony over the valley. In 1886 an event occurred that would have been unthinkable a generation earlier. When Laird, Norton and Company tried to drive their logs on the Clam River, they were forced by legal action to suspend the drive. Local farmers claimed their lands were inundated by logging dams and demanded compensation. Although there was little farming activity apparent from the river, nearly twenty-five settlers claimed damages. With the high-water level rapidly passing, the company had little choice but to satisfy the claims as quick as possible or face the prospect of having their winter's cut be hung up until next year. The "settlers" proved to be tough negotiators, with one securing a $900 payment—much more than his land was even worth. Land speculators bedeviled loggers further by buying the rights to likely dam sites on tributary streams and then holding the land until their price was met.[79]

For lumbermen who chose not to negotiate with their litigious neighbors, the cost could be high. In 1903 the Chengwatana Dam, at the outlet of Cross Lake, was mysteriously dynamited in the middle of the night. The dam was the most important one on the Snake River. It was built in 1848 when the area was wilderness. But by the time the dam needed reconstruction in the 1870s, the lumbermen had to share the area with farmers who complained that the annual log drives flooded their fields. Little was done to address these complaints in the years leading up to the explosion. The explosion hit the operations of lumberman James McGrath hard. He had forty-five million feet of logs ready to go down the Snake, only to be suddenly left without the water to do the job. His costly solution was to take the logs by river as far as Pine City, pull them out of the water, and ship them by rail to Stillwater. There the logs were dumped into the St. Croix and rafted down the Mississippi to the waiting sawmills.[80]

The Failure of Government Regulation
of the St. Croix Pinery

While disputes such as these became more common as the population of the North Country became denser, it is nonetheless striking how little legal and legislative action inhibited the activities of lumbermen. Legal issues were almost always restricted to local courts, even on a border river like the St. Croix. Legislative regulation by the states of Minnesota and Wisconsin was largely restricted to the granting of charters for dams or the passage of laws designed to facilitate the lumber business. The regulation of log marks and the establishment of inspection districts for the scaling of logs were both state regulations that provided a structure conducive to the expansion of the lumber industry. There was remarkably little interest on the part of either Minnesota or Wisconsin in conserving the forests of the St. Croix Valley. The states used neither taxation policy nor their investigative authority to try to promote greater efficiency in forest operations. Instead, tax policy actually promoted the "cut and get out" tactics of the loggers by presenting hefty tax bills to the owners of uncut forestlands. Two prevailing cultural assumptions inhibited a more creative role for government in forest management. The overwhelming majority of Americans believed that the appropriate disposition of land was private ownership and that the highest use of that land was for farming. Most residents of the Upper Midwest believed that the lumber boom was only a temporary phase of settlement that would in time yield to new and greater opportunities.[81]

Aside from the occasional involvement of the U.S. Army Corps of Engineers in navigation issues, the federal government was even less of a factor in regulating St. Croix logging. As the largest landowner in the valley, the U.S. government proved a poor steward of some of the finest forestlands in the nation. For at least a generation, from the 1837 Sioux Treaty to the Civil War, pine timber on federal lands was there for the taking. Large-scale operators like Isaac Staples purchased land to secure their long-term futures, but men without access to eastern capital had the option of stealing public timber as a means of establishing themselves in the industry. In hushed tones over steins of lager beer, loggers debated how extensive was the larceny, but no one doubted its existence. Finally, in 1855 Thomas A. Hendricks, the U.S. commissioner of public lands, issued a circular directing public land offices to collect the value of all depredations brought to their attention. Typical of federal efforts of that era, Hendricks provided no means to investigate thefts. The General Land Office often relied upon the trespassers themselves to report how much timber was taken. Typical was the July 1859 case of two Hudson, Wisconsin, men arrested for timber stealing. The only reason they were brought to trial was that they had fallen out with a third accomplice who "out of pure maliciousness" turned informant on them. The Wisconsin Legislature protested such prosecutions on the grounds that theirs was a frontier state and such logging should be allowed "in view of the necessities of the people." In twenty years only $150,000 was collected in all of the lumber districts in the United States. Federal legal expenses required to collect that money topped $50,000.[82]

A more determined effort to deter timber thefts was launched by the General Land Office in 1877 at the behest of Carl Schurz, the independent reformer who served as Rutherford B. Hayes's secretary of the interior. Schurz sent special agents to several of the most active logging regions with the power to investigate trespasses and to seize illegally cut logs. A wave of panic swept over the industry when Colonel E. A. Pratois arrived in Minnesota to begin his investigation. In May the federal agents were at the Taylors Falls land office "ascertaining the amount of trespassing on the St. Croix River and its tributaries." There was great fear that "a good many prominent lumbermen will be found to have their hands in the grand scheme of denuding the public domain of its valuable timber." The St. Croix lumbermen loudly proclaimed their innocence and complained that the real abuses had taken place on the Chippewa and Upper Mississippi rivers. "That any of the first families of Stillwater were engaged in this business is preposterous," complained the *Stillwater Lumberman*. "But Lord, how those Minneapolis and St. Cloud fellows did steal!" In the end, the agents determined that 81.7 million feet of pine logs had been plundered from public lands, just between 1868 and 1876.[83]

One single case of trespass found that thirty-five million board feet, with a dollar value of over $75,000, was stolen. All told there were twenty-nine cases brought to trial in Minnesota because of Schurz's probe. Unfortunately, thanks to the U.S. Congress, the prosecutions never lived up to the buildup of the investigation. The House refused to increase the Department of the Interior's total investigatory budget above its paltry $12,000, which made it impossible to build cases for prosecution, even though the penalties from those cases would have returned hundreds of thousands of dollars to the federal government. Schurz tried to drum up public support for broad changes in U.S. public land policy. He called for an end to preemption rights and homestead privileges for forest lands on the grounds that the lands were not suited for family farms and that current policy encouraged very valuable public lands to be privatized at next to no return to the public. He advocated managing forestlands for sustained yield by keeping them in public hands and leasing cutting rights. Such a policy would have created the equivalent of the National Forest System in the 1870s. Unfortunately, neither Minnesota nor Wisconsin was ready for reform. "I found myself standing almost solitary and alone," Schurz complained. "Deaf was Congress, and deaf the people seemed to be." Actually, Congress was worse than deaf. Representatives, slavish to the interests of the timber industry, passed the Timber Cutting Act, which made it legal for homesteaders and mining companies to cut public domain timber. More importantly, the act gave St. Croix lumbermen facing prosecution or investigation a giant loophole. Anyone who was found to have stolen timber from the public domain in the past, or accused of doing so in the future, simply was required to pay a fee of $1.25 per acre. The prospect of any future investigations of timber fraud vanished with this legal absolution for past sinners and the shameless invitation to "go and sin some more."[84]

The Schurz investigation was greatly resented not only by the lumbermen of the valley but also by the people of the St. Croix Valley. As an earlier investigator had lamented, "Local sympathy sealed up sources of information." Yet the image painted by the investigation of the lumber baron plundering public lands would linger. It would later resonate quite strongly in the valley when a lack of timber forced once-busy mills to close. Then public ire would be directed against Weyerhaeuser and Laird, Norton and Company. This is ironic because the big downriver timber companies did not become deeply involved in the valley until after the vast majority of the Upper St. Croix forest was no longer in the public domain. In fact Laird, Norton and Company was never involved in a single trespass case in Wisconsin, nor did it resort to filing fraudulent homestead claims. On the other hand, the names of some of the richest of the original Stillwater lumbermen are on homestead applications in Burnett County, Wisconsin, including Isaac Staples and Frederick Schulenberg. By the 1880s the forest was privatized by lumbermen, who controlled thousands of acres, and by railroads, which, due to the land grants offered by state and federal legislation, owned hundreds of thousands of acres more. These large corporate owners tried to protect their holdings from the type of depredations that had previously been directed against public lands. Rather than being timber thieves, the Weyerhaeuser syndicate were themselves victims of fraud. The job of protecting corporate lands from theft largely fell to men known as timber cruisers.[85]

The job of the timber cruiser was to estimate the amount of board feet of lumber that was present in a tract of standing timber. Many of the first cruisers were former government township surveyors. One of these was Albert C. Stuntz, who traveled all over the Upper St. Croix Valley in the 1860s and 1870s. Sometimes he traveled alone; on other occasions he was assisted by Ojibwe canoe men. He camped out in the woods in all seasons, although when he came across a logging camp he was grateful for its hospitality. In 1864 he visited a logging camp on the Namekagon River. "Laid over at Mackey Bros. Camp in Sec. 27," he wrote in his diary. "Nothing to read or amuse myself about camp. Life is carried on as usual Some mending Boots and Fiddling Some reading Some Sleeping & Some Talking." A handful of large lumber companies maintained a salaried cruiser; the holders of railroad land grants usually required a staff of such men. There were, however, a number of experienced land lookers based in Stillwater who were available for assignment as needed by any lumberman. Accuracy and integrity were crucial requirements for timber cruisers. A company might base its entire winter logging program upon their reports or use them to determine the price at which it would buy or sell land. Timber cruisers were most often the first to discover a timber trespass, and their familiarity with what went on in the forest over many years, plus their knowledge of tract books, made them the most effective investigators of the crime. During the 1880s William M. Croom, timber cruiser for Laird, Norton and Company, discovered numerous signs of "old cuttings" in what the company had thought was virgin St. Croix Valley pine forest. Little could be done to seek redress for depredations committed in the indefinite past, but cruisers were kept busy assessing the value of ongoing trespasses. Although these abuses were very common, by the 1880s they were generally the result of error, not larceny. William Sauntry made a practice of firing foremen who "cut a round forty." He told one of his employees, "A man who'll steal for you, will steal from you." Most trespass cases were the result of errors made in laying out boundaries or the result of deep snow obscuring the lines between tracts. Damage claims were generally settled promptly and amicably, short of court action.[86]

Fire in the Forest

On September 1, 1894, the acrid smell of smoke was heavy in the air around the sawmill town of Hinckley, Minnesota. That was nothing unusual for the town's twelve hundred residents. For more than a month the threat of fire hung over the town as the dry swamps and cutover lands of Pine County smoldered, unchecked by fire crews or precipitation.

In fact, no rain had fallen for more than a month. For weeks, trains traveling between Minneapolis and Duluth had been delayed by the blankets of smoke that obscured the track. While the situation was worrisome to the people of the St. Croix Valley, it was not unique. August and September were the driest months of the year, and fires often broke out amid the cuttings. In the course of logging, swampers cut away all branches from the logs. These slashings were left where they fell and over the course of a season would become dry and very combustible. Fires frequently broke out in such terrain, but rainfall usually put them out before they could cause much damage. As morning turned to afternoon in Hinckley, the sky darkened and the air became still, as if before a thunderstorm. But there would be no rain that day.

Hinckley was a mill town on the northwestern edge of the St. Croix Valley. It was located on a small tributary of the Kettle River known as Grindstone Creek. It was thanks to the Lake Superior & Mississippi Railroad that it owed its existence, however, not its meager creek or the Kettle River. The railroad had sped the rate of cutting all along the western fringe of the valley. Instead of loggers merely working their way up toward the headwaters of streams such as the Snake and Kettle from the St. Croix River, the harvesting of pine also proceeded from the railroad toward the river. This two-pronged attack on pine plains of the region ensured that, in spite of the density and quality of its timber, the area was stripped of its trees faster than any other part of Minnesota. By 1894 the town of Hinckley was surrounded by vast stretches of combustible cutover land. As the summer fires spread unchecked from one field of slashings to the next, they merged to form one great, unrestrained storm of flame, surging to the east then to the west, directed only by the whims of the wind.

Incredibly, there were several Hinckley residents who had lived through the terrible Peshtigo Fire in northeastern Wisconsin twenty-three years earlier in 1871. That blaze, caused by a similar set of circumstances, killed more than one thousand people in spite of the fact that the residents of Peshtigo had both a river and Lake Michigan available as a place of refuge. As they saw the smoke darken the sky of Hinckley, these veterans, like those who knew no better, continued to work at their jobs, but surely they must have given some thought as to how they would save themselves if holocaust again stalked them.[87]

All thought of work was cast aside a little after noon when word came over the telegram that the town of Pokegama, just nine miles to the south, had been engulfed in flames. As word spread that most of Pokegama's inhabitants had been burned, people at last awoke to their own danger. It was too late. Almost as soon as they entered the street, they saw an ominous black cloud boil up on the horizon and quickly spread its shadow over Hinckley. The volunteer fire department barely had time to deploy to the edge of town when the monster fire struck. A wave of scorching, overwhelming heat swept over the firefighters as building after building broke out in flames. In an instant, the town was lost and people fled to save themselves. The foolish sought valuables; the prudent gathered up loved ones and immediately fled to the train station. With Hinckley transformed into an island in a sea of flames, the steel rails were the only way out. Hastily a train was assembled from a collection of cars sidetracked at the village. By the time several hundred panicked people filled the train, the heat was so great that the paint on the passenger cars was beginning to blister. The train sped through the blazing forest until it reached the town of Sandstone, where it warned the inhabitants of the coming conflagration. The firestorm, however, was hard on their heels, and within minutes of the train leaving Sandstone, that town was destroyed, with the loss of forty-five people. When the train reached the Kettle River, the trestle bridge was in flames. If the Hinckley refugees could get across the river, they would be able to outrun the fire; if not, they would be trapped and killed. The engineer opened the throttle, and the train made it across just before the bridge collapsed into the inferno.[88]

For those in Hinckley who missed the train, horror and death stalked them. Many fell dead in midflight, killed by the "suffocating choking gases" that preceded the flames. A contemporary reported:

> Dogs, cats, chickens and stock were stricken instantly and died in their tracks without serious burns. In one instance a man was stricken down, but not burned enough to destroy his clothes, and in one of his pockets was found a small leather purse in which were four silver dollars welded together in one solid piece.

Those killed by the gas fell "in the twinkling of an eye." Others suffered the ravages of fire. More than ninety discreet piles of gray ash, in human form, were later found along a railroad embankment. An equal number who had sought refuge in a partially flooded gravel pit nearby survived. Another group of about two hundred people ran up the track of the Lake Superior & Mississippi Railroad. As the flames gained on them, the slow victims perished one by one. Most of them, however, managed to keep running until they were rescued by a train. The train was unable to outrun the flames, and it caught fire, but not before engineer James Root unloaded his terrified passengers near a bog, where they sought refuge.[89]

The great forest fire of September 1894 became known as the Hinckley Fire. It wiped out the Minnesota towns of Hinckley, Sandstone, Poke-gama, Mission Creek, and Partridge and killed 413 people. But it was not restricted to Minnesota. The conditions that caused the fire—dry weather, combustible cutovers, and high winds—existed across the Upper St. Croix Valley. On the Wisconsin side of the valley, a wave of fires swept up the track of the North Wisconsin Railway. Fires born by bad logging practice consumed little mill towns that had lived off the harvest of the forest. Phillips, Wisconsin, was consumed in July 1894, while the town of Barronett was burned on the same day as the Hinckley disaster. At the time of the fire the manager of the Barronett mill was in Minneapolis holding forth before a group of lumbermen how his mill was in no danger because it was surrounded by cleared ground. He was brought up short by a telegram that informed him the fire had leaped cleared land and consumed the mill. The refugees of Barronett no sooner found shelter in Shell Lake than flames surrounded that town. Although more than fifty buildings burned, Shell Lake was able to save its mill and the lives of its citizens. Among the other mill towns devastated that fire season were Comstock, Benoit, Marengo, and Mason. An estimated 1.4 million acres of pinelands and cutover were consumed by the fire.[90]

Although the 1894 fires were the worst that swept over the St. Croix Valley, they were neither the first nor the last. Forest and brush fires had been a regular feature of life in the valley from the time logging developed as a large-scale commercial enterprise. During the 1870s and 1880s virtually all residents of the valley understood that if there was a dry spring, a bad fire season was likely to follow. In 1877, for example, a lack of snowmelt had hampered the spring log drive. Water was so low in the river that in Polk County boys were able to wade across to Minnesota. By July massive fires raged across the valley. The *Stillwater Lumberman* estimated that "fire on the upper St. Croix has destroyed more timber than was cut last winter." Nothing would be done to restrain such wild fires. In 1879 fires ranged all around Grantsburg, Wisconsin. Only when buildings in town were threatened did "all the men and boys" turn out to fight it. The fires in the hinterland were ignored until "the rains put most of them out."[91]

A common misconception of these massive fires was that they burned up thousands of acres of pine timber. This occasionally happened with very intense fires. The 1877 fire on the Upper St. Croix and Namekagon generated heat "so terrific" that it burned "out all traces of stumps." Usually a forest fire passed through standing timber fast enough that it only badly singed the trees. A fire-ravaged forest was reduced in value, but it still boasted most of its board feet. The trouble was that fire killed most of the trees. This meant that the charred timber had to be harvested within a year or be severely degraded by insect damage. Forest fires reduced an asset that was rapidly appreciating to one that had to be immediately liquidated.[92]

The construction of Nevers Dam indicated that the big lumber companies that now dominated the valley were building for the future when they moved into the St. Croix. The Weyerhaeuser syndicate had discussed cutting back on forest operations due to the fallout from the Panic of 1893. With the market for building materials depressed, they had thought it best to keep their pine in the forest. The fire, however, forced their hand. "It is too bad to have such timber as that wasting," lamented the head of Laird, Norton and Company, and he ordered a massive salvage operation. The 1894 fires may have destroyed five hundred million board feet of timber, but equally damaging was the accelerated pace of logging that followed and the long-term impact of scorched earth for future forest growth.[93]

The loss of life that accompanied the Hinckley Fire necessitated some type of political response. In both Minnesota and Wisconsin ineffectual bills were passed establishing local fire wardens.[94] But the fires continued throughout the 1890s and into the 1900s. The big downriver mill owners would have liked to have held their pinelands for better prices, but they were unwilling to force changes on the way logging operations were conducted. They left the actual cutting of timber to local contractors whom they squeezed into operating at a low profit margin. These small businessmen were not inclined to take on the extra cost of piling the branches and brush left behind after logging and conducting a controlled burn. Although a generation of experience taught them better, they left behind the fuel for future forest fires. Farmers who purchased logged-over land were confronted with acres of slash that could be removed economically only one way—by fire. A farmer working alone on his homestead lacked the ability to contain a blaze once it began. He merely doused his cabin with water and waited for the fire to stop of its own accord. Hundreds of fires set in this manner swept over the upper valley each year between 1890 and 1910.

The Last Days of the Lumber Frontier

Each logging season and each fire season took its toll on the forest resources of the St. Croix Valley. Reports of the logging industry's demise, however, were greatly exaggerated throughout the last years of the nineteenth century. Year in and year out, logging continued to dominate life along the St. Croix. While it was broadly recognized that the valley's timberlands were nearing exhaustion, there remained an immutable aura of permanence about the seasonal cycle of the industry. Life on the St. Croix was locked in a familiar, comfortable rhythm, a three-part harmony of the long winter logging season, the dramatic spring river drives, and the sawdust summers of mills humming at full capacity.

To be sure, the names and faces of the lumberjacks changed. The men from Maine and Ireland gave way to Swedes, Poles, and Finns. The name *shanty boys* also faded from the scene, replaced by the less jaunty *lumberjack*. Although it is a name we associate today with colorful images of burly men in red woolen shirts, there was a more ambiguous understanding of a lumberjack at the time. On one hand, they were lauded as men of the frontier, tough enough to challenge nature in her own domain. Yet, in the minds of many valley residents, a lumberjack was "a new sort of animal," crude, dirty, and potentially violent. At the beginning of the twentieth century, with the frontier all but destroyed, the lumberjack was often seen as the lowest form of migratory worker, a hobo without a home. There was little Paul Bunyan romance in the way woods workers were viewed. Some immigrants and cutover farmers settled into the life of a woodsman with a relish and stayed at it as long as there was work. For most, however, life in a logging camp was a way station on the road to Americanization and better-paying jobs. Finnish lumberjacks expressed their distaste for life in a lumber camp with the following verse:

> A wretched home, this cheerless camp;
> And "finer people" sneer, make cracks:
> "You ruffians, bums,
> bearded lumberjacks!"
> Our wages are the rags we wear,
> Our scraps of food no one digests.
> Our beds are bunks
> And fleas our only guests.

Finnish lumberjacks sometimes brought Old Country experience with an axe or saw to their job; others, such as Poles or Hungarians, merely brought a strong back and an empty stomach.[95]

Many cutover farmers had no choice but to seek woods work to raise the cash needed to build their farms. Carl Kuhnly, who grew up on a Burnett County farm in the 1890s, remembered: "The first three winters Dad worked in the logging camp. He would walk about ten miles to come home on Saturday evening, then back again on Sunday evening." His

Figure 10. The Goar and Stinson log-driving crew on the Clam River, Polk County. This picture was taken in 1902, a time when the once dense St. Croix forests were reaching the point of exhaustion. Wisconsin Historical Society, WHi-6806.

mother had to care for both the farm and six kids during the long winter. Families who lived farther from the camp had an even more difficult time. One farm wife remembered her husband worked at a camp twenty miles away. "The Camp could just as well have been a thousand miles away. There were no roads, only snow-buried trails. I never saw my husband till spring." Late in life she bitterly asked the question, "Why was I left alone to carry the whole load while my husband was away destroying the forest for a big lumber company who never knew we existed?" Another "old timer" underscored it was necessity that drove the men into the woods. "The family owed bills at the store for winter groceries and some of the money was used to pay these bills."[96]

The hard-core lumberjacks were men who had spent most of their lives in the woods and who intended to finish their days felling trees. They had initially taken work as lumberjacks for the same reasons of necessity as the immigrants or cutover farmers, but they had long since given up any dream of another life. They liked the freedom of having limited responsibility. While in a camp they slept where and when they were told. When

it was time to eat, the food was put down in front of them. The comrade-
ship of a lumber crew provided pleasant association with men of similar
disposition without the demands of family. A hobo lumberjack cherished
the freedom to quit a camp at any time, collect his pay, and go—perhaps
because the cook was bad, or his bunkmate snored too loudly, or he did
not like the foreman, or simply because he wanted to go to town for a
drink. Similar to sailors on a ship or soldiers in barracks, the logging camp
was a predictable, structured, male world governed by well-established,
easily discernible rules of behavior. The bulk of these lumberjacks for life
were specialists, such as sawyers. In fact, it was not uncommon for sawyers
who worked well together on a crosscut saw to team up and work together
for years at a time. A competent foreman took the measure of his crew
fairly quickly and worked to keep in camp the skilled and hard-working
jacks. Often men would become regulars with a foreman, year after year.
All lumberjacks shared in common, however, the ability to put up with ill-
lit bunkhouses reeking of sweat, bunks with only a "snort pole" between
occupants, and dirty clothes infected with lice.[97]

Stillwater was a prime destination for woodsmen when spring came,
and their arrival was viewed with a mixture of curiosity and dread. In 1856,
the morning after a group of lumberjacks painted the town red, the editor
of the *St. Croix Union* sought to lecture the woolen shirt brigade:

> Our town is now filling up with lumbermen. They have been cooped up all
> winter in the woods; and when they get down here they feel like having a
> sort of jollification. We would have them understand that we are their
> friend; we know how to sympathize with them; for at one period of our life
> we were engaged in a humbler calling than theirs. . . . It is not the dress, or
> the occupation, that makes a man; but the motives and principles by which
> he is actuated. But while we are a friend to the lumbermen, we are also a
> friend to law and order. Gentlemen, take your fun, but let it be innocent
> and harmless.

Such condescending editorials had little impact on the young men with
full pockets and high spirits. Besides, as a rival newsman lamented, the
business community profited handsomely from the jacks: "There is always
room and welcome for a new saloon or strychnine whisky depot." A resident of Marine later recalled that Stillwater was "a good place for a rough-
and-ready lumberjack to spend his money," with "about 25 licensed saloons," and plenty of "women of easy virtue were also available." After a
few five-cent shots of whisky, many lumberjacks were ready for a visit to
"Red Nell," the leading madam of the lower river. Together with "Perry the

Pimp," Nell operated one of several well-frequented bordellos across the river at Houlton, Wisconsin.[98]

As the lumber industry spread to the upper reaches of the valley, new "whoopee-towns" were developed. Barron, Wisconsin, although one would hardly know it today, was reputed to be a rough town during the 1880s when lumberjacks frequented it each spring. In 1880 the small town of Chandler, Wisconsin, boasted fourteen saloons and two gambling houses "hastily nailed together of rough boards or logs . . . giving evidence that few, if any, of its citizens went there to stay." A common saying in the North Woods a century ago was that "the three roughest places in the world are Hayward, Hurley, and Hell." Certainly in the 1890s, Hurley, a town on the Wisconsin–Michigan border, deserved its ill fame with the wall-to-wall saloons and flophouses along Silver Street. Hayward operated on a smaller scale. It was only a village of about one thousand residents and even in its roughest days it never had more than seventeen saloons. In fact, the "good people" of the town voted on at least one occasion to make the town dry, although that prohibition did not last long. Hayward's worst blind pigs and brothels were located away from the main town, easy for lumberjacks to find but not flaunted in the face of decency. "No lie about it," recalled one former lumberjack, Hayward was "one of the wildest little towns in the state." Men who had too much to drink could be found on almost every corner "bucking up." For years after the lumber trade had slackened, the proprietors of Hayward saloons delighted in pointing out to gullible tourists bullet holes and caulk marks from loggers' boots on the floors of their rundown bars.[99]

While there may be considerable local color in tales of wild "timber-beasts" cavorting in saloons, modern readers need to remember that hundreds of lumberjacks, perhaps the majority, departed for home straight from the logging camps, or after having only a single drink with their winter comrades. For men waiting for a train, a saloon was one of the few places in most towns where a dirty lumberjack would be welcome. Most saloons put out "free lunches" for men who would buy a drink. Plenty of lumberjacks lost their entire paycheck on one wild spring spree, but these tended to be the hobo lumberjacks—men who lacked, or chose to neglect, family responsibilities and had few ambitions in life other than doing what they pleased, as long as the money lasted. These men took pride in their physical prowess in the woods and their capacity for alcohol in town. Their life followed a pattern of long periods of hard work and short periods of riotous living. Sherman Johnson, who grew up in the valley, recalled a friend who went to work as a lumberjack at the age of fourteen. "He then made the logging and harvest field circuit for 14 successive seasons, each year spending his money for liquor and women." Only when he gave up logging was he able to turn his back on the bottle. More than a few who stayed on that path ended up as old men shaking through delirium tremens.[100]

Logging never really ended in the St. Croix Valley. Just as white pine yielded to Norway pine, so too did jack pine and cedar forests become the focus of the lumberjacks. Hemlock, which Isaac Staples regarded as scrap, became a very important forest product during the 1890s. The editor of the *Bayfield Press* proved an accurate prognosticator when he wrote in 1883, "While laying no claim to being a possessor of the gift of prophesy the writer would, nevertheless, hazard the opinion that ten years or less hence will see this once despised timber take high rank in the markets of the land." Unlike the buoyant white pine, hemlock and hardwood would not float for long and could not be driven via the St. Croix to market. Logging railroads and steam haulers were the only way to transport the logs and bark from the forest. The harvesting of hardwoods and hemlocks in the region climaxed in the 1920s, after which time the few remaining sections of virgin forest were nothing but a handful of wood lots. The Edward Hines Lumber Company of Chicago was one of the few big lumber companies to operate in the valley in those later years. In 1902 they purchased remaining forestlands of the North Wisconsin Lumber Company, including the big mill at Hayward. The Hines Company operated a logging railroad that ran spurs all through the upper Namekagon valley.[101]

The final phase of logging in the St. Croix Valley featured the rise of the rubber tire lumberjacks. It began in the 1930s and continues to this day, featuring small-scale logging on selected private wood lots or public forests. During the bleak Depression years the backwoods townships of the valley were very hard hit. Many a cutover farmer kept food on the table only by harvesting cordwood from tracts of "weed trees"—second-growth stands of poplar and balsam. Only after several decades of professional management of public forestlands did logging begin again in earnest through government timber sales. By this time logging camps were a thing of the past, as lumberjacks commuted to work in their automobiles. Heavy equipment operators replaced the physically demanding work of a sawyer or swamper. The name *rubber tire lumberjack* was spawned by the diesel-powered skidders used to harvest wood and reflects the impact of such equipment on the lumberjack's physique.

Through the long twilight of the logging industry in the St. Croix Valley, from the days of river driving to railroads to truck trails, the one consistent feature was the lumberjack. As long as there was work to be done in the woods, there were men willing to take up what has always been one of the riskiest jobs, in terms of personal injury, in America. In 1970 Fred Etcherson, who cut pine at the turn of the century and scraped by through the Depression hauling pulpwood, speculated on what drew men to work in the forests. Although most lumberjacks just barely made enough to live on, he was confident that they were happy with their lives because they "didn't care to get rich." Eighty years later, he tried to explain his personal motivation:

> I just love to be in the woods. I'd just love to go down there in these woods right now. Lay down and go to sleep, I love to be in the woods. Wonderful place to be.

Cutting trees gave Etcherson a chance to develop a love of the forest, and he spent his life destroying that which he loved.[102]

The Impact of Logging on the St. Croix Valley

Like the fur traders before them, the lumberjacks' embrace of the St. Croix Valley transformed it. They had turned the forest—a place they all valued, many appreciated, and a handful loved—into lumber, a utilitarian if prosaic commodity. The volume of lumber produced by this single valley was staggering. During the peak year of 1890, the St. Croix Valley, as either logs or lumber, produced 450 million board feet. The total production between 1840 and 1912, if loaded onto standard log cars, would have required 2.2 million rail cars. As a single train such a span of cars would be long enough to reach across the continent more than six times. The transfer of this wood, from where nature intended it along the Upper Mississippi valley to the treeless region to the south and west, made possible the agricultural settlement of the Great Plains. The majestic white pine of the St. Croix lived again—in some cases still lives—as homes, barns, corn cribs, fence posts, and doors and in products ranging from the support beams in great public buildings to lowly outhouse seats. While the establishment of grain farms on the plains was not in itself an unmixed ecological benefit, in the balance the loss of a vast forest for the gain of a breadbasket was a trade nineteenth-century Americans would have been pleased to accept.[103]

Masked behind the balance between the Upper Midwest's loss and the Great Plains' gain is the enduring impact of the logging frontier on the St. Croix Valley. The sudden, dramatic loss of the valley's forest was an ecological change unrivaled since the last descent of the glaciers. The vast plains of old-growth white pine, an area exceeding four thousand square miles and boasting trees two to three hundred years old, have never been replaced. White pine had dominated the pre-settlement forest because of its ability to adapt to a wide range of conditions. However, intensive logging and forest fires destroyed the natural reseeding mechanism of the forest. Well-meaning but misguided efforts to reseed white pine led to the introduction of an Asian tree disease known as blister rust that devastated white pine seedlings and led to the elimination of most efforts to replant the forest's most valuable and beautiful tree. A generation of hardy immigrants broke their lives trying to follow the axe with the plow on the cutover lands. Only a persistent handful, blessed with a patch of rich soil, survived. The homesteads of the rest are today lost amid succession forests of poplar or plantations of jack and Norway pine. The myth of the upper river as a land of inexhaustible forest resources was quickly replaced by the myth of the region as future agricultural cornucopia—each myth burdened with tragic consequences. Much of the Upper St. Croix is again a forest, but it is not, nor can it ever again be, a wilderness. Instead, it is a curious mix of the failure of agriculture and the success of sylvaculture, as much a product of human design as a Kansas wheat field.[104]

Logging introduced urbanization to the St. Croix Valley. Milling and transportation concentrated the harvest of wood on specific, reoccurring locations. Initially these were waterpower sites on the lower river such as Taylors Falls and Marine. Eventually most of the energy of the logging frontier focused upon Stillwater, and it grew to a city of more than a dozen mills and thousands of inhabitants, the majority of whom were beholden to the forest for their livelihood. After the Civil War a new pattern of town development followed the blueprint of the steel rail. A string of new mill towns sprouted along the Lake Superior and Mississippi Railroad and along the North Wisconsin Railway. Hinckley, Minnesota, and Hayward, Wisconsin, each came to symbolize the success and failure of the logging frontier. These hinterland towns did not displace Stillwater's importance as the principle funnel through which the bulk of the pine flowed. Pine remained largely oriented to the river, but once the softwoods had been cut, the railroad towns became the focus of hemlock and hardwood production. These towns and much of the cutover countryside turned their backs on the St. Croix River.[105]

The St. Croix River was left vastly changed by the logging frontier. What had been in 1837 a wild river, disturbed only by a handful of Ojibwe fish weirs, had become one of the most controlled and manipulated river systems in America. There were between sixty and seventy gated dams in the St. Croix watershed and uncounted numbers of splash dams, hastily constructed of brush and earth. When combined with the loss of forest cover to logging and frequent brush fires, the dams left as their legacy a river that flowed much less clear and whose banks were more prone to erosion. The habitat of brook trout and other native fish that favored clear, cold waters was gradually destroyed. The strong current of the Upper St. Croix River and the flushing action of the multitude of dams sent waves of turbid water to the lower river. Where the current slackened, the sand and earth suspended in the river settled into bars and shoals. Where steamboats easily navigated in the 1840s, commercial vessels repeatedly were grounded in the 1880s. Even when the loggers did not hold back water at Nevers Dam, dredging and wing dams were necessary for boats to effectively navigate between Stillwater and Taylors Falls. In addition to all of the silt and sand sent downriver, the lumbermen infringed on the St. Croix at Stillwater with extensive landfills. The mill owners had created more than ten acres of new waterfront land either by accidentally creating the conditions for mudslides or by consciously trying to increase their river frontage by dumping massive amounts of slabs and sawdust into the St. Croix. Such annual deposits further clouded the water.[106]

While logging as a business continues and will continue to linger in the valley in the twenty-first century, the logging frontier ended in 1914. In that year the boom at Stillwater, the great net of wood and chain that captured and sorted all of the pine driven on the St. Croix, handled its last log. It was a demise that had been long expected. More than a decade earlier, William H. C. Folsom, one of the valley's first pioneers, who had lived and prospered long enough to become its first historian, observed:

> The business has been a wonderful one; it has enriched many; it has furnished and is still furnishing a means of livelihood for thousands but is going rapidly and like the sands in the hour glass that keeps running, ever running on, its day will soon come. And then what?

Most of the men of Folsom's generation had come as young men from New England to make their fortunes in the woods. Reflecting on their lives before marble-clad hearths, in the comfort of homes paneled with finely grained wood, they took satisfaction in their accomplishments. The white pine boom had lasted long enough to see them into plush retirement or honored internment as founders of prosperous communities.[107]

Younger men were left to ponder the question posed by Folsom: "Then what?" Many men cast their lot with the business of logging, not the valley of the St. Croix. From the boss logger of the river Frederick Weyerhaeuser to a modest lumberman like William Veazie, many a man who made his fortune on the St. Croix gambled he could make another in the rainforests of the Pacific Northwest. Some who stayed moved into farming, which by 1900 supported more people in the valley than logging. Others looked to tourist excursions, manufacturing, or mining to be the next boom for the valley.

The men who prospered in the lumber boom left behind ravaged forests and splendid Victorian homes. Visitors to Stillwater can today see the homes of Roscoe Hersey, the partner to Isaac Staples, and that of Captain Austin Jenks, who made his fortune rafting St. Croix timber down the Mississippi River. The lumber barons of Stillwater had the financial means to build in whatever style struck their fancy, and they did so with the intention of erecting not only a comfortable home but a monument to all that they had accomplished in their lives on the frontier. John McKusick arrived from Illinois in 1840 and stayed on in Stillwater, eventually founding its first sawmill. His brothers Jonathon, Ivory, and Noah joined him in the Minnesota lumber business. Today the Ivory McKusick house in French Second Empire splendor stands in Stillwater as an example of how well the family did. Albert Lammers celebrated his success in the lumber industry by building elaborately with wood. In 1893 he chose the Queen Anne style, with its elaborate millwork and hand craftsmanship, for his new home. Like the McKusick, Jenks, and Hersey houses, the Lammers mansion is on the National Register of Historic Places. William Sauntry was not one of the founding generation of St. Croix lumbermen, but he did so well through his association with Weyerhaeuser that he was able to join the elite in 1891 with his own fine residence. It was not, however, a place where Sauntry lived out a prosperous and contented retirement.

The young men of Stillwater could take little comfort from the fact that their fathers had done well. On the frontier, social mobility moved in two directions, and the challenge for those who stayed in the St. Croix Valley was to find a path to profits that did not lead to the played-out pineries. Among those who stumbled in pursuit of illusory new ventures was William Sauntry, the most successful of the second generation of St. Croix loggers. Although he lacked formal education and polished manners, Sauntry had convinced Frederick Weyerhaeuser to trust him with management of some of the timber trust's biggest projects, from the boom company to Nevers Dam. In the years that followed those coups, Sauntry demonstrated again and again his mastery of the business of logging. He directed the Ann River Logging Company, the large and multifaceted firm that cut the bulk of the remaining pine on the St. Croix. There was a swagger and what one historian called an esprit de corps about the lumberjacks who worked for the Ann River Logging Company. In 1891, for example, they showed off their prowess by loading a sled with a mountainous 31,480 board feet of logs and then hauling it over their own ice-rut roads for one mile. Sauntry was the hard-driving, tireless, and inspirational leader of the company. When the Ann River outfit cut its last log, Sauntry invested his sizeable fortune into a variety of mining ventures. He was in a new field, however, and his energy and drive only plunged him deeper into losing investments. By 1914 he had lost all that he had won from the forest—money and reputation. His splendid house on Fourth Street in Stillwater became just another asset to be wagered on an increasingly bleak future. When even his old associates from the Ann River Logging Company turned their backs on him, William Sauntry purchased a revolver and put it to his head.[108]

That same year in Stillwater Frank McCray, the master of the St. Croix River Boom, hopped onto the last pine log to ever enter the boom. Workers watching from the cribs and log channels sent up a hollow cheer. More than thirteen billion board feet of logs before that, in 1856, a much spryer Frank McCray had guided the first log through the Stillwater boom. To mark the occasion the lumbermen invited all of their old employees to the boom company boarding house for a farewell feast on the banks of the river. The old timers slapped each other on the back and told again the stories of their youthful antics and the epic scene of a river of logs. The St. Croix River had remained an important logging stream much longer than any of its Lake States rivals, longer than Michigan's fabled Tittabawassee or Muskegon, longer than Wisconsin's Chippewa River. Yet the era opened and closed within the course of one man's working life. "It makes one sad to realize," a veteran of the logging era later wrote, "that a great industry has absolutely faded, like a mist before the sun, largely because of the greed and hurry and lack of foresight of the generation that is gone."[109]

Back in 1837, during negotiations with the U.S. government, the Ojibwe had proposed not to sell their lands, but to lease them to the Americans. The Ojibwe were aware that it was the lumbermen's lust for St. Croix pine that was pushing them off the land. "It is hard to give up the lands," lamented Chief Flat Mouth. "They will remain but you may cut down the trees and others will grow up." The Ojibwe proposed a lease of sixty years. Although the American negotiators brushed their offer aside, the Ojibwe had rather accurately predicted how long the lumber frontier would last. They missed the actual ending of logging by only seventeen years. Yet when the St. Croix Boom closed, it was not the Native people of the valley who inherited the deforested lands. New people from old lands across the ocean were already re-imagining the St. Croix as a cutover cornucopia, a North Star of opportunity.[110]

3 | "The New Land"

Settlement and Agriculture

The unregulated free market of the logging era left the Lower St. Croix River choked with logs and silt and partially stripped of forests. The Upper St. Croix was a moonscape of a land denuded of much of its flora and fauna. Guilt over this environmental disaster, however, was not in the psychology of nineteenth-century lumbermen or state and local officials. Many predicted that the farmer would follow the lumbermen as the forests receded, and the entire North County would be turned into an agricultural paradise. New England Yankee farmers readily responded to this opportunity for cheap land, as did a variety of European immigrants. Many of these new settlers came with no greater expectation than to acquire a piece of land to farm for their families. The St. Croix Valley's excellent water routes, however, connected the region to national and international markets. With access to markets came the lure of prosperity. Railroads came to complement the region's water highways, and technological innovations in agriculture led to increased efficiency. Its agricultural bounty, like its timber resources, made the St. Croix an integral part of the Midwest economy that supported growing industrial cities like St. Paul, Chicago, and St. Louis. Although eastern or European metropolitans might regard the Upper Midwest as an isolated frontier, St. Croix Valley farmers felt very much connected to the fluctuating markets of the city. Pride in their importance to a growing nation and optimism in their region inspired the farmers who followed in the loggers' wake.[1]

Dividing the Valley

Agriculture changed the St. Croix's natural landscape by creating the first permanent settlements in the valley. Since the Erie Canal had opened in 1825, the Great Lakes and Mississippi valley region became accessible to East Coast entrepreneurs and pioneers. Settlers by the tens of thousands began migrating westward.² Federal surveying teams began penetrating the Wisconsin frontier in the late 1830s. They first began in the populous mining region of southwestern Wisconsin. From there land offices opened in the southeastern portion of the state and moved north and west. As land was made available, the Wisconsin frontier gave way to farms with amazing speed. By 1845 federal land sale offices sold nearly three million acres in the territory.

The problems in settling the St. Croix Valley, however, were its remoteness and lack of governmental presence and authority. Until 1838 the land belonged to the Ojibwe. Legal jurisdiction for matters concerning soldiers, fur traders, and the like fell under territorial government authority. The St. Croix Valley first came under the jurisdiction of the Indiana Territory and then the Illinois Territory. After 1819, legal jurisdiction over the valley was granted to Crawford County, Michigan, which in 1836 became Crawford County in the Wisconsin Territory. The land under the jurisdiction of Crawford County, Wisconsin, included the entire western portion of the Wisconsin Territory east of the Mississippi River and north to Canada. Its county seat was in Prairie du Chien. Anyone living east of the Mississippi in the Wisconsin Territory had to travel there for any legal transactions. In 1840 only 351 non-Indian people were found by U.S. census takers to live in the "Lake St. Croix District," which ran from St. Croix Falls to the Ojibwe Mission at Pokegama.³

The early settlement of the St. Croix Valley was inextricably tied to the founding of the Twin Cities of Minneapolis and St. Paul. The colorful Joseph Renshaw Brown, the former soldier, fur trader, and lumberman and the future farmer, storekeeper, and government official, played an influential role in how the valley would be settled. In his quest for prime real estate, Brown hedged his bets by laying claim to land along both the St. Croix and Mississippi rivers, which created a bitter rivalry with Fort Snelling occupants. The eventual consequence of the Brown–Fort Snelling rivalry for the St. Croix River Valley was the dispersal of commercial and government activities along the river rather than the creation of a major urban center at the more logical junction of the St. Croix and Mississippi Rivers near Prescott and Point Douglas. Brown was initially interested in sites for ferry landings with the potential for future towns. He knew once the Upper Mississippi was opened for settlement, prospective homesteaders would flock to the area and would need places to disembark, temporary accommodations, provisions, and access to interior lands. Brown made a claim near Fort Snelling. It seemed like a perfect location to build a northern Davenport, Galena, Dubuque, Peoria, or even Chicago! It was at the northern reaches of navigation on the Upper Mississippi, since narrow channels and sandbars limited access to the Falls of St. Anthony. It was also the only site available for settlement in that vicinity, since Fort Snelling was on the northeastern shore of the Minnesota River and the western shore of the Mississippi. The southern shore of the river was still in Indian hands. Because settlement was also heading up the St. Croix River, Brown made three claims there. One was at the mouth of the river near present-day Prescott, another was at the head of Lake St. Croix where he had a warehouse to supply his upper river trading posts, and the third was on the lake at St. Mary's Point near the small voyageur and "half-breed" settlement.[4]

Settlement also depended upon the presence of governmental authority to ensure the legality of land claims, property rights, and law and order. In 1839, therefore, Brown made his first foray into politics when he sent a petition to the Wisconsin Territorial Legislature for a permit to run a ferry from his claim across the Mississippi to the Fort Snelling reserve. He also won a position as justice of the peace. Brown's victory helped make legal matters more convenient for settlers in the St. Croix Valley, who used to have to travel to Prairie du Chien.

Brown also set up a whiskey depot on his claim across from Fort Snelling that quickly became a popular recreation site for lonesome soldiers short on entertainment. After one drunken spree that put nearly two-thirds of the men in the guardhouse, army commanders aimed to put an end to Brown's house of libation. It seems, however, that Brown's whiskey shop was only a pretext for their claiming the best ferry landing between Fort Snelling and the Falls at St. Anthony. St. Anthony Falls was obviously a choice spot for waterpower. Several officers from Fort Snelling, including the commanding officer, dabbled on the side in land speculation. If the area around the falls was to flourish, these would-be entrepreneurs needed to eliminate competing commercial and settlement sites such as Brown's. They too hedged their bets by claiming a prime landing spot at the mouth of the St. Croix River at present-day Prescott. Not only was Brown's whiskey shop in the way, but so too were old fur traders, refugees from the Red River colony in Winnipeg, and various French Canadian vagabonds. On the pretext of military necessity, fort commanders decided to extend the fort boundaries, ousting all nonmilitary personnel. Major Joseph Plympton claimed these hangers-on were using up more than their fair share of fuel wood and their horses and cattle overgrazed public lands. He ordered Lieutenant James L. Thompson to mark out new boundaries for the fort that included what is now the Twin Cities. About 150 squatters were told to leave. In October 1839, in a show of support for Fort Snelling's actions, the secretary of war sent an order to the U.S. marshal for the Wisconsin Territory to remove the settlers immediately and use force if necessary. The order, however, was misdelivered and delayed for months.[5]

In the meantime, outraged settlers formed a citizens' group and selected Brown to present a petition against the military reserve extension to the Wisconsin Territorial Legislature in Madison. They hoped the civil government would stop the extension of the military holdings into land intended for civilian settlement. In November 1839 Brown accompanied Ira Brunson, the representative for Crawford County and deputy marshal, to Madison, where Brunson submitted the petition to the territorial government. By December the territorial government passed a resolution against the military reserve extension to the east bank of the Mississippi and notified the secretary of war that the military was preempting land that was under the civilian control of the Wisconsin Territory without its consent. By March 1840 the dispute reached Congress. The War Department's influence in Washington, however, led to the petition's death in committee. Fort Snelling commanders immediately forced the squatters out of the military reserve.[6]

Fort Snelling's land grab had important consequences for settlement in the Upper Mississippi River valley. Many of the evicted squatters moved upriver to St. Anthony Falls to join a small group of earlier settlers. This settlement grew into St. Paul and Minneapolis—the region's center of immigration and business, not Fort Snelling's site at the confluence of the Mississippi and the St. Croix rivers, which remained under military control. The St. Croix River Valley was bypassed in the process. This proved to be a blessing and a curse for the valley. The region would lack a cohesive, unifying economic and political center, but in the future this would preserve its bucolic charms.[7]

Brown was not dissuaded by his defeat and pressed another citizens' petition on the Wisconsin Territorial Legislature in the winter of 1839–40 requesting that a new county be formed out of northwestern Crawford County with a county seat south of the fort on the Mississippi. The Wisconsin Territorial Legislature was aware of increased settlement in its northwest region and was ready to entertain the idea of a new county. However, loggers from the Marine and St. Croix lumbering companies submitted a counter-petition to the legislature to create St. Croix County with the county seat at Prescott. Brown quickly recognized the greater viability of a proposed county centered along the St. Croix River rather than the Mississippi, but he was determined to prevent Fort Snelling officers from being beneficiaries of a county seat at Prescott. Instead of submitting his own petition, Brown sought a compromise with the lumbering interests to find a new county seat elsewhere.[8]

The lumbermen wanted to have all the timberland on the St. Croix and Chippewa rivers fall within the new county's boundaries. Brown wanted locations for future ferry landings, town sites, and farmsteads for arriving immigrants. By January 1840 a compromise bill passed the legislature establishing St. Croix County. Its southern boundary was set at the Porcupine River—now Rush River—on Lake Pepin. From its first fork its boundary went in a straight northeasterly direction to the Hay Fork of the Red Cedar River and then directly north along the Bois Brule River to Lake Superior, and westward along the lakeshore to the Canadian border. Its western boundary was the Mississippi River.

In 1842 Brown put forth a resolution to Congress to have the public lands north of the Wisconsin River surveyed as well as appropriated for a new military road from Fort Howard, located on Green Bay on Lake Michigan, to Fort Snelling via Plover Portage on the Wisconsin River and Dacotah on the St. Croix. Travel would be cut from five hundred to two hundred miles by avoiding the water route along the Fox and Wisconsin rivers. Both measures passed. Brown thereby enhanced the prospect of settlement for the north territory as well as increasing the value of "Joe Brown's claim."[9]

The early days of the county seat at Dacotah were modest, as the settlement was made up primarily of Brown's relatives. While many people passed through the settlement on their way to the pineries, few chose it as a permanent settlement. With no finished lumber available, the first "court house" was made from tamarack logs with mud plastered in the chinks to keep the wind and cold at bay. While the county commissioners met here, David Irwin, a judge from the Green Bay district court, was appalled on his first visit in June 1840 by its primitive conditions and lack of formalized proceedings. He was quoted as saying that he would never again go to that "God-forsaken spot." Even when he was accused of neglecting his duties, Irwin refused to go back to Dacotah. The Wisconsin Territorial Legislature was then forced to amend the St. Croix County legislation and give Crawford County legal jurisdiction there. Commissioners were allowed for convenience's sake to set up offices at Red Stone Prairie on the Mississippi River, now Newport, Minnesota. Joseph Brown even abandoned Dacotah in 1843 for Grey Cloud on the Mississippi. By 1846 Dacotah was all but a ghost town.[10]

The beauty and resources of the St. Croix, however, did not depend upon the likes of Joe Brown to attract settlers. By the 1840s the St. Croix

River Valley as well as the Upper Mississippi River valley had emerged from a mysterious country of wilderness and Indians to a well-known area mapped out and described by explorers. Eastern developers and potential settlers had already begun migrating here. The first towns were lumber centers, such as Stillwater, Marine, St. Croix Falls, Osceola, Hudson, and Arcola. Other settlements appeared near fords in the St. Croix that were most conducive to shipping, such as Afton, or places that had been Indian trading centers, such as the Dacotah mission, which became Newport. The old trading post Crow Wing gained importance when the Winnebago/Ojibwe agency was located across the Mississippi River from it.[11]

Settlement and agricultural expansion, however, was hindered by the lack of roads and the collapse of the county government. The county had no taxing system. By 1841 St. Croix County business was back in Prairie du Chien. By 1843 county commissioners stopped meeting entirely. County Clerk William Holcombe, however, sought to resurrect the moribund county government. In 1843 he petitioned the Wisconsin Legislature for a variance to permit the clerk to act as sheriff so elections could be held under his supervision. The office of judge probate was also revived. Yet even with these changes, most legal matters still had to be presented in Prairie du Chien. In 1845 Holcombe and other St. Croix citizens asked the legislature to relocate their county seat. They argued that since St. Croix County now had more people than Crawford County, it should not be subordinated to a government with a smaller population. The legislature agreed and granted St. Croix citizens the right to vote on a new county seat, yet they never held an election.[12]

The lack of roads and the log-choked river made it impossible to ship agricultural products out of the valley. Economic activity in the rudimentary settlements along the St. Croix, therefore, was limited to serving the local economy mainly through barter. The first farmers were simply loggers or mill workers who saved enough of their wages to claim land for a homestead and then sent for their wives and children. They found raising livestock was the quickest entrée into farming, as they could sell their meat and dairy to local timbermen. Once a flourmill was built in Afton, farmers grew grain to supplement lumberjack diets with pancakes and biscuits. They also collected seasoned wood from cleared land to sell to steamboats that plied the St. Croix.[13]

Settlement along the St. Croix depended upon its citizens getting control of and representation for the valley in Madison and Washington. The Wisconsin Legislature, with its sights on statehood, also was frustrated by the lack of governmental organization in the county. In 1846 it seized the initiative and selected Stillwater, the most important logging center on the Upper Mississippi, as the seat of county government. In that year Stillwater rivaled St. Paul with approximately the same number of permanent families, ten and twelve, respectively, and three to five stores. The territorial legislature also wasted no time in building the first road from Stillwater to St. Paul that year. With its northwestern county back on track, the Wisconsin Legislature began its quest for statehood and applied to Congress in 1846. By 1845 Wisconsin counted 155,000 residents in a mid-decade census, exceeding requirements for statehood. Wisconsinites proceeded to elect delegates for their December convention. Statehood seemed quickly assured.[14]

Wisconsin's northwestern boundary, however, became a hotly disputed issue that had long-lasting consequences for the St. Croix Valley and its residents. When St. Croix County was created, it encompassed the entire St. Croix Valley. Its geographic remoteness and economic uniqueness created a culture apart from other settlers in the Wisconsin Territory. Settlers saw their interests and identity as different from the farmers and miners in the south and eastern portions of the territory. Most St. Croix men were lumberjacks who came from Maine or other New England states. They bypassed most of Wisconsin and its typical path of settlement by heading directly up the Mississippi from Illinois to the St. Croix River Valley. Their acquaintances were old fur traders or soldiers who became town-site speculators and lumber entrepreneurs. Settlers here realized that this far northern country had no chance to secure any major public institutions, such as a capital city, a major university, or a penitentiary. St. Croix residents felt their best interests politically and economically would best be served by separating themselves from Wisconsin and forming a new territory, and so they proposed a boundary near the Chippewa River. However, the Wisconsin Territorial Legislature intended to keep possession of all the rich timberlands, prairies, rivers, and lakes, including the growing city of St. Paul, and to extend the boundary to the Mississippi River.[15]

The St. Croix Valley's fate also got caught up in the growing sectional crisis of the United States. When the issue of Wisconsin statehood was brought before Congress in 1846, expansionists such as Senator Stephen Douglas of Illinois proposed a third option for the territory's western boundary at the St. Croix River. They argued that a state with its boundary extended to the Mississippi would be too large to manage and by creating another state the North would gain a Free State advantage over the South and the threatened expansion of slavery. When the enabling legislation passed Congress, the politically powerful expansionists won. Wisconsin's boundary was designated at the St. Croix River.[16]

Statehood-enabling acts, however, were generally considered recommendations, not binding acts. Wisconsin's constitutional convention had the right to consider Congress's actions regarding their boundary proposal. Many Wisconsinites were appalled at the prospect that they would lose so much territory in the northwest. St. Croix County residents feared that their river valley community might be divided; yet they were encouraged by Congress's willingness to create a new territory. Therefore, much was at stake on what would happen at the state constitutional convention. The citizens of St. Croix County faced the daunting task of trying to convince both Congress and most of Wisconsin that the boundary should run further south than either of them wanted. They elected William Holcombe, who was one of the original founders of St. Croix Falls Lumbering Company and who also had interests in steam boating and land speculation, to represent their interests.[17]

Holcombe skillfully made the case that congressional expansionists were right that a state with its boundary at the Mississippi was too large, and that another state would work to the North's advantage in national politics. He also argued that Madison, nearly three hundred miles away, was too far and remote a location of government. Echoing the democratic philosophy of Thomas Jefferson, Holcombe added that St. Croix County residents were self-reliant frontiersmen who needed and wanted a government that was close by and accessible in order to exercise their democratic rights effectively. The St. Croix River Valley with a boundary from present-day Winona to the western edge of Michigan's Upper Peninsula, he proposed, should form the nucleus of a new territory made up of people with common interests in lumbering and business. Stillwater, "the hotbed of St. Croix separatism," planned to be the capital. A divided valley, he argued, would "alienate the interests of society, perplex the trade and business of the river, and retard the growth of the settlement."[18]

If Holcombe had made his case to Congress, he might have made some converts. By arguing, however, before the Wisconsin constitutional convention that the state should give up more land than even Congress proposed as well as access to Lake Superior, his amendment never stood a chance. However, Holcomb did win a compromise. The convention did agree that the valley should not be divided even at the cost of some territory for Wisconsin. It proposed a new boundary from Lake Superior south to the Mississippi, which ran approximately fifteen miles east of the St. Croix. Holcombe and his supporters seemed to have victory close at hand. If Wisconsin voters accepted the proposed constitution, the St. Croix Valley would not be part of the state. The only opposition in the state to the loss of territory came from Crawford County. Residents in Prairie du Chien feared they would be left on the fringe of Wisconsin and politically marginalized in a state dominated by eastern interests. The proposed constitution, however, also foundered on other issues. There were provisions in it to ban bank chartering, which reflected the typical western suspicions of outside moneyed interests controlling their destiny. The convention also had written a constitution that was fairly liberal for the time. Married women were given property rights, and Negroes suffrage. Conservative Badgers, even some in St. Croix County, roundly rejected the document in an April 1847 election.[19]

In 1847 Wisconsinites held a second constitutional convention. It was smaller, more politically balanced, and more representative of the sentiments of the territory than the first had been. In the interim, political influence along the St. Croix shifted to the "Bostonians"—a group of eastern capitalists that included Caleb Cushing, Rufus Choate, and Robert Rantoul Jr. In 1845 these men had formed the St. Croix and Lake Superior Mining Company. They planned to mine copper on the Upper St. Croix and to develop timber resources and waterpower at St. Croix Falls as well as at the Falls of St. Anthony. This syndicate felt their economic interests were not compatible with farmers, merchants, and lead miners in the southern half of the state. The inclusion of a ban on bank chartering in the first constitution convinced them that pioneers were in general suspicious of eastern investors. Controlling their own territory would give them the tax breaks they felt were needed to promote their economic goals. They *had* to get a boundary favorable to their interests. Some confidently predicted Cushing would be the first territorial governor with Stillwater the new capital. The Bostonians and others in St. Croix County hoped the new body's more pragmatic members might be influenced to turn over more land to the valley than the previous compromise. They selected a new delegate, George W. Brownell, to represent them. Brownell was a geologist and mineralogist who came to the St. Croix Valley in 1846, where he discovered lead and earned a living as a newspaper editor. He was also Cushing's agent.

The new delegate made a dramatic entrance into Madison by arriving on snowshoes after a three-week trek to demonstrate the remoteness of the St. Croix River Valley from Madison. Brownell reintroduced Holcombe's original proposal for the border and made the same Jeffersonian claims for local government. Wisconsin convention delegates, however, were savvier than their predecessors. Perhaps because of the close ties between Brownell and Cushing, they did not buy the argument advanced that the St. Croix Valley was "worthless" to Wisconsin. They realized that the St. Croix Valley had singular natural resources in timber and waterpower. By simply accepting the boundary of the enabling act—the St. Croix River—they could claim at least some of these resources and shore up their statehood quest by assenting to the original act of Congress. Other bolder delegates wanted to claim as much of the Old Northwest Territory as possible. They set a boundary from the first rapids of the St. Louis River near Lake Superior to the mouth of the Rum River down to the Mississippi, thereby seizing the entire St. Croix Valley as well as the St. Paul area. Their proposal swept the convention by a vote of fifty-three to three.[20]

As long as the valley was not divided, many St. Croix residents were resigned to their inclusion in Wisconsin. Others on the west side of the St. Croix River, however, wanted no part of the new state and schemed to create a new territory. Individuals such as Morgan Martin, Wisconsin's territorial delegate to Congress, envisioned a Minnesota Territory and future state that included not only land stretching from Holcombe's original border to the Mississippi, but also land further west into the Louisiana Purchase Territory. The Dakota Indians, however, would have to be approached to sell their land. Martin's fur-trading associates stood to benefit from the opening of land not yet depleted of fur-bearing animals. They also thought the St. Paul area would attract more settlers if the Indians were pushed further west. This faction took their case directly to Congress.

By the spring of 1848 the St. Croix River Valley found itself at the center of a heated national debate that threatened to jeopardize Wisconsin's admittance into the Union. The Minnesota faction found a supporter in Robert Smith of Illinois. Smith argued that the St. Croix Valley was too remote from Madison and that it would be economically unfeasible to divide settlers on the river under two governments. If the valley was left intact and outside Wisconsin, the St. Croix settlers could form the nucleus of another state. Expansionists in Congress did not miss the implications of this proposal for creating another northern state. Other congressmen, however, did not think Minnesota a viable territory because of its small population. Congress also had more pressing issues—in the aftermath of the Mexican War, the acquisition of new territory in the Southwest, and the expansion of slavery west—than reflecting on the remote northern country and potential Indian problems. For the most part, Congress was anxious to remove the Wisconsin-Minnesota issue from the national stage and create another northern state, so it sought a simple solution. It voted to admit Wisconsin to the Union on May 29, 1848, with its northwestern border as specified in the enabling act—the St. Croix River. The valley was separated for political expediency.[21]

While the rest of Wisconsin celebrated its new status in the Union, St. Croix River Valley residents lamented their separation from each other. The river that had initially united them now divided them. They found themselves under different governments and legal jurisdictions that would forever complicate life along the St. Croix. The most immediate problem was that St. Croix County, Wisconsin, did not have a county seat, since Stillwater was across the river. St. Croix County on the Minnesota side had a county seat but no legal authority under it. Wisconsin quickly rectified this situation by choosing a new county seat, initially called Buena Vista, for its portion of St. Croix County at the mouth of the Willow River. In 1852 it changed its name to Hudson. While Congress declared that Wisconsin territorial laws were in force across the river, residents there found themselves with no courts, no law officers, no legislature, and no representation in Congress. They had to politically mobilize themselves once again.[22]

Since it had been the county seat, Stillwater's residents took the lead in convincing Congress to grant territorial status to Minnesota. Word went out to settlers of the region to meet in Stillwater for a convention on August 26, 1848. Among their concerns was that in their current state of political limbo, Congress might not appropriate money for internal improvements. The convention resolved to petition Congress and President James K. Polk for a more clearly defined territorial status. They unanimously elected Henry Sibley of Mendota, Iowa Territory, to present their case. At first Congress was reluctant to seat Sibley but relented when they decided he was a delegate of the remnant Wisconsin Territory. The expansionist Senator Douglas helped him steer through Congress a bill to create a new territory. Douglas's assistance, however, tinged the Minnesota cause with a Democratic Party hue that southerners quickly turned into a sectional issue. Although independent frontiersmen and believers in popular sovereignty, Stillwater politicians were decidedly uninterested in the national implications of Minnesota statehood. But the spirit of Manifest Destiny was on their side, and on March 3, 1849, although lacking sufficient population, the Minnesota Territory was created. Its boundaries matched those of today's state. Minnesota's St. Croix County was renamed Washington County, after the first president, on October 27, 1849.[23]

Although boundaries and governmental jurisdictions were finally settled, the division of the St. Croix Valley had an initially depressing effect on its economy. Residents on both sides of the St. Croix River regretted their separation for years to come. Wisconsinites, with their North Woods identity, longed to be part of the new territory. Minnesotans, in turn, pitied their poor Wisconsin cousins for being so close to them but captive of another state. In his visit to the region in the late 1840s, travel writer Ephraim S. Seymour noticed the disheartening effect of the division of the valley. "Another circumstance detrimental to the prosperity of this place, at least temporarily, is the location of the boundary line," he wrote. "Several of the citizens [of Wisconsin], preferring to unite their fortunes with the new Territory of Minnesota, have removed from St. Croix to some of the thriving towns now springing up in that flourishing territory. The only business now prosecuted at the Falls is that of the sawmills, and incidental business connected with it."[24]

However, by the mid-1840s the federal land office was inundated with pleas that the land along the St. Croix be surveyed. In 1841 Congress had passed the Pre-Emption Act granting squatters the first opportunity to purchase lands they had already settled for the minimum amount of $1.25 an acre. This act was later replaced by the 1862 Homestead Act that granted a free quarter section to any settler who farmed it for five years. However, land along the St. Croix had not been surveyed, and some squatters started to log trees to which they had no legitimate claim. Timber speculators petitioned Congress and President John Tyler to do something about the theft of timber along the river. They feared that any legitimate business enterprise would never take root in the region unless clear title to land was established. Therefore, the General Land Office authorized the opening of a land office in St. Croix Falls in 1848. It was moved to Stillwater in 1849 and to Willow River (Hudson) in 1849. Another legal hurdle to settlement was surmounted, and the opening of legal settlement in the St. Croix could finally begin.[25]

Farmers and the Repopulation of the Valley

In the 1850s the lure of cheap land encouraged land speculators and farmers to seek their fortunes in the West. Nearly one in four Americans was on the move from one state to another. European immigration to the United States began in earnest. Wisconsin was determined to steer as many of these people to the state as possible. Once these migrants reached the Mississippi River, the St. Croix River's natural advantage as a major artery of transportation worked in its favor.[26]

In order to lure prospective settlers, however, the valley had to dispel its reputation as a frozen, barren wasteland. The tactic taken was to emphasize the advantages of seasonality. St. Croix newspapers tirelessly wrote about the North Country's comparable advantages to other regions.

Local newspaper articles were reprinted in such prominent publications as the *New York Tribune,* thus ensuring a national audience. All the leading papers argued that its cooler climate eliminated pests and health threats without sacrificing agricultural productivity. "Minnesota is the healthiest State in the Union," claimed the *Stillwater Messenger.* "Hundreds and hundreds of families are annually driven from other Western States to take up their residence in Minnesota to escape this offensive and troublesome foe [fever, ague, consumption] to the emigrant and his family."27

Not everyone was so easily convinced. Puzzled as to why someone would choose to live in the St. Croix Valley, Seymour interviewed a Marine Mills resident:

> I was impelled by curiosity . . . to inquire of Mr. Lyman why he had left the fertile soil an[d] sunny prairies of Illinois, and wandered off here in the woods, in the northern clime. . . . He stated . . . that he had been severely afflicted with the ague in southern Illinois . . . [and] had become nearly broken down by the disease. Many of the young men of his neighborhood came up to work at the Marine Mills, and he noticed that all returned with recruited health, although suffering with ague at the time of their departure.

Mr. Lyman, therefore, decided to seek relief in the North Country. When he arrived, he was so weak he said could scarcely walk from the landing to the boarding house but "was immediately restored to the enjoyment of excellent health."28

"As to the Agricultural capabilities of Minnesota as compared with those of Illinois," the *St. Croix Union* pointed out to its readers, "Minnesota is a far better country for the producing of some agricultural staples than Illinois," particularly oats, wheat, Irish potatoes, and garden vegetables. While the paper admitted that the growing season was shorter, it claimed that crops matured more quickly here due to its richer soil, nighttime rainfall, and warm sun. "The vegetable productions of a northern climate are generally superior to those produced in a lower latitude. For this reason Minnesota wheat, corn, potatoes, etc. will doubtless command a premium in the Southern markets." The North Woods were also a hunter's paradise teaming with fish, fowl, and game, and its forests and marshes were abundant in wild rice and cranberries. "We have used this kind of rice a number of times" wrote the *Hudson Star,* "and believe it to be richer and better than the southern rice, and equally wholesome."29

The soil "is rich with decayed vegetable matter, yet owing to the large proportion of silica which it contains it neither bakes in dry weather nor becomes very muddy in wet. Minnesota is almost entirely exempt from that intolerable nuisance—mud!" In an era before paved roads and when fields were manually plowed and sowed, mud produced by rain or melting snow was a major problem in New England. Minnesota's and northern Wisconsin's other advantages compared to the prairie states below them was that they were one of the best timbered and one of the best watered and water-powered regions in the country. Access to transportation was another major advantage.[30]

Given the challenge of wintering in northern Wisconsin and Minnesota, North Country settlers could not ignore the reservations expressed about the weather. Many residents tried to minimize its harsher realities by emphasizing the positive side to winter snows. "Sometimes we have good sledding for 100 days in succession," the *St. Croix Union* bragged. "The farmer, or lumberman can make all his arrangements for hauling on the snow, with a perfect assurance that they can be consummated." Humor, too, proved a useful coping mechanism. "It is an established fact," wrote the *St. Croix Union,* "that [Minnesota] is not further north than the North Pole, and that men can winter here without becoming congealed like mercury when it is subjected to cold about 400 below zero." And, "Here we have snow, snow, snow everyday; it remains with us like a true friend." The *Stillwater Messenger* could not help but gloat after eastern newspapers reported that frost had arrived in Connecticut by September 9, "three weeks earlier than Minnesota." The tinkling of sleigh bells on a bright, brisk winter day prompted many to claim that "the winters are the most delightful season of the year," and would make others "envy Minnesota life."[31]

Another advantage Minnesota boasted of to attract settlers was that it still had plenty of land that could be claimed under the pre-emption law. By 1855 federal surveying teams had not finished their work. A squatter could still claim choice land of up to 160 acres without having to first purchase it. He could farm it, thus making an exclusive claim as well as have the opportunity to sell his crops and raise money to purchase the land at $1.25 per acre before it was put up for sale. Under this system the average settler was able to beat out land speculators who purchased large tracts of land and held onto them until values increased and a profit could be made without any improvements made to the land.[32]

Land speculators were also a feature of the Minnesota frontier. Simeon P. Folsom, a land agent from St. Paul, advertised his three thousand acres near White Bear Lake in St. Croix Valley newspapers. He claimed his company aimed to sell its land to "actual farmers" and would provide them with "prices and terms that will make them the cheapest lands in that section." Tracts of land were not to be sold in parcels less than 160 acres.

Banks also tried to make a profit out of the popular pre-emption. If a squatter had not made the money necessary to purchase his land when it came up for sale, the Banking Office of C. H. Parker and Co. in Stillwater gladly provided "reasonable terms."[33]

It was recognized early on that the St. Croix's geographic position placed it in a prime position to distribute its produce locally and nationally. Throughout the 1850s many logging encampments still imported much of their food from states to the south, such as Iowa, Illinois, and parts of Wisconsin, and local boosters such as the *St. Croix Union* pointed out that for "those who desire to settle on lands [in the St. Croix] . . . the pine lumber interest will furnish them a market for surplus produce." In addition to local demand, however, was a growing awareness that the towns and cities down the Mississippi would demand St. Croix surplus. The numerous navigable streams kept communication to the outside world open for seven months each year. "The publication of these and other facts . . . characterize Minnesota as a competitor for Western Emigration," claimed the *St. Croix Union*. In 1850 Seymour noted that demand from new manufacturing towns on the Mississippi would lead to the St. Croix being "dotted with farmhouses, and enlivened with the songs of multitudes of cheerful and thriving husbandmen" to meet their needs.[34]

While newspapers of the St. Croix Valley did their best to promote the North Country, Wisconsin's state legislature decided more aggressive tactics were needed to lure settlers. In 1852 it established the position of commissioner of emigration with an office in New York. This office distributed thirty thousand copies of a pamphlet written by John Lathrop, chancellor of the University of Wisconsin in Madison, describing Wisconsin's singular features. Half of these pamphlets were sent to Europe. The remaining pamphlets were distributed to ships and hotels in the East. In the pamphlet, Lathrop tried to dispel fears potential settlers had of a hostile, northern environment by claiming that Wisconsin's climate and topography compared well with New York and New England. It was perhaps even milder and more beautiful. Its soil was richer. It was certainly more healthful than the East Coast with its diseases and epidemics and free of the hot, stagnant, malaria-infested lands of Illinois and Indiana. It had more timber than these prairie states. In 1853 traveling agents hired by the emigration office posted advertisements in more than nine hundred newspapers throughout the Northeast and Canada. Once settlers made it to Wisconsin, the office provided personal assistance. It also worked with emigration societies, foreign consuls, shipping lines, railroad companies, and freight handlers. By 1854 the emigration office opened operations in Quebec.[35]

Competition between states for immigrant settlers was fierce. Although it was still a territory, Minnesota refused to be outdone by Wisconsin. In January 1855 the governor of the territory requested that the Minnesota Territorial Legislature establish an emigration office. "We need not stop to inquire why it is that thousands of our fathers, brothers, and friends can content themselves to stick to the worn out and comparatively barren soil of the old States, rather than seek a home in this invigorating and healthy climate, and fertile soil," the governor asserted to the legislature. "They will soon find out our facilities for wealth and comfort, when we take steps to advertise them." Upon the governor's request the legislature opened an emigration office in New York City to provide "correct information of our Territory, its soil, climate, population, productions, agricultural, manufacturing and educational facilities, and prospects." The governor complained that he received numerous inquiries from other states about Minnesota winters and whether it is so cold "stock freezes to death." The governor also asked the legislature to prepare "a brief well written pamphlet giving the facts."[36]

St. Croix newspapers heartily endorsed these efforts. "The emigrant whose purpose is to find a home in the west, on his arrival at our eastern ports," wrote the *St. Croix Union,* "must hail with heartfelt joy the man who can give him reliable information in regard to any portion of our unoccupied lands, and who can instruct him what route to take in order to reach these lands." The paper expressed approval at the selection of a native of Switzerland who was fluent in many languages to be the territory's emigration agent and that he took the trouble to visit the St. Croix Valley to survey the land before he set out for the East. "We understand he is much pleased with this section of our Territory," wrote the *St. Croix Union.* "We may expect a large immigration the present season."[37]

These efforts to recruit migrants paid off. The 1850s were exciting years for the St. Croix Valley and marked the emergence of an agricultural economy beyond a subsistence level. Newcomers arrived regularly by steamboat, and in 1856 a sixty-mile military road was completed from the Point Douglas area northward to Sunrise, providing easier travel upriver. Between 1849 and 1863, Hudson conducted a "land office" business, especially in the middle years of the two decades. In 1849 it sold 9,097 acres. From 1854 through 1856 over 500,000 acres of land went into private hands. By 1863 the rush had peaked but was by no means over.[38]

The climate did profoundly affect agriculture. The St. Croix Valley is in what is known as the Wisconsin tension zone that divides the state into

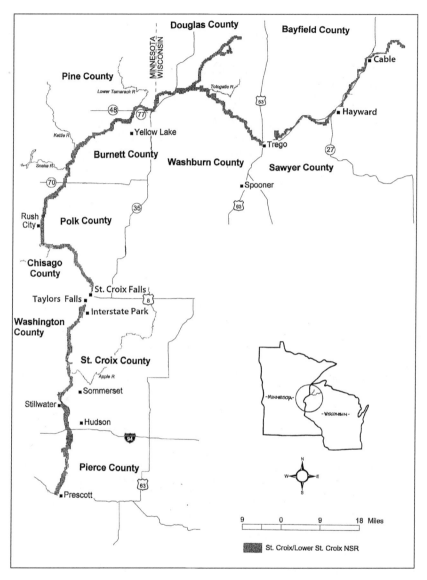

Map 6. Settlement and agriculture era, St. Croix National Scenic Riverway (Source: National Park Service).

two distinct floral and vegetation regions. In the northern zone, the climate is generally marked by cool, dry, continental arctic air masses from Canada. Its winters are longer, colder, and snowier. In the southern zone, the climate is influenced by the interaction of Pacific air, warmed and dried from its passage over the Rocky Mountains, and warm, moist, tropical air from the Gulf of Mexico. Its summers are longer and warmer, and snowfall varies. In this tension zone is a mixture of vegetation common in the southern part of the state—such as prairie, oak-savanna, and southern-hardwood forests—and vegetation common in the northern zone—such as boreal forests, conifer-hardwood forest, and pine savanna.[39]

The first farmers in the St. Croix Valley faced the challenge of discerning the quality of the soil in this geologically and climatically varied landscape. Those who arrived first had the advantage of choosing prime prairie land along the Lower St. Croix. There they found rich, virgin soil that produced bountiful harvests. In 1853 one settler noted that the "steady tread of the immigrant land looker" wore down roads where none had been before. The delta region was the first area pioneers saw before heading upriver. It did not take long for would-be farmers to recognize its potential. In 1854 a guide pamphlet titled "Description of Pierce County" described the Prescott area as composed of terraces of limestone and sandstone deposits that created "a beautiful prairie." By 1856 the wheat harvests were bountiful enough to encourage the town of Prescott to erect its first flourmill and begin exporting its agricultural surplus.[40]

Just north of the delta on the Wisconsin side of the St. Croix River was another vast stretch of prairie land that began near Hudson and extended to the Willow River, up to the Apple River on the north and down to the Kinnickinnic River on the south. On August 23, 1848, the surveying work was completed, and many squatters then made their claim. On the first day of filing, three men, Louis Massey, Peter Boucha, and Eleazer Steves, claimed nearly all the land fronting Lake St. Croix. These Yankee settlers were taken by the resemblance of the St. Croix to the Hudson River valley in New York and had the new town christened "Hudson."[41]

Hudson's steamboat landing became an immigrant port of entry for the St. Croix Valley. Newcomers from many different countries disembarked there for the rich prairie lands to the east of the town and made Hudson's population very ethnically and religiously diverse, with a Methodist, Baptist, Congregational, Episcopal, Presbyterian, and Roman Catholic church. In 1854 the *St. Croix Union* commented, "It appears to contain a very intelligent, industrious, and enterprising population, whose principal aim appears to be, to make Hudson, *the* town of the St. Croix valley." The

Stillwater paper, however, added, "With the exception of *Stillwater,* they will doubtless succeed." The Wisconsin Legislature's chartering that year of a railway line from Lake St. Croix through Superior to Bayfield no doubt provoked the paper's defensiveness. Many people along the St. Croix had speculated that Hudson would become the "metropolis of the west." Stillwater's actions to keep itself out of Wisconsin had backfired on this issue. The Badger State's legislature could hardly be expected to promote a Minnesota town as the terminus for its railroad.[42]

Church and community played a role in the settlement of the back-country prairie land as illustrated by the stories of a devout Episcopalian from New York State named Varnum Maxon. In 1846 Maxon claimed land near Cedar Lake for the Episcopal Church. He returned to New York and recruited fourteen families who arrived by wagon train in 1856. In 1854 two brothers, Thomas and Trueworthy Jewell from Massachusetts, came west. They were taken by the beauty of the countryside that reminded them of their old home in New England. They bought seven hundred acres of land near present-day Star Prairie and offered free lots to anyone back home.[43]

Irish immigrants made their way to this farming frontier in 1855. They chose homesteads along the east fork of the Kinnickinnic River, near present-day Town of Pleasant Valley. Lawrence Hawkins led the party of eighteen who began their journey back in County Galway, Ireland, in 1852. They had stopped along the way in Connecticut and then in Madison before finding their way to the St. Croix Valley, forming the nucleus of an Irish farming community.[44]

German immigrants also carved out a place for themselves in the St. Croix Valley. In 1851 Haley and Nicholas Schwalen disembarked in Hudson, having come all the way from Hunsfeldt, Germany. They were looking for a large tract of land to support a small community of German farmers. They selected a site approximately six miles southeast of Hudson and then returned to Germany. The Schwalens organized their family and friends into an emigration party and sailed for America. By 1852 they landed in Racine, Wisconsin, took the water route from Chicago to Galena, and then went up the Mississippi River. Once they reached Hudson, the entire group had no choice but to live in one building until individual families could file their claims and build their own homes. They survived their first years by selling their products, particularly butter, to the townspeople in Hudson. The "German Settlement," as it became known, eventually prospered. "The soil is the best I have seen in America," wrote H. H. Montman to his parents in Germany. "They have birch trees just like in Germany. The climate is very much as it is at home."[45]

The ethnic diversity of this prairie region was further enhanced by the arrival of Dutch immigrants to the area near present-day Baldwin. The Dutch organized a whole community to emigrate together across the Atlantic. The first Dutch pioneers reached the St. Croix Valley in 1857 after a long, hard, overland journey. Another Dutch group took trains to the Mississippi and then tried to make it the rest of the way to Hudson by river. Although technically spring when they arrived in April, the Mississippi north of Winona was still frozen over. The group continued by rail to St. Paul and then made their way by foot to the Baldwin area.[46]

French-Canadian settlers added to the ethnic mix of the St. Croix Valley. They were attracted to the prospect of larger farms than the ones they had along the St. Lawrence River. In 1851 Joseph and Louis Parent chose a spot along the Apple River. Their enthusiasm for the location was communicated in their letters back home, and soon a stream of French-Canadians made their way to the St. Croix Valley. By 1890 nearly two hundred families of French-Canadian descent lived along the Apple River up to the town of Somerset.[47]

The land further upriver from Stillwater displayed some features common to the tension zone. Most of it was heavily wooded or marsh and swamp land with smaller patches of clearings. The land was very rich and fertile with glacial sediment, but breaking ground for farming in this region was an even greater challenge since the land had more trees to clear. One of the few areas that had stretches of prairie was near Osceola on the Wisconsin side of the St. Croix. "Osceola [has a] . . . beautiful situation, commanding a fine view of the river in both directions for miles," wrote the *St. Croix Union* in 1855, adding, "There is no place that had such prospects to become ultimately a large and flourishing place of business." The town became the county seat for Polk County, and also had a good stretch of river bottomland for a steamboat landing. Also in the *St. Croix Union,* one resident wrote, "I can without any hesitation, pronounce it the best landing from Point Douglas to the Falls. . . . The land within two miles of this place is all taken up in farms and under good cultivation, and had hitherto produced abundant crops." Farmers were also lured to the area by the prospect of selling their products to local lumber companies. According to the *St. Croix Union,* "Corn is worth $1, and at this time the loggers are offering seventy five cents for oats and cannot get them at that as they are scarce. For the want of farmers up here the loggers have to buy their corn, Oats, Flour and Pork, below." Osceola became one of the first wheat-producing areas in the St. Croix Valley. Osceola Creek provided an ideal site for a water-powered flourmill that was built in 1853. However, the settlement of Polk County was much slower than the lower river because of the difficulty of breaking ground in this more wooded area. The population did not take off until after 1866.[48]

In the 1850s wheat began to rival logging as a valley export. The St. Croix River gave settlers an advantage not enjoyed by all farmers in Wisconsin. Until railroads penetrated into the hinterland of the frontier, many farmers could not participate in commercial agriculture. While many farmers in more remote regions depended upon local markets and bartered goods, St. Croix farmers were able to engage in a national market almost from the time they broke ground. Wheat became their first commercial crop, and it was ideal for a pioneer farmer. Unlike other crops, which required careful cultivation or large start-up costs like animal husbandry, wheat was easy to grow. Unlike corn, which required more refined breaking of the root-packed prairie sod to allow its own deep roots to grow, the soil for wheat only needed minimal preparation before sowing. Tree stumps did not even have to be removed. Farmers could ignore the wheat crop until harvest and in the meantime spend the rest of their time clearing more land and fencing it in. Wheat was easily stored in private warehouses. Farmers were issued wheat "receipts," "tickets," or "certificates," which they could then use in local stores. "As good as wheat" was a common expression on the frontier, which implied that it was often used in lieu of money. The *Prairie Farmer,* published in Chicago, celebrated the virtues of wheat production, claiming, "It pays debts, buys, groceries, clothing, lands, and answers more emphatically the purposes of trade than any other crop.[49]

The Willow, Apple, and Kinnickinnic rivers and their tributary creeks offered prime locations for mills to turn wheat into flour. It did not take long for commercial flour mills to appear. Caleb Greene and Charles Cox built the first mill in this area on the Willow River in 1853–54. They called it Greene's Paradise Mill, and *Paradise* became its brand name. The Bowron brothers followed suit by building a second mill near the confluence of the Cedar Lake Creek and the Apple River. In 1855 Horace Greeley, the renowned editor of the *New York Tribune,* took a trip up the St. Croix River and proclaimed, "The cry is Wheat!! Wheat!! . . . Every steamboat goes down the river with all the wheat on board she will take, and a couple of wheat laden barges fat to her side."[50]

Wheat was not the only export from the St. Croix Valley during these years. Cranberries grew wild in the abundant marshlands scattered throughout the woodlands. It did not take long for settlers to realize the potential profit from harvesting what grew naturally and cultivating it to enhance their harvest. The first experiments with systematic cultivating began in the late 1840s near Stillwater. Travel writer Seymour noted, "the soil and climate of this region are so well adapted to their culture . . . it is not unreasonable to presume that its culture may, hereafter, become so general as to render it a prominent article among the staples of Minnesota." The berries fetched a good price. "We hear of [cranberries] selling at from $3 to $4 per bushel," wrote the *St. Croix Union* in 1855. The paper explained the advantages of cultivating cranberry vines. "One acre will yield from 200 to 400 bushels each season. This would be better than corn, or wheat or almost any other crop." By 1859 the St. Croix cranberry trade exceeded $10,000 with five thousand bushels of the fruit harvested, according to the *Stillwater Messenger*. Wild blackberries also had a ready market. "These berries are found in great abundance in the valley above us," the *Stillwater Messenger* reported, "and are of a very fine quality."[51]

The Swedish Frontier

On the Minnesota side of the St. Croix opposite Osceola stood another forested region with a series of beautiful, pristine lakes known as Chisago Lakes. When pioneers made it to Taylors Falls, they reached the head of navigation on the river. There they disembarked into the most splendid scenery along the river. The colorful Dalles rock formations soared from 50 to 250 feet straight up. While land along the river was either bought up by speculators or under the control of logging interests, much of the inland lakes region could still be squatted on or bought at the government rate of $1.25 per acre. Within a decade the area attracted scores of Swedes who would eventually make this region the largest Swedish-speaking rural area outside of Sweden. This nearly exclusive dominance of one ethnic group was unusual not only for the St. Croix Valley but for the entire country.[52]

What made this concentration of Swedes possible was that the Chisago Lakes was off the beaten path for westward migration. Most pioneers preferred the prairie and hardwood forests to the south. Coniferous forests to the north were known to have stony, acid soils with little organic materials. The Chisago Lakes area, however, was in the transition zone between the hardwood forests in the south and the coniferous forests of the north in what is called the "mixed forest." Most settlers assumed the land was infertile. They were wrong. Glacial sediment had enriched the soil, and it was a handful of Swedish immigrants who first realized this.[53]

In 1850 Carl Ferstrom, Oscar Roos, and August Sandahl claimed a forty-acre site near Hay Lake. While they chose to move on the following year, they sold their farm to another Swedish immigrant, Daniel Nilson. His home formed the nucleus of the growing Swedish community. These pioneers grew corn, potatoes, and rye their first year. In 1842 Eric Norberg came to America and explored the Minnesota Territory after the land office was set up on the St. Croix in 1848. In his travels, Norberg met Gustaf Unonius, a bishop of the Swedish Episcopal St. Ansgarius Congregation in Chicago. Unonius was very devoted to assisting Swedish immigrants find a good life in America and was constantly on the lookout for better opportunities. When he heard Norberg was heading north, he asked him to send back a report. In 1851 Norberg wrote to Unonius that Minnesota would make a better settlement for Swedes than Illinois. He wrote, "West of Taylors Falls . . . there are many lakes, streams, and rivers, and a better place for a large settlement I have hardly ever seen." He urged Unonius to send Swedish immigrants here. In June 1851 the first group of Swedish farmers arrived by steamboat and promptly cut a road from Taylors Falls through the woods to the lake region. In this more remote area they took advantage of the pre-emption law and claimed this uninhabited land. The resemblance to their homeland was appealing and comforting. Within five to six years nearly all of the government land in the Chisago Lakes area was claimed, almost entirely by Swedes. This migration from Sweden eventually encompassed the entire region between the St. Croix and the Mississippi Rivers. Through letters to their homeland or to other Swedish American communities, they began one of the most exclusive chain migrations to one area in the United States.[54]

This Swedish community remained isolated for years due to its remoteness, lack of transportation networks, and continued dependence on a common foreign language. Rudimentary farming skills also hampered their access to a market economy. "We were on the whole a poorly selected company as none of us was skilled in any trade," wrote Oscar Roos. "None of us could cut hay to feed our oxen." Most did not have draft animals and lived at a subsistence level, living off the wild game and fish in the surrounding woods. They grew white and brown beans, hay, and rutabagas. Later on they began to grow potatoes, corn, small grains, and garden vegetables. The men and older boys often hired themselves out to logging companies in the winter for some cash. Once they were able to acquire a cow, making butter and cheese became a means to barter for store-bought products, such as sugar and coffee. Chickens and hogs came later. Their first customers were the logging companies. It took many years before they joined the wheat export economy.[55]

Despite their aloofness, the Swedes were well regarded. "The Swedes are a moral class of people," wrote one observer. "They are very industrious and strictly honest. They attend to their own business, and let the balance of mankind attend to their own." While preferring to speak their own language among themselves, the Swedish community made English a priority for their children, which pleased Americans. "We learn with pleasure that English is taught to all the children . . . it behooves them to instruct their offspring so as to fit them for the community in which they live."[56]

By the late 1850s agricultural settlements penetrated as far north as Sunrise. "Its advantages in an agricultural point of view are first rate," wrote one settler. "It has an excellent range for cattle, and abundance of timber of every sort . . . excellent bottom, meadow and upland." The local pineries provided a profitable market. "For doing business with the lumbering interest this point has a decided advantage on account of its proximity to the pineries." Farmers here could sell their goods at "Stillwater prices" and take the added advantage of the "transportation" markup without having to transport goods upriver and still have competitive prices.[57]

The St. Croix Valley also prospered by catering to disembarking migrants at steamboat landings who planned to head further west. After the Sioux treaties were signed in 1853, Minnesota's southwest territory opened for settlement. Pioneers crowded steamboats that plied the Mississippi, the Minnesota, and the St. Croix rivers. In 1855 the *St. Croix Union* wrote, "The immigration to Minnesota the present season bids fair to be immensely large, exceeding by many thousands that of any preceding year. . . . Capt. Smith estimated the number of the emigrants . . . to be at least four thousand." Towns along these waterways hustled to assist and provision these sojourners before they continued their westward trek on new government roads. The bustling activity of the Taylors Falls and St. Croix Falls area became so great that in 1854 prominent residents formed the St. Croix Bridge Company, and by 1856 the bridge was completed. It "was the first bridge that spanned the St. Croix and Mississippi rivers."[58]

The population along the St. Croix had grown so fast that in 1853 the Wisconsin Legislature created two additional counties out of St. Croix County—Pierce to the south with Prescott as its county seat and Polk County to the north with St. Croix Falls its seat. On the Minnesota side of the river, the influx of settlers prompted a movement to create Chisago

County north of Washington as early as the fall of 1851. By January 1852 the first county commissioners' meeting was held at Taylors Falls, the new seat of government for Chisago County.[59]

Land Speculation and Growing Pains

In 1856 John Bond expressed the optimism of the time when he wrote in his territorial guidebook, "The immigration to Minnesota is composed of men who come with the well-founded assurance that, in a land where Nature had lavished her choicest gifts—where sickness has no dwelling place—where the dreaded cholera has claimed no victims—their toil will be amply rewarded, while their persons and property are fully protected by the broad shield of law." This rosy picture of the North Country led to a "real estate mania" in the valley. Yankee land speculators bought up land along the river. There were high expectations that the timber resources and waterpower would transform the region into a great manufacturing center. In 1855 the *St. Croix Union* wrote that from Taylors Falls began "a succession of falls and rapids six miles in length, creating one of the most extensive water powers in the North-west, and easily controlled. With the materials for manufacturing with which the St. Croix Valley abounds, there must spring up here manufacturing cities which may surpass Lowell, Nashua, or Lawrence [Massachusetts]."[60]

Land that had been claimed under pre-emption and bought for $1.25 per acre sold for $5 an acre after it had been "improved" with a small lean-to and a small tract of broken sod. Dozens of towns were platted, many before they were even surveyed. Buyers both in the St. Croix Valley and in the East eagerly bought the land. Eastern capitalists deposited large sums of money into St. Paul banks that in turn made loans of 3 percent to land investors in the valley. In some of the more settled towns, it was said that a parcel of land that sold for $500 in the morning could fetch $1,000 by evening. All it took was a fast-talking salesman who could convince an unwary buyer that he could hear a train whistle in the not-too-distant future. As a local commentator put it, "Every settler felt himself a prospective millionaire, and the public imagination soared high with greedy hope." Even remote settlements like Sunrise encouraged speculators. In 1856 a local resident wrote in the *St. Croix Union* that "the advantages of this point as a place for investments . . . [add to] the confidence of everyone owning property here . . . , I will venture to say, [its value] has increased, is increasing, and ought *not* to be diminished."[61]

St. Croix Falls, however, did not experience this rush of development. The town was situated in a picturesque setting near waterfalls that dropped fifty-five feet in a six-mile stretch that never froze in the winter. This feature made it an ideal location for year-round water-power. Ironically, it was this asset of St. Croix Falls that ignited the "greedy hope" that arrested its development. In the 1850s a bitter lawsuit between Caleb Cushing and William Hungerford kept either from fulfilling their economic ambitions. When it was finally resolved in Cushing's favor in 1857, a financial panic hit the country, and Cushing, back east, lost touch with his Wisconsin interests. His lack of attention continued throughout the Civil War. Cushing and Hungerford "unitedly accomplished the ruin of their town," wrote William H. C. Folsom. Another hindrance to pioneering in the St. Croix Falls area was, of course, its dense forest land. It did not help matters that the land office moved further downriver to Stillwater in the Minnesota Territory and then to Hudson in 1849.[62]

Although giddy with hope for the future prosperity of the St. Croix Valley, the communities here did not escape the growing pains experienced by the rest of the country. By the mid-1850s many Anglo-Protestant "nativists" were alarmed at the large number of foreigners, especially Catholics, entering the country. They felt the essential free and democratic character of the country was threatened by the highly centralized, authoritarian Roman Catholic Church. Catholics had not been schooled in independent, democratic practices, they argued. Their church required slavish devotion, and their parochial school undermined the public school—the backbone of a democratic society. The American Party, or the Know-Nothings, aimed to restrict immigration by limiting office holding to native-born Americans and to restrict citizenship to those with a twenty-one-year residency. This platform was aimed at states like Wisconsin that had generously offered the vote to foreign-born persons who lived in the state for one year—a first in the country. By the end of the decade, however, the nativist movement foundered on the growing sectional crisis over slavery, and the Know-Nothing Party disappeared from the national stage. Some of its members were absorbed into the new Republican Party that muted nativism under its slogan of "Free Soil, Free Men, Free Labor." However, the damage was done in Wisconsin. The state legislature was forced to disband its entire emigration agency.[63]

Nativism reached even the raw frontier communities along the St. Croix River populated by New England Yankees. After an election in 1857 the *Stillwater Messenger,* a Republican paper, printed derisive comments about Irish voters. The paper complained that they voted as a "solid phalanx" in soiled, shabby clothing. The Democratic *St. Croix Union* angrily responded:

> The Messenger has unwittingly showed the cloven foot. The great majority of the black Republicans hate an Irishman much worse that they do the devil himself, and the Messenger . . . publicly proclaims what is privately taught by their leaders. They hate an Irishman, and some of them are determined that the Irish, and Dutch, and French, and all foreigners, shall be deprived of the right of suffrage . . . *Our* opinion is, that any foreigner, if he be naturalized, or has declared his intention to become naturalized, has as good a right to vote as a native born.[64]

Protestant immigrants, apparently, were acceptable. In 1860 the *Stillwater Messenger* described Scandinavians as "hardy, industrious, frugal and honest people" who were "succeeding well" in Minnesota. Of the Germans it wrote, "The enterprise of the German people who had come among us, and the success which has almost invariably rewarded their laborious and indefatigable efforts . . . have had the effect to induce a large emigration of that worthy class of people among us." But bitterness toward Irish Catholics still lingered in the paper when it wrote, "The emigrants expected from Ireland this season, are said to be of a superior class from those usually found among us, belonging mainly to the agricultural class, educated, and pretty generally Protestants, and possessed of some money and means."[65]

The valley, however, needed immigrant labor too much to risk offending the foreigners. The *St. Croix Union* argued for a more pragmatic approach to the immigrant issue:

> Strike out what the Irishman has done for America, and the country would be set back fifty years in the path of progress. Corn would grow where the Erie canal bears the freight of millions of fertile acres; the lumbering coach would take the place of flying trains on ten thousand miles of railroad. . . . Hundreds of millions of dollars could not purchase from the American people the property and advantages that have absolutely been bestowed upon them by Irish labor . . . it is an essential element in American thrift and progress and we could not lose it for a month without recurrence of chaos.[66]

Despite this anti-immigrant sentiment, Wisconsin's emigration office had laid the basis for the state to become one of the most ethnically diverse in the country in the nineteenth century. However, the recruitment of new immigrants would for the time be left in private hands. "American letters" from friends and relatives who made it to Wisconsin enticed their European counterparts to make the journey themselves. Churches also played a critical role in "chain migration," where letters, money, immigrant guide books, and even prepaid ship fares were sent to the old country to bring the next group over to start a new life on the Wisconsin frontier.[67]

Sectional issues and the collapse of the Know-Nothings were perhaps less important to the decline in nativism in the St. Croix Valley than economic issues. In 1857 a natural downturn in the national business cycle provoked a general panic as holders of banknotes rushed to their banks for redemption. Many banks temporarily suspended specie payment, and others completely failed. An economic depression followed as businesses throughout the North folded. While most of Wisconsin's banks remained solvent, it was among the states worst hit by the economic downturn. For two to three years coins and reliable banknotes were extremely hard to come by. Wisconsin businessmen could not pay their eastern creditors. Loans were impossible to obtain. In many areas the economy reverted to the primitive system of bartering goods and services. Federal land sales plummeted.[68]

The panic reached into the St. Croix frontier, since logging companies and land speculators had financial ties to the East and saw its markets for lumber collapse. Cash and credit quickly evaporated. Land agencies folded. Several banks in the valley, such as the Hudson City Bank, the St. Croix Valley Bank, the Farmers and Mechanics Bank, and the Chisago County Bank, were forced to close their doors. Wildcat currency was refused, and barter became the only means to exchange goods. In November 1857 the *Stillwater Messenger* made a plea for farmers to keep the economy solvent by selling the wheat they had been withholding from the market in the hope of fetching higher prices later. "There cannot be a greater mistake than for the farmers to hold on to their grain in hopes of higher prices, rather than sell at present prices and pay their indebtedness to country merchants and other creditors," the newspaper wrote. "For in the present unsettled state of the money market, capitalists will not invest money for wheat in store, except at very low rates . . . a considerable rise in the price of wheat is very dim, and it is more likely to go lower than higher." It argued that it was also a matter of "justice and common honesty . . . [to] sell their crop and pay their debts. . . . It is their neglect to do this, which is the main cause of the money pressure," the *Messenger* asserted. "The city merchant cannot pay his debts in New York or Boston, because the country merchant cannot pay because the farmers . . . are indebted to him for one, two, and sometimes for even three years' purchases. And here is the root of the whole evil."[69]

The financial panic also brought the land speculation frenzy to an end. Antipathy toward speculators had been building for the last few years. Certainly the *Cushing v. Hungerford* affair was the most glaring example of speculators single-handedly interfering with "progress." Cushing became the embodiment of the evil nonresident speculator in the St. Croix Valley for years to come. Residents from Marine Mills complained as early as 1855 that speculators were responsible for the slow growth of the town. "The most pernicious of all, which has, perhaps, caused the sluggish growth of this town," wrote one settler, was "the grasping, griping propensity: 'Get all you can and hold on to all you get,'" of land speculators. Land in the valley was "principally owned by non-resident speculators," complained one resident, which "renders its agricultural development almost impossible. This deplorable landed monopoly is the only barrier this portion of Minnesota meets in its onward progress." Many of the towns that had been platted with high hopes and expectations never materialized. Land speculators in Marine Mills were forced to unload their monopoly on town lots "on very favorable terms." Folsom wryly noted that the "town" of Drontheim in Chisago County, which had been platted in 1856, was "still a brush and swamp plat" in the 1880s. The "town" of Chippewa in Chisago County, which was platted the same year, Folsom facetiously noted, "makes a fair farm."[70]

The financial panic also changed the course of agricultural settlement and development along the St. Croix. Hundreds of people, who had been reluctant to get into farming because of the insects, drought, and prairie fires that plagued the first farmers, abandoned towns for the land when money and employment evaporated. By the spring of 1858 the amount of land brought under cultivation in Minnesota had doubled. In Marine Mills the change in the economy finally brought to the town the completion of a flourmill that it bragged was "superior to any other in Minnesota."[71]

The experience of Irish immigrants to the St. Croix provides a good example of how an economy that went bust pushed them into farming and respectability. Wisconsin's eagerness for internal improvements and economic development of its North Country acted as a recruiting agent for immigrants to the state and the St. Croix Valley. Soon after the Wisconsin Legislature chartered the Lake St. Croix, Superior, and Bayfield railway to extend through Hudson, Congress relinquished the land for the right-of-way, and the people of Hudson raised $47,000 in subscriptions for the railroad. When work began, a large colony of Irish railway workers came to Hudson. They joined their compatriots who had come to work on the river in the logging industry. When the financial panic struck, many Irish loggers and railroad workers lost their jobs. Many moved out onto "the prairie" and joined the small Irish farming community in Erin Prairie on the Willow River. In 1860 nearly all of the heads of families in Erin Township were born in Ireland. Erin Prairie became a prosperous farming community and dispelled the negative stereotypes of the Irish as "barroom loafers" and "ignorant."[72]

While national demand for lumber eventually helped the St. Croix Valley make some recovery from the panic, agriculture played a significant role in its economic turn-around. St. Croix farmers, anxious to make quick cash, continued to grow wheat. Cheap land, good soil, and access to a national market made this possible. The St. Croix and Mississippi rivers, of course, were the main highways for agricultural export, particularly to states to the south. By the late 1850s railroads enhanced their connections to eastern markets. In 1857 a railroad reached Prairie du Chien, and another reached La Crosse in 1858. Wheat was then shipped to the Port of Milwaukee and east through the Great Lakes.[73]

The St. Croix Valley's wheat boom was possible because its soil was rich and fertile when the soil in the lower Midwest and Great Lakes region was exhausted by single-crop agriculture. A contributor for the Ohio *Farmer* admitted to the *Stillwater Messenger,* "It is already shown that Ohio can produce crops that are better for her soil and climate than wheat; one good crop in three years is about all we can expect." In southern Michigan, southern Wisconsin, Indiana, and Illinois, "the result is no better." The scientific thinking of the time argued that wheat grew better above forty-five degrees latitude and corn grew better below that. "When the country has had a few more years of cultivation," the Ohio *Farmer* advised Minnesotans, "there will appear in the New York market, flour barrels with Minnesota brands, and that it will be of superior quality." Railroads, the *Messenger* argued, were key to reaching eastern markets via Lake Superior. "That accomplished, the time will not be far distant when Minnesota flour will find European markets."[74]

By May 1859 the *Stillwater Messenger* could brag, "The past season is the first season that Minnesota has been a produce-exporting State." One steamboat leaving Hudson was laden with 2,377 sacks of grain and potatoes. Another packet shipped out 3,500 sacks. The bountiful harvest of 1860 "will long be remembered as a very propitious and fruitful season," the *Messenger* boasted. "One great feature of the soil and climate of our favored Minnesota, is the great certainty of producing a good crop every year. There has not been a general failure of any crop since the first settlement of the country."[75]

In many respects, though, wheat production was a symptom of the poverty of frontier life because of its ability to command cash with minimal investment in labor, time, and material. Few people in the St. Croix Valley paid any heed to other regions with declining yields. Many immigrants with little means did not have the luxury to think long-term. They had to survive in the present, even if that meant compromising the fertility of the soil in the future. Many migrants from eastern states were creatures of habit who left the East because of soil exhaustion and moved on rather than adopt better farming techniques.[76]

The seriousness of mono-agriculture could be seen in the central part of the state that went from producing twenty to twenty-five bushels of wheat per acre to only five or six bushels just a few years later. In addition, in some years limited snowfall in the winter left soil moisture so low that it led to crop failures in the early 1850s. These crop failures prompted an agricultural reform movement in Madison. In 1851 legislators and interested citizens organized a state agricultural society and sponsored the first state fair and cattle show that fall in Janesville. The society also encouraged the formation of county-level societies. Through state and county fairs, agricultural reformers hoped their exhibits and demonstrations would encourage farmers to adopt scientific farming practices, such as crop rotation and diversification, use of fertilizers, stockbreeding, animal shelters, raising hogs and sheep, the best grasses for hay, and producing better butter and cheese.[77]

Some counties in the St. Croix Valley began to heed the warning. St. Croix County organized an agricultural society in 1857, and Pierce County followed suit in 1859. In 1855 the first annual fair of the Minnesota Territorial Agricultural Society was held in Minneapolis. Although Washington County showed its agricultural bounty at this fair, its subsistence economy prevented its establishing a society until 1871. Minnesota newspapers reported on the successful cultivation of other crops. "Onions are among the vegetables which luxuriate in our soil," proclaimed the *Stillwater Messenger.* "Several specimens measured 13½ inches in circumference." As to potatoes, "Minnesota produces this vegetable in perfection." Tomatoes here grew "in greatest abundance," some even weighing "a trifle of two pounds." The exorbitant price of fruit brought up from Illinois prompted the local newspapers to encourage farmers to take up fruit growing. "That several excellent and healthy fruits can be raised here, we have not a particle of doubt," chided the *St. Croix Union.* "We do therefore trust that our farmers will set out more fruit trees. . . . It is a duty they owe, not only to themselves and family, but to posterity." The paper ran subsequent articles on how to plant an orchard.[78]

The Civil War Years in the St. Croix Valley

Before the St. Croix Valley could relish the prosperity of the wheat boom or plan for the future, sectional conflicts between North and South ignited the Civil War. Many other issues, seemingly on the surface unrelated to slavery, were tied to sectional issues. In response to this new political situation, Wisconsinites of all political stripes met in Ripon, Wisconsin, and formed the Republican Party. Although former Whigs dominated the party, they minimized their former anti-immigrant and temperance crusades. Enough disaffected Democrats and Free Soilers felt they found a viable alternative to the "conspiratorial slave-power" Democrats and joined the new party. Wisconsinites and residents of St. Croix County moved into the Republican camp. By the 1856 presidential election, most Wisconsin counties voted Republican, including Pierce, St. Croix, and Polk counties. The only holdouts for the Democrats were Irish and German Catholics, and other non-English-speaking immigrants with the exception of Scandinavians.[79]

It was against this political backdrop that the Whig/Republican *Stillwater Messenger* and the Democratic *St. Croix Union* of Stillwater fought their respective local battles. National issues also crystallized the values of St. Croix Valley residents. These rugged pioneers switched to the political party that promised to safeguard their homesteads and independent labor—the Republican Party. When war broke out between the North and the South in April 1861, however, there was not an enthusiastic response to the first call to arms in Wisconsin except among Anglo-American Protestants. Foreign-born immigrants were reluctant to commit themselves to the cause of the North. German and Irish Catholics had trouble forgetting the nativist sentiment among many Republicans and would not put their lives on the line for a country whose host population disdained them. While many ethnic groups did not want to see slavery spread, few cared about the abolitionist cause. Given its high immigrant population, Wisconsin did not contribute as many men to the Union in proportion to its population as Illinois, Michigan, Indiana, Ohio, Minnesota, and Iowa.[80]

While statistics for the Wisconsin side of the St. Croix are not readily available, Washington County in Minnesota took great pride in its wartime contribution. "Every citizen of Washington county should feel proud of her war record," wrote the Republican *Stillwater Messenger*. "Compared with other counties, she is not behind any in patriotism: while she distances nearly every other, in proportion to population, in the number of troops furnished for suppressing the rebellion." Of its total population of 6,770, there were 980 men who fought for the Union. The *Messenger* pointed out that this figure did not include the number of men who were enticed into Wisconsin regiments where they could receive generous bounties. These bounties were attractive to pioneers as a means to get out of debt or purchase a farm.[81]

On the home front the war brought prosperity to the North. By 1863 Wisconsin farmers enjoyed a boom like never before. Wheat and other agricultural products were in great demand, and Wisconsin's and Minnesota's rich, virgin soil allowed their farmers to cash in. Wartime profits and the shortage of men stimulated the mechanization of farming. A mechanical reaper could replace four to six men in the field. While reapers were fairly common throughout most of Wisconsin before the war, St. Croix pioneers could now afford them. Threshing machines had been rare in the state before 1860, but increasingly more and more farmers began using them.[82]

These labor-saving devices also had the effect of spreading wheat-growing mania. Since wheat was already easy to cultivate, labor-saving machinery made it possible for one farmer to bring even more land under cultivation, and labor scarcity diminished the possibility of turning to more labor-intensive crops that required more vigilant hoeing and tending. Proximity to the St. Croix River also kept farmers in the valley tilling wheat since this gave them access to the national demand for wheat.[83]

As early as August 1861, the *Stillwater Messenger* was able to remark about the countryside, "Four years ago, but one farm had been opened. . . . Now scores of them, in a high state of cultivation, are to be seen, where there was either unbroken prairie or forest." The rattle and clatter of reapers made "merry music" as they harvested the "golden sheaves" of wheat. "The farmers of Minnesota have abundant reasons to rejoice," exclaimed the Stillwater newspaper. "Far removed from the desolating track of war, and with the granaries filled to overflowing, they have great cause for the most profound gratitude." The wheat bonanza was added proof that farming in Minnesota was not only possible but also profitable. In 1861 wheat sold for about 50 cents a bushel. By 1866 it fetched more than $1.50. The *Messenger* published an example of wheat fields turning into gold: "In 1863, J. W. Treager purchased thirteen hundred acres of unimproved land in

Washington County . . . for which he paid $10,000. . . . In the summer of 1863 he broke seventy-acres, upon which he raised a crop in 1864. That crop was sold for sufficient to pay for the land upon which it was raised, for breaking and fencing it, and all the expense of raising, harvesting and marketing the crop, and $1,100 besides." Farmers in Washington County who reported wheat yields of twenty-four bushels per acre abundantly demonstrated the richness of the soil in the St. Croix Valley for growing wheat. The war itself, of course, stimulated this demand for wheat. American farmers also benefited from the misfortunes of Europe stemming from the Crimean War. However, war also brought about inflation, and farmers' expenses for new machinery and other goods also rose. Many farmers took on heavy debts to expand their operations. Thus their dependence upon wheat continued at the expense of diversification.[84]

The Farming Frontier Moves up the Valley

When the Civil War ended in 1865, Wisconsin still had nearly ten to eleven million acres of unsold land, which amounted to approximately one-third of the entire state. Most of it was considered valuable for its timber rather than its suitability for farming. In 1866 the federal land office opened up the last big tract of land in Wisconsin, amounting to nearly 6.5 million acres. Most of it was located in the Chippewa pinery, but the Upper St. Croix also went up for sale. A total of 7.5 million acres in agricultural scrip was sold. The holders of this scrip came from all over the country, and most buyers bought tracts of land of at least eight hundred acres.[85] In this period, however, the average settler to the frontier and the St. Croix Valley faced stiffer competition for land. While land was still available through the public land office under the Homestead Act or even pre-emption claims, most of the best land went into the hands of speculators. Farmers had to pay more for this land up front. Homestead land not only had the $10 fee and no title to it for five years, but it also might have wetlands that needed to be drained, have poorer soil, or be off the beaten path with little access to markets before railroads and graded roads were built. If they were lucky, some newcomers were able to purchase farms that were already developed by a previous owner.

Caleb Cushing, the persona non grata of St. Croix Falls, made the largest purchase, 33,000 acres, in Polk County. In 1868 Cushing had helped organize the Great European American Emigration Land Company and served as its president.[86] This company was incorporated in New York

State with a million dollars of capital from investors in New York, Georgia, Wisconsin, and six other states with branch offices in Stockholm, Hamburg, and Liverpool. They directed immigrants to *their* land, not state-owned land.

Cushing's big schemes, however, often failed because he often selected incompetent or unscrupulous agents to execute his plans. The general manager of the Emigration Company, Henning A. Taube of Stockholm, was one of them. The Stockholm office provided an elaborate prospectus on how to reach St. Croix Falls, the cost of getting there, employment prospects in logging, and the opportunity to buy good farm lands from the company. The first group of 125 Swedes reached the St. Croix in early summer 1869. However, confusion over titles made lands not available for purchase until October. Taube then sold certificates of title to the land before the immigrants arrived. However, colonists found they could not exchange these certificates they bought from Taub in Sweden and Prussia for land in the St. Croix. Taube dumped the problem on Cushing. To his credit, Cushing honored the certificates and refunded the settlers' money. They all found land outside St. Croix Falls. Cushing resigned as president and trustee of the company. All the other trustees did as well, and the Great Emigration Company closed.[87]

Cushing still owned land in Polk County and continued to buy more. By 1875 he owned 45,000 acres in the county, and he finally found a reliable agent in J. Stannard Baker. Baker took control of the Cushing Land Agency in 1874 and quickly turned the enterprise around. "His appearance," wrote Alice E. Smith, "marked the end of thirty years of mismanagement resulting from absentee landlordism, controversial claims, lack of policy, and negligence." Baker put an end to trespassing on Cushing's land for logging, wild hay harvesting, and cranberry picking, and kept Cushing's tax rate low. His agricultural lands, sold to both lumbermen and farmers, finally began to produce a return after thirty-three years of investment.[88]

Articles placed in Madison newspapers lured a colony of Danes to Cushing's lands near the town of Luck. West Denmark, as the settlement was called, stretched over three townships. "These hardy Scandinavians were very thrifty and industrious," commented Harry D. Baker, son of J. Stannard Baker. Their only goal was "to get homes and [they were] willing to go into the wilderness and cut down the heavy hardwood timber and build their log houses and clear small patches of land from which to raise crops for the necessities of life. . . . They . . . were a wonderful asset in the development of this heavy hardwood timbered country." They were far from the St. Croix River, which limited their opportunities to join in the export of agricultural products and isolated their communities until the railroads came through. The virgin soil had the characteristics of the northern end of the tension zone. It was mostly black loam with a subsoil of clay or gravel. While it produced good crops for small grains, vegetables, hay, and corn, farmers found it unreliable. They became dairy farmers much sooner than other settlers further south and claimed to have established the first cooperative creamery in the state of Wisconsin.[89] Cushing eventually sold the rest of his land to William J. Starr of Eau Claire. These heavily timbered lands were logged off and eventually became farms.

However they were able to obtain land, settlers and immigrants continued to be attracted to the St. Croix Valley. Between 1865 and 1873, newcomers to Wisconsin arrived and followed the paths of settlement that were laid out by the first wave of settlers. Virtually the same ethnic groups continued to migrate to the state. The difference in settlement from the antebellum years was that large tracts of public lands were often bought up before anyone took up residence. These newcomers were also not solely dependent upon the waterways to reach their destinations. Railroads took them deeper into the North Country. By 1870 the population of Wisconsin increased to over a million inhabitants, and it kept its position as having one of the largest immigrant and foreign-born populations in the country.[90]

The state of Wisconsin renewed its efforts to recruit foreign emigrants. In 1867 Wisconsin established an unpaid board of immigration. Its main strategy was to distribute a thirty-two-page pamphlet written by Increase A. Lapham that described every facet of the state, including its location, resources, educational institutions, churches, system of government, rights of citizens, the Homestead Law, and routes to the state. The cheapness of land in Wisconsin relative to other areas was particularly stressed. Besides its English version, this pamphlet was translated into German, French, Welsh, Swedish, Norwegian, and Dutch. The governor appointed a three-person committee in each county to compile a list of names of family and friends in the Old Country to whom to mail the pamphlet. The state of Wisconsin thereby institutionalized the pattern of chain migration that had begun in the 1850s. Pierce, St. Croix, and Polk counties continued to attract Irish, Swedes and Norwegians, some Germans and English, and a few Danes, along with the native-born Americans from eastern states.[91]

In 1871 the Wisconsin Legislature created an elective commission with a full-time office in Milwaukee and a part-time agent in Chicago. The first elected commissioner, the Norwegian-born Ole C. Johnson, sent pamphlets directly to government and emigration agencies in England, Germany, Belgium, Denmark, and Norway. Johnson also provided the extra perk of free travel for women, children, and elderly men on the Milwaukee & St. Paul Railway to their Wisconsin destinations.[92]

Minnesota joined the competition to recruit immigrants. In 1867 the state established its own board of immigration. Its secretary was Hans Mattson, a Swedish immigrant and pioneer. He also served as a land agent for the St. Paul and Pacific Railroad. Mattson journeyed to his homeland to recruit prospective settlers for the Minnesota frontier. "He was a modern Marco Polo returning from fabulous lands beyond known horizons," wrote historian Theodore Blegen, "and he never ceased to describe his chosen state as a land of milk and honey." Minnesota's Immigration Board also sent out pamphlets in Norwegian, Swedish, German, Welsh, and English to other recruiting agents back east and to Europe. Railroads offered cheap fares for immigrant families and provided temporary shelter for them when they arrived. Ever sensitive to its reputation for cold weather, Minnesota made more exaggerated claims of its health benefits for ailments such as ague and consumption than it had in the 1850s and minimized other sicknesses common on the frontier, such as diphtheria and typhus. This strategy worked well. In 1879 the *Rush City Post* wrote that the night train brought in nearly "three-hundred Swedes who were on the way to locate in this and Burnett counties. . . . Mr. [Charles] Anderson went over to Sweden sometime early in the spring for the purpose of bringing over a ship load and he succeeded well." These Swedish immigrants made an enthusiastic impression on the *Post*. "A better more healthy and well dressed lot of foreigners never landed in Rush City before. . . . They make the best of citizens."[93]

Railroads: Regional Rivalry and Growth

As the population grew and more farms appeared in the post–Civil War period, railroads became the focus of economic expansion for entrepreneurs, businessmen, and farmers. It was clear by the 1870s that the waterways had reached their limit of development. The Upper Mississippi River was not always navigable around the Rock Island rapids. Its tributaries in Wisconsin were also undependable. The Wisconsin River had constantly shifting shallows. Logs and lumber rafts choked the St. Croix and other rivers. In an era of horse- or oxen-drawn vehicles, public roads were not a viable access to markets. The railroads represented the most modern and efficient mode of transportation.[94]

In 1856 and 1864 Congress had set aside nearly four million acres of the public domain in Wisconsin for railroad construction. The railroad companies that were granted land under the 1856 act abandoned projects after the Panic of 1857 and much of the land was forfeited. In the 1864 Railroad Act, Congress renewed and enlarged the land grants. One of the three grants was for a northwest route from the St. Croix River or Lake St. Croix to Bayfield on Lake Superior. Another route was to connect the St. Croix to Tomah, approximately forty miles east of La Crosse. The third was to take a north-central route. Congress had stipulated that the first two railroads must be completed in five years and the last within ten years. By 1869, however, the St. Croix & Lake Superior Railroad Company had not laid any track along its proposed route from Hudson to Bayfield, thereby forfeiting its claims. In 1871 Congress proceeded to select the North Wisconsin Railway Company to build the line, with the assumption that the same land would be granted to it. However, once the issue was reopened in Congress, Minnesota and Wisconsin citizens fought over the route to Lake Superior. Once again the old decision to divide the state of Wisconsin from Minnesota at the St. Croix turned the respective sides into competitors rather than cooperators.[95]

Citizens of Minnesota had envisioned a railroad route from St. Paul to Duluth that would give the Twin Cities access to Great Lakes shipping without having to pass through Wisconsin or Illinois. During the Civil War, Minnesota governor Alexander Ramsey argued the line was essential to national defense. A line to the North Country where the lumber industry was expanding was also an incentive. In 1864 Congress had passed the Lake Superior and Mississippi Railroad grant, and the Minnesota legislature allocated the remaining land necessary. The line, however, was not immediately built. When the St. Croix and Lake Superior Railroad Company folded and the grant dispute reached Congress, Minnesotans rallied to defeat the plans for a rival railroad in Wisconsin. The people of Duluth were adamantly opposed to Bayfield becoming a rival port and railroad terminus. They argued before Congress that they had a better harbor than Bayfield, and that the area surrounding Duluth was better suited for a larger city. Congress agreed, and by 1871 the Lake Superior & Mississippi Railroad opened for business. Duluth became the third largest city in Minnesota behind the Twin Cities and the reigning city on Lake Superior, not Bayfield, Wisconsin.[96]

As the Twin Cities became the North Country's metropolis, national railroads were built to the west of the St. Croix. Stillwater still intended to benefit from these transportation improvements. In 1867 many prominent men of Stillwater, Taylors Falls, Marine, and Baytown organized the Stillwater & St. Paul Railroad. Within a few years they had raised enough capital to begin building. By December 1870 the line was completed. Stillwater's economy boomed as a result of this connection, and settlers were lured to the area by the 63,850 acres of government land grants available for sale. By 1878 the Stillwater & St. Paul line became part of the Northern Pacific Transcontinental system. Since it was at the head of deep-water navigation on the Mississippi and had rail access to the Twin Cities, Stillwater fully expected to "be more than ever a prominent wheat market."[97]

Northern Wisconsin suffered from Minnesota's transportation expansion and from its restrictive constitution that prohibited public funds for internal improvements. Investors in this region were only interested in timber. Most farmers preferred prairie land, or at least wooded areas with oak openings, to the dense forests of the North Country. Land-grant railroad companies had to build the line first before they got title to the land. Railroads had a tough time raising capital to build lines in northern Wisconsin. In 1868 and 1869 railroad lobbyists tried to get a constitutional amendment passed that would allow the state to give financial support for railroad construction. However, the more populous southern counties that already had access to railroads defeated it. Local railroad promoters also had bad luck with eastern capitalists who preferred to invest in the less risky consolidation of railroads and the more profitable business of monopoly building than in new construction into Wisconsin's deep woods frontier. Nonetheless, by 1873 the West Wisconsin Railroad reached Hudson, where a bridge was built across the St. Croix to the Minnesota side.[98]

St. Croix Valley residents felt helpless to change this situation. The region was so sparsely populated that timber and sawmill owners were able to defeat any local tax increases that might adversely affect them. While railroad transport facilitated timber extraction, the lumber companies were still able to make use of the superb waterways. Railroad lobbyists, however, generally found a sympathetic ear in Madison. While the Wisconsin Legislature could not grant money, it did create new counties to isolate the timber factions. Northern Wisconsin's counties above an east-west line from Marinette on Lake Michigan to New Richmond on the west were all created with less than five thousand people by 1880. Burnett County had been set off from Polk County in 1856, but it was not organized until 1865. Bayfield County was created in 1868.[99]

By 1873 the North Wisconsin Railway Company had won the right to build a line from Hudson to Lake Superior. The West Wisconsin line, however, also coveted the route and made sure that the legislature put so many conditions on the line that Alexander Mitchell, the president of the Chicago, Milwaukee & St. Paul Railway and a Milwaukee banker, rejected it. Before the West Wisconsin Railroad could celebrate its coup, the Panic of 1873 and the ensuing depression of the 1870s put an end to the railroad boom. The St. Croix to Bayfield line was not completed until 1883. Construction of this line from St. Croix to Superior required a bridge to cross the Namekagon River at Trego. The line was eventually brought into the Chicago & North Western system.[100]

When St. Croix railroads skirted the eastern side of the valley, many towns along the river complained they were left out of the new prosperity that was developing in the railroad towns to the east. "We are shut out from the world of travel and business, by not being on the line of an important railroad," lamented the *Polk County Press*. "Our river is of great benefit to us, but that serves us not more than one half of the year, and some years not as much as that,—especially when the logs blockade us for two months of the best part of the season."[101]

The St. Croix River, which had been so essential to the development of the region, no longer was. The key to agricultural and related economic development was access to a railroad. So despite the linkage of the St. Croix River at Hudson to Lake Superior, many towns further upriver were convinced more railroads were needed. In 1884 the *Polk County Press* asked, "Will a Railroad Pay?" Its answer was definitively "yes." The paper pointed out that two-thirds of the wheat produced in the county was hauled to Stillwater because farmers could get a better price there. "Why? Because in Stillwater there is a competition in the market, more buyers and a higher price is paid." The reason for more buyers was because "the town had the shipping facilities of three railroads, and will soon have a fourth or fifth. These facilities give that city mills to grind the wheat, and a demand for it. . . . With a railroad there would be a demand for Osceola wheat in Stillwater, St. Paul and Minneapolis." Osceola's mills did expand, and farmers got the same price for their wheat in town as they would in hauling it the nearly twenty miles to Stillwater.[102]

In the early 1880s a new era of cooperation in railroad building began between Minnesota and Wisconsin. The Lake Superior & Mississippi Railroad did not fulfill all expectations. The Port of Duluth did not readily give the Twin Cities access to national markets. In winter, ice hampered lake traffic. There were also complaints that subsequent summer freight rates were too high—having to make up for winter losses. By the 1880s the Twin Cities had become a major milling center with enough economic and political clout to consider running a line from Minneapolis–St. Paul across northern Wisconsin and Michigan's Upper Peninsula to Sault Ste.

Marie—the entrepôt between Lake Superior and Lake Huron, as well as the proposed terminus for a Michigan railroad. "It is imperative to all our interests," wrote the *Minneapolis Journal,* "that a railroad system be constructed which will place us at least on an equality, both summer and winter, with all other shipping points in this vast region."[103]

This proposal created great excitement in the St. Croix Valley. "It will be an entirely new road, crossing the river somewhere between Stillwater and St. Croix Falls," the *Polk County Press* enthusiastically exclaimed. Wisconsin was eager to have the line built since it would travel through the heart of its hardwood and softwood northern forests, as well as penetrate into its iron regions. Financing the road through Wisconsin, however, was fraught with the usual fiscal problems. The *Polk County Press* was undaunted and encouraged its readers to support the project, which was expected to come down the Apple River, cross the St. Croix River, and proceed down to Stillwater. "This great public improvement presents itself in close relationship to the people of Polk County," the paper wrote. "It is proposed by its incorporators to cross the territory of this county, and our people. . . . They propose to develop this section with a great trunk railroad: they will ask in return but a small amount of assistance from us, when compared to the benefits the road will bring to us." The *Press* proved prophetic as work began on the Minneapolis, St. Paul & Sault Ste. Marie Railway, or Soo Line, in 1886 and went from the head of the Osceola prairie into Osceola itself, with a branch line up to St. Croix Falls. From there it ran downriver to the mouth of the Apple River, where it crossed the St. Croix to Stillwater.[104]

"The coming of the railroad has nearly doubled the price of land," rejoiced the *Polk County Press,* "and those who wish to buy should do so soon, as property in Polk county will never be worth less than it is to-day." In August the people of Osceola celebrated. "We are in the world," proclaimed the *Polk County Press.* "Connected with business and commercial life." After thirty-five to forty years of waiting for a railroad, the people of Osceola were giddy with excitement. Their winter hibernations were over as was their exclusion from the growing national market both as sellers and consumers. "Old men and old women leaped for joy; young men and maidens gaily tripped the streets," an observer noted when the track was completed. "Everybody was happy . . . even the sick smiled, and those who did not smile must have been nearly dead." The arrival of the railroad to Osceola brought with it a new grain elevator and warehouse in anticipation of bountiful harvests of wheat grown on the prairies of Osceola.[105]

From Wheat to Dairy Farming

The transportation revolution stimulated some farmers to turn to dairying. This quicker, more efficient form of transportation made it possible for farmers to break their reliance on wheat. Astute farmers found they could now ship fresh dairy and animal products to urban areas. Dairy farming proved to be a more reliable source of income, and it relieved the ecological problems of single-crop production. Farmers' lives were also transformed. Equipped with extra cash and cheaper transportation, they had access to the growing consumer economy that had developed in the late nineteenth and early twentieth centuries through catalogues, regular mail service, and the possibility of train trips to the city for shopping and entertainment.[106]

The switch to dairying, however, was gradual, as many farmers continued to believe in wheat despite the warnings. In 1860 St. Croix County led the state in wheat yields. Pierce County had also emerged as a significant wheat-producing region. St. Croix County reported almost no livestock in the county save for draft horses. Oats were still cultivated, but even potato production was down, as were the number of swine along the St. Croix, all this to make way for wheat. While other river counties began showing some interest in raising sheep, St. Croix and Polk counties displayed no interest in these woolly creatures. In 1868 the State Agricultural Society lamented in its *Transactions* that wheat was still the most highly valued crop in Wisconsin.[107]

The reluctance to give up on wheat along the St. Croix was in part because virgin land was still being brought into cultivation in the 1860s and 1870s, and the decline in soil fertility was not rampant. Between 1870 and 1890, the ten-county stretch of land from La Crosse to St. Croix County gave the state a 25 percent increase in farm acreage in Wisconsin. Burnett County also joined the ranks of wheat-producing counties in the state. The land values in this region went up with railroad service, but so too did taxes. Farmers, therefore, needed to make more money to pay their property rates. Hence, they relied on wheat as long as they could.[108]

The wheat-growing era in the St. Croix Valley, however, had its limits. Agricultural prices in general began to decline in the 1870s and into the 1880s. To make up the loss, farmers had to expand their operations. While sowing wheat and watching it grow was relatively easy, harvesting it was a problem. Ripe wheat was easily damaged by heat, wind, hail, and pounding rain and had to be harvested quickly. Horse-drawn reapers and binders were essential. While mechanical farming equipment worked well in the prairie lands of the valley, the hillier terrain in the old forests was less accommodating to machines. In addition, the new roller-mills that began to dominate in the milling centers in the Twin Cities preferred hard spring wheat varieties rather than softer winter wheat. Spring wheat, however, was more susceptible to disease and forced many farmers to rethink their approach to farming.[109]

It was the knowledge, experience, and enthusiasm of migrant New Yorkers that ushered in the dairy industry in Wisconsin. Earlier in the century, New York had undergone the same crisis in farming. When wheat farming had exhausted the soils in the Empire State, farmers were forced to search for alternative means of support. By 1860 New York dairy farmers had successfully made the transition to dairy products. To sell broadly, farmers found they had to improve the quality and quantity of their cheese and butter. New York dairy associations began to insist on more sanitary processing, better salt, truth in fat content, and improved packaging. Consistency in quality, improved packaging, and an increase in quantity created a national and even international market for New York dairy products.[110]

By the time Wisconsin's wheat era had played out in the 1870s, the New York dairy industry was a well-developed model. Many a New York dairy farm boy migrated to Wisconsin, attracted by the relative cheapness of the land, feed, and labor. New Yorkers usually took the lead in building cheese factories, in organizing breeder associations, and in uniting the efforts of the state's dairy farmers. In 1872 they helped found the Wisconsin Dairymen's Association. Many of its presidents and executive officers hailed from the Empire State and contributed much-needed expertise to the fledgling organization. Through dairy manuals, articles in local newspapers, and public meetings, farmers' apprehensions about the factory system were assuaged. It was hard to argue with the fact that factory-produced cheese was of a consistently higher quality than any produced at home. Factories also could buy supplies in bulk at cheaper costs. The main challenge was getting the milk to the factory before it spoiled. The neglected state and county road system and the vagaries of Wisconsin weather discouraged many farmers from participating in the factory system, but not necessarily in the viability of dairy farming. Farmers also wanted to retain rights to the whey produced from their milk for cattle and pig feed. Many issues needed to be resolved before dairying became widely accepted and practiced throughout the state.[111]

Wisconsinites were the fortunate beneficiaries of the Morrill Land Grant Law that helped establish the University of Wisconsin in Madison. Its College of Agriculture was created in 1866 after a farmers' convention, sponsored by the Wisconsin State Agricultural Society, put pressure on the state legislature to establish an agricultural school in Madison. Scientific agriculture proved to be critical to the development of the dairy industry. Initially, it was difficult to find experienced faculty as well as attract students. Therefore, in 1886 the college began short courses for farmers and in the following year began a winter Dairy School. It was the first of its kind in the country and trained and certified students in the making of cheese and butter. Within a few years it produced enough graduates to man the factories throughout the state. By working closely with farmers and factory workers, the faculty and researchers at the College of Agriculture were able to develop practical and scientifically sound knowledge for dairy farmers. The college also initiated an extension division, published its research, and regularly hosted farmers' institutes. "Scientific men and practical farmers occupied the same platform," wrote historian Joseph Schafer, "with the result that science was more closely controlled by experience and experience definitely guided by science. No other feature in the history of agricultural advancement . . . had been so resultful [sic] in developing mutual respect and confidence between the farmer and the man of scientific learning."[112]

In 1878 the Minnesota Dairymen's Association was founded. The Minnesota Butter and Cheese Association followed in 1882, establishing new scientific innovations as well as the idea of participant-owned businesses from creameries to elevator companies. The co-op movement succeeded beyond anyone's dreams. By 1921 the hundreds of co-ops that had developed joined together to form the Minnesota Cooperative Creameries Association.[113]

However, St. Croix River counties clung tenaciously to wheat. Even in 1885 Osceola bragged that its "wheat market is now as good as any in the valley." But the valley could no longer avoid the problems associated with wheat farming. In 1879 St. Croix County held first place in wheat production in the state of Wisconsin. Ten years later it had dropped to forty-fifth. Pierce County took tenth place—down from third—and Polk County fell to twenty-second place. This did not mean farmers immediately abandoned wheat growing. They instead sowed other grains, such as rye and barley, with their wheat. By 1899 St. Croix County had moved up to second place in wheat growing. Pierce and Polk counties held third and fifth place, respectively. Most of the state, however, had shifted to dairy farming, thereby reducing competition. Yet it demonstrates the reluctance to give up what was seen as an easy cash cow. St. Croix County even expanded its milling operations.[114]

Astute farmers, however, realized wheat yields and profits would not continue. In the fall of 1883 State Senator James Hill, who represented Polk County, urged farmers who attended the Barron County Agricultural Fair to begin crop rotation if they intended to continue raising wheat. Reliance on one crop was "ruinous to the success of any farmer, as it robs the soil of its plant food," Hill explained. "Worn out lands are brought into a higher state of cultivation by seeding to clover and ploughing it under." The natural complement of crop rotation was livestock farming. Cattle could graze in the clover fields and eat and rest in hay as well as provide their own natural fertilizer to the fields. Root crops, such as turnips and potatoes, were also natural fertilizers in addition to providing nutritious slop for pigs.[115]

Still, the change to dairying in the St. Croix Valley proceeded slowly due to its expense and the long time it took clover to thrive on exhausted soil. Since most farmers did not own stock, they had no manure. Therefore, many had to purchase commercial fertilizers before they could count on pasture grasses to naturally fertilize the soil. Then they also had to buy milk cows. Gradually the number of cows increased through the 1890s, but there was no comparable increase in the growth of factories to process milk products. Farmers also had to adapt the crops they did grow to meet the nutritional needs of their animals, especially by growing hay and corn.

Those who were without the financial means or unwilling to commit to the intensive work of dairy farming migrated out of the St. Croix Valley. This began another trend that affected the ethnic composition of the region. Yankee farmers had been the most devoted to wheat and were more likely to move to the broader, fertile prairies of central and western Minnesota, where they could use machines more effectively. Immigrants and their children were more likely to stay. German farmers in particular were more attached to the land they homesteaded and were willing to take over exhausted Yankee farms.[116]

In 1885 William A. Henry, dean of the College of Agriculture at the University of Wisconsin, began his crusade to persuade farmers to switch to dairying when he began a lecture circuit of the newly inaugurated Farmers' Institutes in the St. Croix Valley. He visited Luck, Milltown, and Osceola that winter. Henry urged farmers to diversify their crops rather than rely upon just one, particularly wheat. He pointed out that wheat cultivation was expanding not only in the Dakotas but also in Canada. "With all this region just developing," he argued, "why should farmers on the high priced lands of the St. Croix valley, persist in ruining their lands with wheat exclusively?"[117]

By 1886 Polk County revived its agricultural society and held a county fair. "It is a gratification to once more witness an agricultural display in Polk County," wrote the *Polk County Press*. The next Farmers' Institute, held in Osceola in November 1887, was a big success. More people came than were expected, filling the hall to capacity at every session. Lecture and discussion topics included "Silo and Ensilage," "Reproduction of Animals," "Dairying," "Swine Husbandry," "Horse Breeding," and "Buttermaking." The interest and excitement of the Institute encouraged Polk County farmers to organize a Farmers' Club.[118]

At a Farmers' Institute in River Falls, Wisconsin, growers were told how farming practices affected property values. Farmlands in dairy counties were worth $45 to $55 dollars per unimproved acre. Farmland in St. Croix County was valued at $34 per acre, while farmland in Sheboygan County was valued at $84 dollars per acre. "In St. Croix, that produces more wheat than any other county of the state, the lands are valued less than one-half of that of the dairy counties," related the *Polk County Press*. "In Pierce county, engaged in the same business nearly as largely as St. Croix, the lands are of nearly the same value. . . . Wheat raising has cost us largely in the most valuable elements of fertility of the soil."[119]

The Farmers' Institute directors strove to make their sessions practical and helpful. For the 1887–88 winter program, dairy machine makers brought their latest developments and demonstrated modern creaming techniques throughout the state. Oil and butter tests were also done from different mixes of milk in order for farmers to be able to actually see the differences that resulted from various methods. Some farmers were dubious of these institute lecturers, whom they considered "theoretical, rather than practical farmers." The *Polk County Press,* however, whole-heartedly supported these educational opportunities and pointed out to their suspicious readers, "No man is employed in institute work who is not himself a practical farmer, the owner and tiller of a farm, who is carving out success by the intelligent application of brain and muscle to his occupation." The paper urged farmers to attend these institutes, arguing, "No state in the union is progressing so rapidly in its agricultural industries as is Wisconsin, and the progress is largely the result of the teachings in these institutes."[120]

Institute workers were tireless in their promotion of their knowledge and findings. "They travel from place to place, frequently reaching sections of country not traversed by railroads," lauded the *Polk County Press,* "carrying with them workers skilled in the theory and practice of agriculture. . . . They scatter . . . information . . . they awaken enthusiasm: they incite farmers to discuss their business; to compare methods of work; to improve their homes and lift themselves out of the ruts into which they have fallen."[121]

Despite the enthusiasm of the press and the standing-room-only crowds at Farmers' Institute sessions that continued through the 1890s, change took place slowly. Farmers who clung to raising grain were on their way to the poor house as far as William Henry of the University of Wisconsin was concerned. Dairying had brought farmers in Sheboygan County riches. In one bank, he claimed, farmers had over a million dollars in deposits. The land along the St. Croix was better, he argued, and city and lake markets were easily accessible. "They can just as well make their land worth seventy-five dollars per acre, and salable at that price as to have it in its present condition." The *Polk County Press* concurred, "The fact that every farmer here who had tried it has succeeded, ought to satisfy everybody."[122]

While the railroad was the key to the transformation of Wisconsin from wheat to dairying, road building became the next logical transportation revolution for the dairy industry. Railroads might be crucial to getting finished products to urban markets, but first farmers had to get their milk from the farm to the factory. Bad weather made travel on rural roads arduous to impossible, especially in the winter and early spring. Oddly enough, the Wisconsin Dairymen's Association and the Wisconsin State Agricultural Society did not agitate for roads. The University of Wisconsin's Farmers' Institutes only began to raise the issue in the late 1880s. What accounted for this reluctance to push for better public roads was not disinterestedness, but farmers' fear that they would lose their ability to pay for roads through their labor. They also wanted to control where the roads would be built. Otherwise, they might find themselves with a heavy tax burden and few roads that served their needs. State and county officials, they believed, were more likely to build roads that linked cities and towns rather than farms and villages.[123]

What stimulated farmers' enthusiasm for road building was the prospect of obtaining free mail service. In 1891 the U.S. postmaster general proposed rural free delivery for all Americans. While this policy was not implemented until the 1930s with the New Deal, farmers realized Uncle Sam would not be able to show up at their doorstep if he had to travel along muddy, rutted, or washed-out roads. Within a few years a coalition of farm leaders, village merchants, people in the mail-order business, and recreational cyclists began to push for roads. In 1893 the Wisconsin Legislature did away with the old local road districts and turned over authority for road building to township boards. Folks in the St. Croix Valley were reassured that they would not be in danger of losing their property to tax delinquency

Figure 11. Pierce County calf clubs on parade. County fairs and 4-H clubs played an important role in promoting dairy farming in the Lower St. Croix Valley. Wisconsin Historical Society, WHi-42448.

since the law still allowed them the "opportunity to work out their road taxes." While it had a modest beginning, road reform began in Wisconsin, and the dairy industry was a chief beneficiary.[124]

Between 1890 and 1920, dairying became Wisconsin's single largest industry, and the state became the number one producer of butter in the country, a position it held until 1950 when Minnesota and Iowa pulled ahead of it. By then Wisconsin had established itself as a leader in cheese production. Wisconsin's efforts were aided by developments along the St. Croix. By 1901 Polk County operated sixteen creameries and four cheese factories that also made butter. Pierce County was close behind with fifteen creameries. St. Croix County also showed gains in dairying.[125]

By the early twentieth century, the Wisconsin shore of the lower St. Croix Valley had successfully adapted to the sweeping changes that affected the U.S. economy since the Civil War. It met the challenges of the opening of lands in the West and the closing of the frontier, the growth of a network of railroads that created a national and international market. The people here became good custodians of this rich land. They restored its fertility and introduced a relatively stable and productive use of that land.[126]

The "Wisconsin Idea of Dairying" transformed the state and the St. Croix Valley. This "idea" was the result of the willingness of the University of Wisconsin and public agencies to work in concert to advance dairying in the state. The Dairymen's Association, the College of Agriculture, and elements within the Republican Party in many ways foreshadowed Wisconsin's Progressive movement where in twenty years the state "became a laboratory for scientific experiment, teaching, and legislation." This progressive approach to solving Wisconsin's farming problems also spared the state much of the political turmoil and class antagonisms of the Granger movement in the 1870s and Populism in the 1890s in which farmers pitted themselves against big corporations and monopolies. In later years Theodore Roosevelt called the "Wisconsin Idea" a lesson in scientific self-help. It made Wisconsin the Dairy State.[127]

The Minnesota side of the St. Croix, however, was much slower to adapt to dairying. The self-contained Swedish communities in Chisago County and the northern part of Washington County were slower to respond to outside pressures for change, whether economic or social. The Swedish Lutheran Church was diligent in "maintaining the Swedish language and culture" and "insisted upon tight discipline among their parishioners." Swedish was spoken as a primary language among immigrant children and grandchildren into the twentieth century. Farming was more of "a way of life" for many of these settlers. A farm "was a place to live and to raise a family," rather than a commercial venture. Many of the farms in the area around Marine began as part-time ventures. The farmers raised crops that the lumber camps wanted, and then they worked in the pineries during the winter to supplement their income. When winter jobs in the woods diminished, their style of farming would not support a family. With few resources to farm on a larger scale, many of these men drifted into the Twin Cities for wage work.[128]

Although the Chisago County Fair was organized in 1891 to encourage diversified farming, change was slow. Even in 1897 wheat was still a major crop, and a new milling company was opened in North Branch, Minnesota. During World War I the mill operated twenty-four hours a day in order to keep up with demand for its Model Home flour. The problems associated with wheat, however, did not escape farmers here. As a local saying went, "Rye after rye, and you'll have bread 'til you die; but wheat after wheat, and you'll have nothing to eat." Rye and potatoes, therefore, became their cash crop. Center City farmers found the soil good for potatoes and harvested from 150 to 300 bushels an acre. The abundance of potatoes required six warehouses to store them. A starch factory made use of the surplus. North Branch claimed to be the "Hub of the Potato Belt." Harris, Minnesota, however, claimed to be the "Potato Capitol of Minnesota," since potato buyers from St. Louis and Chicago made that town their headquarters. Between August 1 and January 11, 1912, a total of 527 carloads of potatoes, at an estimated value of $300,000, were shipped from North Branch.[129]

Potato growers benefited from the inventions of a German immigrant named F. Splittstoser, who invented potato diggers, sprayers, and planters. Splittstoser established a factory in North Branch to manufacture his equipment. Before commercial fertilizers were available, however, potato farmers ran into the same problem wheat farmers did—soil exhaustion. More savvy farmers used their years of cashing in on potatoes to build up their dairy herds. Through the early twentieth century, issues of the *Stillwater Weekly Gazette* and the *Minnesota Farmer* circulated among the interested farmers along the St. Croix River and encouraged better farming practices. Chisago farmers began to consider dairying more seriously. Creameries had begun to appear in the county in the 1890s, but unlike their Wisconsin counterparts, all failed. The Rush City Co-op, established in 1907, became the first successful dairy venture. By the 1920s several cooperative creameries were in operation throughout the county.[130]

Washington County also had its wheat boom years in the second half of the nineteenth century. Flour mills sprung up in Stillwater in the 1870s, but soil exhaustion, competition from western lands, and Rocky Mountain locusts forced farmers here to find other alternative uses of the land. The county revived its agricultural society in the 1870s. Crop diversification, rotation, and scientific farming principles took root.[131]

The development of an agricultural market economy in the St. Croix Valley served several purposes. The landscape was permanently changed from giant white pine forests with oak openings to a scene of dairy barns with silos towering over pastures. Dairy cattle replaced the wild elk, moose, and bear. The separate ethnic enclaves along the river towns pursued the same economic ends in the switch from wheat export to dairying in order to improve their material lives, and in the process they became American farmers.

Farming the Cutover

The settlement and development of agriculture along the Upper St. Croix River had a much more torturous history than the lower river. The eighteen northern counties in Wisconsin became known as the *cutover*. These included Burnett, Washburn, Sawyer, Douglas, and Bayfield counties along the St. Croix and Namekagon rivers. The southern St. Croix River counties were technically not included in this designation,

Figure 12. Riddled with stumps and raved by fire, the devastated land of the northern Wisconsin cutover was a difficult challenge for the homesteader. Photograph by the Farm Security Administration. Wisconsin Historical Society, WHi-31545.

even though major portions of them had been "cut over." Stretching from the northwest corner of Polk County and running along the St. Croix and Namekagon rivers up into Douglas and Bayfield counties is a region called the Pine Barrens. It was composed of "coniferous forests and open expanses of sweet fern and grassy barrens." The first road up the Namekagon, the "State Road" that ran from New Richmond to Ashland, was not cleared until 1877. This limited settlement possibilities. These counties were also the most densely forested of the St. Croix Valley and did not have the rich soil, oak openings, or prairies of the lower river. Roads were only gradually built into this part of the valley. The Hayward-Cable road was built near the railroad in 1892 and turned into a "county highway" the following year. Other logging and tote roads penetrated the forest. By the turn of the century, commercial logging firms had blazed their way through this region leaving in the wake of the lumbermen's ax stump fields, brush, tree limbs, and discarded logs.[132]

Although it was a truism in the nineteenth century that farming would follow lumbering as part of the onward growth of civilization, Burnett County newspaperman Edward Peet expressed the general contempt residents had for lumber barons who claimed there would be this march toward progress but did little to bring it about. Peet complained, "The interests of the lumbermen and others were in an opposite direction and more was done to keep settlers out than was done to bring them in." But he noted optimistically, "The day of lumbering is passing away. . . . The scarcity of timber from the lumberman's standpoint and an over supply from the standpoint of the tiller of the soil are one and the same thing. It is just this condition that is now turning the attention of farmers to Burnett county."133

Farmers had not been unknown in the cutover. After all, lumbermen had to eat. Loggers, however, did not want homesteaders who would demand the usual amenities of settlement, so they initially created corporate farms that supplied them with wheat, potatoes, vegetables, beef, and hay. The first independent farmers in the area were lumberjacks. In winter they worked in the logging camps, and in summer months they grew foodstuffs to supply the camps. Their farm sites were chosen more for their closeness to winter employment than for their agricultural potential. Burnett County had its share of winter lumberjacks and summer farmers.134

By the 1890s, when the American frontier was declared closed and the best western lands were taken, many Wisconsinites began to take a second look at cutover lands in the state. "There now remain only comparatively small areas suitable for settlement by pioneer farmers in this country," William Henry related to the *Polk County Press* in 1895, "and today Wisconsin offers the largest body of good agricultural land for such settlement possible to any state or territory in the Union. . . . Where ten years ago we could not have stopped the migratory crowd intent on reaching the plains of the west, many are now ready to hear of what Wisconsin has to offer in an agricultural way." Other observers also noted this trend. People from "Texas, Kansas, and the Dakotas and Minnesota," related one cutover resident, "have come back and taken up their homes permanently here."135

The first pioneers ventured into southern Burnett County. In 1850 St. Croix Falls was "regarded as the dividing line between savage and civilized life. . . . Beyond that point," wrote Seymour, "white traders and others have Indians wives; and the entire population, with few exceptions, is Indian or half-breed." The pressures of white settlement were soon felt. In 1856 the county was created out of Polk County, and the Homestead Act of 1862 brought in Norwegian and Swedish immigrants as well as some settlers from eastern states. Civil War veterans with homestead credits also made their way here. "A homestead of 160 acres of land can be taken inside of railroad limits," announced the *Burnett County Sentinel* in 1879. "Any settler who has taken 80 acres before can take an additional 80 provided it lies adjacent to the first. He can also sign back to the government his 80 acres and take another 160 in another place, the time he lived on the first applying on time he must stay on the last entry." Many pioneers eagerly responded to the offer. "There is quite a rush for homesteads in the vicinity of the last twenty miles location of the North Wisconsin railroad," wrote the *Taylors Falls Journal* in May 1879. In 1880 the *Burnett County Sentinel* announced, "Emigrants are coming thicker and faster. A load comes up almost everyday. Most of them are settling in towns Eureka, Luck and Sterling." Some of these settlers were disenchanted farmers from southern Minnesota. "Grasshoppers are the cause of their leaving," explained the *Burnett County Sentinel*. "Year after year their crops have been destroyed by these pests until they had become sick of the country and they concluded to leave."[136]

Burnett County was rich with lakes, swamps, marshes, bogs, and wet meadows. In 1850 the federal government passed the Swampland Grant Act that gave Wisconsin approximately three million acres of federal wetlands, a good part of which held valuable timber. This land was desirable because it could eventually be drained and turned into productive farmland. Because loggers often trespassed on state land harvesting its valuable timber, the state felt compelled to dispose of its lands quickly and let private owners police their own lands. These lands were sold through public auction with the usual pre-emption rights. The quick sale of land, however, often resulted in it being sold below market value or being bought up by land speculators. In 1871 Wisconsin governor Lucius Fairchild asked that state lands be withdrawn from the market in order to appraise their value accurately. This suggestion, however, went unheeded until the turn of the century. At that time Burnett County still had 35,000 acres of state-owned land.[137]

Optimism abounded that business and agriculture would thrive here. Those who settled in the southeastern section of the county joined in the wheat craze of the 1870s and 1880s. "The farmers are very busy harvesting," declared the *Burnett County Sentinel*. "Nearly all the wheat and oats in this section are ready to cut; and from what we have seen and heard, it will be fully up to the average yields of former years. Corn never was looking better." Approximately 879 acres of wheat were cultivated in the county, with an average wheat yield of fifteen bushels an acre, "which ought to keep the people of the county in flour, instead of importing as has been done before." In 1879 the *Burnett County Sentinel* wrote about the "Flattering Prospects" that the county expected to enjoy with the building of a railroad from Grantsburg to Rush City on the Minnesota side of the St. Croix

River through St. Cloud. "The road taps not only a region abounding in pine and other timber, but also traverses one of the finest agricultural regions of [Minnesota] and Wisconsin."[138]

As with earlier homesteaders in the lower St. Croix, the weather was a major concern of prospective settlers. Promoters of the North County optimistically spoke of the healthy but bracing quality of the climate. The shear abundance of wood for fuel, it was argued, made the winters more tolerable here than in warmer but fuel deficient areas. "Our winter storms are sometimes bad but they are very mild affairs compared with the prairie blizzards," commented Ed Peet. "No person who has ever experienced a prairie blizzard had any fault to find with our winters. Hot, scorching winds are unknown and drought seldom causes even a partial crop failure." Peet, however, also expressed the naive and even quack scientific thinking of the times by claiming the climate was changing because of the denuding of the forests. "The climate is also changing," he wrote. "The winters are milder and not so long . . . and other unfavorable conditions will change." The expectation was that the climate would become more like that of southern Wisconsin.[139]

New settlers marveled at the fertility of the soil when new growth bloomed so quickly in areas loggers had cleared recently, especially if fires had swept through and had burned it. Poplar, aspen, and white birch "seeded thick as grain in a field, and grew fast." Blueberry and raspberry bushes thrived, as did the deer that feasted on them. Wild hay and pasture grasses, such as timothy, flourished. This new growth only encouraged the notion that the cutover would make a great dairy region. "The highest scientific agricultural authority says that in no place in the world is there any better grass for butter and cheese making than in Northern Wisconsin," proclaimed Ed Peet. "All over this region grasses and clover thrive. . . . It seems to follow as naturally as a second growth of timber." Even William Henry predicted that northern Wisconsin would become one of the great cheese producing regions in the country.[140]

Burnett County farmers benefited from the proximity to the agricultural changes happening in areas further south along the St. Croix. The Burnett County Agricultural Society began holding annual fairs in 1877 and enjoyed its connections with the Farmers' Institutes. "Our farmers who used to raise hay and oats for the logging camps, and many of them who used to be half farmer and half logger," wrote Peet, "are now devoting all their energies to the farm and modern methods are being adopted. They are going into dairying and stock raising and a manure heap has now a value in it."[141]

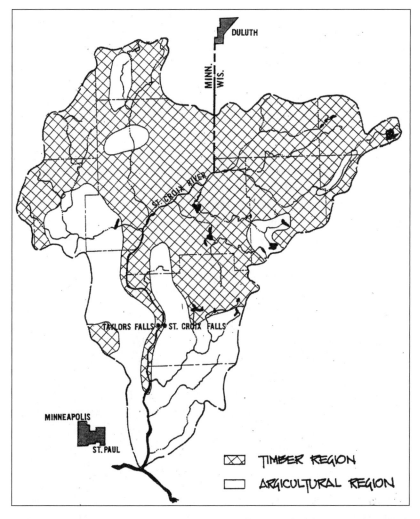

Map 7. Timber and agricultural regions, St. Croix Valley watershed (Source: National Park Service).

However, there was no escaping the problems associated with excessive wheat growing and the limitations on agriculture in poor cutover soil. By 1883 it was reported, "Wheat does not turn out as well on the barrens and away from the timber as expected, but in the hardwood it is fully up to the average." By 1888 the *Burnett County Sentinel* lamented that "cinch bugs will make harvesting and threshing of wheat unnecessary this fall. Wheat is turning yellow and looks dead throughout the county." In the following spring Grantsburg voted to purchase a carload of seed wheat for farmers whose wheat crops had failed for the past two seasons. It was not, however, a handout. Farmers were expected to sign a note agreeing to pay the town back the principal outlay with interest in their 1889 taxes. Some farmers simply gave up.[142]

Farmers in Burnett County also cultivated livestock feed such as oats, corn, barley, and rye. When wheat failed, farmers began to grow potatoes as their new staple crop. In 1890 a starch factory was planned, and by 1894 the *Burnett County Sentinel* noted, "Potatoes [were] coming in quite lively," and the starch factory was receiving culls for grinding. By 1895 the *Rush City Post* reported that farmers shipped ninety cars of potatoes from that city. The *Burnett County Sentinel* complained that the Grantsburg area farmers then could only get five empty cars out of the fifty to seventy cars that they needed. Schools were closed for two weeks in October because most of the students were picking potatoes.[143]

Burnett County's abundant wetlands provided another cash crop—cranberries. Before 1870, Indians and frontier settlers brought them to market in their canoes or wagons. Ten to fifteen bushels a day was considered a good haul. State and federal governments, who owned these marshy lands, considered them fair game for public plunder.[144]

Commercial cultivation began when lumberman A. N. Badger of Oshkosh, Wisconsin, was scouting out pinelands in the St. Croix Valley in 1872 and discovered its marshlands, including the "Big Meadow" north of Grantsburg. He was familiar with the experiments of businessmen in Oshkosh with cranberry cultivation and lured them here, where they entered claims for five thousand acres of what they considered the most valuable marshlands north of Grantsburg. The first two crops they gathered from nature's bounty netted them $10,000 in profit.[145]

Not content to sit idly by and let nature take its course, these entrepreneurs began to make improvements. Since it took about five years to get an acre of marsh into commercial production, cranberry growers had to have a good deal of initial capital for investment. "Men without capital can not engage in the business," warned the *Burnett County Sentinel*. Ditches were dug to drain the land when necessary, and then dams were built across these ditches to flood the marshes when more water was needed. Flooding the marshes killed off wild grasses and "other encumbrances" and allowed more cranberry vines to flourish. By 1875 forty miles of ditches and ten miles of dams were constructed. At harvest time these companies employed local farmers, their families, and Indians. This necessitated building sleeping and cooking accommodations for the pickers and dry-houses.

Local newspapers cooperated by putting out the cry for pickers. "Cranberry picking will commence next week," announced the *Burnett County Sentinel*. "A good many persons will be employed as all the berries will be picked by hand, no rakes used. 75 cents per bushel will be paid for picking." Large growers usually provided music and even a dance hall to attract pickers.[146]

Burnett County encouraged the development of the cranberry trade by lavishing money for roads and bridges and even looked into the feasibility of constructing a bridge across the St. Croix River. The purpose of the bridge was to connect the cranberry farms to the Lake Superior & Mississippi Railroad on the Minnesota side of the St. Croix at Pine City. By 1875 Marshland, Minnesota, became the headquarters for several companies owning marshes and cranberry farms in the area.[147]

"'How's cranberries,' is all the cry now!" wrote the *Burnett County Sentinel* in 1876. In that year the Buttrick and Gill harvesters expected "to get 1,000 barrels of cranberries from their marsh this season." However, the manipulation of water levels and intensive cultivation of cranberries also presented environmental problems. Worms began to infest the cranberry marshes. The only way to kill off these pests was to flood the marshes, but if the area became short of water, only Mother Nature could solve the problem. Manipulation of water levels also left the berries exposed to damage from freezing and thawing. In the fall of 1879 leaking dams and seasonal drought conditions left cranberry marshes uncovered for the winter. "These marshes need snow on them the whole winter to protect them from the frost. . . . Half of the vines on the marshes . . . were winter killed."[148]

The forest fires that often ravaged cutover regions also threatened cranberry marshes. The first reported fire affecting them came in April 1877. "The cranberry marsh owned by Andrew Ahlstrom of this village, and the marsh of Albert Bugbee's, were burned over last week," reported the *Burnett County Sentinel,* "preventing them from bearing berries this year." The marshes did rebound somewhat despite this dismal prediction. "From all reports, the cranberry marshes are doing well considering the damage that was done last winter and by fires last spring," noted the *Sentinel.* "There will be no big crop gathered of course from any of the marshes, but it will be the largest crop that was ever gathered in this county." By 1887 Wisconsin produced 52,567 bushels of cranberries of which 18,000 came from Burnett County.[149]

The 1894 Hinckley forest fire burned approximately four thousand acres of cranberry land. Afterward there was an attempt to drain fifteen thousand

Figure 13. Cranberries being loaded into a washing machine, Washburn County, Wisconsin. Cranberry farming began in natural marshes, but cultivation for commercial production required altering the landscape with dikes and ditches. Photograph by G.A. Marquardt, September 1934. Wisconsin Historical Society, WHi-43015.

acres of the marsh for small farms. When this failed, Crex Carpet Company bought the land to make grass mats. It closed in 1914 when other native grasses crowded out wiregrass. Much of the land in the Big Meadows was eventually sold for taxes to prospective farmers. Rather than see these ecological changes as an environmental disaster, the draining of swamp and marshlands was seen as progress. "Man is at work now continuing this drying process and the county is undergoing many changes," wrote Ed Peet. "Within the memory of the present residents of the county, lakes have changed to marshes, marshes to meadow lands and meadow lands have become good dry plow lands. . . . There are but few swamps in the county that will not at some time in the future be reclaimed and turned into profitable farm land."[150]

Large-scale cranberry farming began near Pokegama Lake by the beginning of the twentieth century. S. H. Waterman & Sons invested heavily in the area, expecting to put a hundred acres under cultivation. They carefully selected the best plants and over the years produced "Jumbo Cranberries."

By the 1920s Andrew Searles and his son, C. D. Searles, natives of Wisconsin Rapids, continued the scientific cultivation of cranberries and turned the marshes of Burnett County into a thriving business.[151]

Blueberries also grew wild in the area. Although native to the region, blueberries thrived in cutover lands once its towering trees were gone and sunshine reached the brush. These berries were very susceptible to frost and drought, though, causing the crops to vary from year to year. Pickers eagerly awaited reports on the condition of the blueberry crop in the *North Wisconsin News* that was established in Hayward in the 1880s. In July 1900, forest fires and frost wiped out the crop. However, in 1901 the blueberry crop in the Namekagon valley between Hayward and Spooner was reportedly worth $75,000. During good years "it was impossible to harvest even a small part of the berries available," recalled Eldon Marple, who picked berries in his youth before World War I. "They grew in such profusion and abundance and over such a vast area that only the berry-stuffing bear, deer and birds ever saw most of them."[152]

Blueberries were often so plentiful many families made money picking them through the summer months. Some pickers even camped out in the woods for the season. Buyers shipped the berries out by the carload. Local Indians also joined in the blueberry trade. "During the summer," wrote one settler, "Indians came from seemingly everywhere with wagons and ponies and some walking carrying packs on the backs, to pick blueberries . . . there were as many as 500 camping at Webb Creek." The Indians traded the berries for various groceries such as pork, salt brine, and bologna at the Webb Lake Store. At the end of the summer, the Indians were reported to dress in feathers and beads, play their tom-toms, and dance their tribal dances "day and night for long periods. Settlers could hear their drums for miles."[153]

Although blueberries were sometimes shipped to places like Kansas City, most of the crop from the North Woods was shipped overnight to Chicago on "the Blueberry Line," "where the berries arrived on the morning market, still fresh and with the dew on them." Conductors on the train would sometimes select a choice spot in the Namekagon valley and invite passengers to fill a hat or apron with berries. "Passengers reported that the berries were in clusters as large and heavy as a bunch of grapes," reported Marple, so "a person could eat his fill while sitting in one spot."[154]

Although it had some trying times, Burnett County's future looked promising at the turn of the century. It had a gristmill, starch factory, eight cooperative creameries, and a cheese factory. In 1896 a refrigerated railroad car set off for New York City every Wednesday. By Monday morning folks back East were enjoying Burnett County butter. Interest in settling this Upper St. Croix Valley continued to grow. "We will boom as well as any of the places in this northern section of the country," boasted the *Burnett County Sentinel*.[155]

The University of Wisconsin's College of Agriculture even began to take notice of the cutover and extended its Farmers' Institutes into the region. The state agency also encouraged the establishment of county immigration agencies. Beginning in 1895 Wisconsin held immigration conventions throughout the state to excite interest in the cutover and re-established its Board of Immigration. In 1907 the board was reorganized and operated for the next twenty years. Its main job was to attract settlers to the cutover by sending representatives to fairs, distributing promotional literature, and connecting potential buyers with land dealers. The state distributed fifty thousand copies of William Henry's book *Northern Wisconsin: A Hand-Book for the Homeseeker*. English, German, and Norwegian emigration pamphlets used illustrations from the book. The Board of Immigration also worked in conjunction with the Northern Wisconsin Development Association and the Northern Wisconsin Farmer's Association. The Farmer's Association dispatched "Grasslands Cars" through Wisconsin, Iowa, Illinois, Minnesota, and Nebraska. These converted passenger railroad cars were lit up by large electric signs boldly declaring, "Farm Products From the Best Region on Earth for the Homeseeker." Railroad companies joined in these efforts. Because they needed to sell their land in the cutover, they hauled the cars free of charge until 1906. They were decorated with maps and pictures, and carried samples of farm products."[156]

In 1895 the Wisconsin State Legislature authorized Dean William Henry to research a book on the agricultural resources and opportunities in the state. A good part of the book focused on the more thinly populated areas of the North Country. The land's apparent lack of value prompted many lumber companies to forfeit their lands through tax delinquencies. Part of their reasoning was that if they recruited settlers and sold some of the land while there was still pine in the area, the lumber companies would pay the lion's share in taxes for roads, schools, and other improvements. If they hung onto their land until all the pine was gone, or until new growth could be harvested, they would be paying taxes for years on lands that had no present value to them. Henry was cautiously optimistic about farming prospects in the cutover, however. He candidly wrote, "There is no royal road to farming in northern Wisconsin." Clearing a farm up here would be long, hard work. "The brush, stumps and undergrowth is often sufficient to make one's heart grow faint . . . before the ideal field can be secured," he

Figure 14. Cutover farmers clearing stumps from their land. This photograph appeared in *Northern Wisconsin: A Hand-Book for the Homeseeker*, published in 1896 by the University of Wisconsin College of Agriculture. The state later realized it had been in error encouraging people to settle the cutover, and the university advised replacing farms with forest. Wisconsin Historical Society, WHi-10565.

warned. "No one should make the venture of home building in the new north before he has carefully counted the cost in the beginning and looked clear through to the end." The state of Wisconsin's promotion and Henry's scientific research in the cutover, historian Robert Gough argued, "conferred legitimacy on the idea of the cutover as a promising region for agricultural settlement."[157]

The agricultural prospects of the cutover were also promoted by the *Milwaukee Journal*. On March 12, 1900, the paper ran a special "Northern Wisconsin Edition." Articles discussed the present condition of the cutover, what role Milwaukee could play in its development, and its agricultural resources and prospects. The *Journal's* editors enthusiastically claimed that northern Wisconsin's agricultural and livestock possibilities would make it the "'richest part of the state.'" The cutover even attracted the attention of the U.S. secretary of agriculture James Wilson, who proclaimed to a Chicago audience in 1902 that northern Wisconsin would make a great agricultural region.[158]

Burnett County eagerly joined in the recruitment of settlers and hosted a state immigration convention, which "attracted so much attention through the daily press that the county got a [lot of] good advertising." The county soon formed its own Board of Immigration. Burnett County newspaperman Edward Peet wrote articles and printed speeches and maps in the *Journal of Burnett County* celebrating the progress of settlements of the county and the potential demand for agricultural products. In 1902 he furthered these efforts by writing a promotional pamphlet titled *Burnett County, Wisconsin* in which he extolled the unique virtues of Burnett County for prospective settlers. He claimed the county's farming prospects were as good as in southern Wisconsin. "A crop of timber is a sure index to the soil. Hardwood land is the best agricultural land we have and runs largely to clay soil. . . . White pine . . . is never found on poor land. . . . People who are accustomed to a black soil and nothing else will have great difficulty in believing that crops will grow on any other color. . . . Farm it well and farm fewer acres than you would of clay and you will make money." Peet admitted not all the soil in the county was good, but he claimed it could still be useful. "If you are coming to this county with plenty of money and want to engage in stock farming you are making a safe investment to buy clay land and go into the raising of clover and other tame grasses."[159]

Peet aimed to attract a particular type of homesteader and wrote, in no uncertain terms, whom he preferred. "The best advice the writer of this book can give anyone," he wrote, "is to not attempt to move a family to any place among strangers if the head of the family is unable to make a living where he now is. . . . There must be a disposition to hard work and physical ability to match the disposition. The 'ne'r do well' is not wanted in Burnett County. . . . Those who can help themselves from the start are the folks who are wanted. . . . The natural local conditions will do the rest for them."[160]

The Chicago, St. Paul, Minneapolis & Omaha Railway brought "land-hungry [settlers] from the cities and overcrowded farming areas" to the "stump farms" of Washburn County in the early twentieth century. For the next twenty years, "settlement and fires cleared land until it was hard to believe that this new landscape had once had great forests." Fires were not just started as the accidental aftermath of logging debris catching fire. Many of these pioneers used fire as an easy way to clear the scrub brush to provide pasture for their livestock. "These fires often went uncontrolled and sometimes did not burn out for weeks," recalled an old settler, "leaving little life in their path. . . . Vast areas became prairie with blackened stumps and rampikes to show where the forest had been."[161]

A legacy of the logging era to settlers in the cutover, however, was that they would not be "homesteaders" in the strictest sense of the word. By

1900 nearly all land in the cutover was in private hands except in Burnett County, which had a fair amount of government-owned land because of its wetlands. At the time Peet's pamphlet was published, however, state land had been withdrawn from the market. Prospective settlers had the choice to buy land from the Chicago, St. Paul, Minneapolis & Omaha Railway or from land speculators. Railroad land was limited, however, and many townships in the county did not have any railroad-owned land. The largest amount of land for sale was the "Baker Lands" of the Cushing Land Agency in St. Croix Falls. Most of the land available was in the sparsely settled northern portion of the county. The oldest settlers had only lived there for four or five years.[162]

Unlike their nineteenth-century counterparts, therefore, most settlers purchased land from private landowners rather than claim land from the federal government under the Homestead Act. One of the most notable land dealers was the American Immigration Company (AIC). It was formed in 1906 when nine lumber companies, including Weyerhaeuser-Laird-Norton subsidiaries and the North Wisconsin Lumber Company located in the Hayward area, decided to form a consortium to unload their land. By 1939 AIC had sold 438,000 acres to settlers.

Railroad companies also sought recruits to their cutover lands by sponsoring low-fare "home-seeker excursions." These allowed prospective settlers to survey land before they bought. However, many land agents who met these trains often put the hard sell on unsuspecting homesteaders, who made the mistake of buying "sight unseen." One homesteader remarked in 1919 that land "was being unloaded on unwary people." These agents often made a purchaser include inferior land along with the good. But land values in the cutover were relatively inexpensive, and land agents offered reasonable down payments and repayment schedules. All parties expected land values to go up, and those conventional mortgages could be obtained to pay the balance when it came due. This was true for the first two decades of the twentieth century, when farm values in the cutover generally quadrupled and farm products fetched a good price during World War I. A lender in Polk County boasted there were no foreclosures there before 1920.[163] However, these settlers took on more debt than their earlier counterparts.

Another ploy used to sell lands in the cutover, especially in Sawyer and Washburn counties, was to claim they had mineral deposits. In 1854 the U.S. Geological Survey had surveyed the Penokee-Gogebic Range with the use of a simple compass that found magnetic attraction in the area. "'Mineral lands,' was the bait," recalled one pioneer. "Iron and copper was everywhere! Any area that had magnetic rock formations could be hawked as mineral lands." Timber and railroad interests put pressure on the state of Wisconsin to do more prospecting. The AIC and the Loretta Mining Company, along with others, did some of their own surveying as well. By 1917 approximately one hundred holes had been drilled, mostly in Sawyer County. The results of this extensive drilling revealed no iron deposits but an underlay of Huronian rock that was laced in iron. The best that could be said of the results was that some of the holes provided great artesian wells of ice-cold water. Mining in the cutover of the St. Croix Valley proved another elusive promise of wealth and opportunity.[164]

Turning the cutover into productive farms was not easy. Many heads of families were forced to take up the lumberjack trade in the winter, leaving their wives to care for their families and homestead during the long, lonely winter months. "The first three winters Dad worked in the logging camp," recalled Carl Kuhnly, whose family settled in Burnett County at the beginning of the twentieth century. "He would walk about ten miles to come home on a Saturday evening, then back again on Sunday evening." "Farmers in this country couldn't get any money excepting from the timber in pineries north of here," recalled Harry D. Baker. Therefore, "Farming was very poorly developed with small clearings, log houses, little production, only a few cattle, and very meager living which was supplemented . . . by shooting deer for meat and by fishing and berry picking and making of maple syrup and maple sugar. Not until the development of the dairy industry in the nineties did farming become really at all profitable and the settlers . . . had a very bitter struggle."[165]

To address this problem, farmers were encouraged by the Board of Immigration to turn to dairying. In 1911 B. G. Packer, a lawyer and former secretary for the Farmers' Institutes, became the board's secretary. Packer was responsible for changing the direction of the board's purpose from simple recruitment of settlers to being an advisor to farmers. The Progressive movement, which championed scientific solutions to social and economic problems, had a significant influence on the Board of Immigration and the College of Agriculture. Outside experts became increasingly interested in cutover farmers, who they assumed needed their professional guidance to be successful farmers. "In doing so," Gough claims, "it laid the foundation for the public policies which would later discourage agricultural settlement in northern Wisconsin." Some farmers did take advantage of this assistance and turned to dairying because they realized the limits of the climate and soil and because dairy products paid well. Bayfield County Finns and Danes led the development of cooperative creameries and cheese factories. Between 1910 and 1930, these sprang up throughout the cutover, and hay became the principal crop.[166]

Women played a crucial role in the settlement of the cutover. Although homesteading required a strong back, most men in the cutover found they could not get a farm going alone. Starting a farm here required the labor of the entire family. Father, mother, and children spent many springs clearing the land of rocks, removing brush and stumps, building basic shelter for the family and animals, and digging wells. In this early phase of development, women's contribution was recognized and appreciated by their husbands. Men who could not find a wife usually gave up on farming. Some women even took up homesteading without men. In 1901 several women who had grown up in the sandiest part of Minnesota "and were not afraid of light soil came over to see what the county was like." They were interested in the lands along the St. Croix and Clam rivers. Pleased with what they saw, they took advantage of the liberal Homestead Law that allowed single or widowed women the opportunity to claim land like a male head of household. These women made their filing for land, "went home, did missionary work and sent others to take up land in the same town." Within six months twenty widows and single schoolteachers bought land and started farming in Burnett County "and have already made a good showing."[167]

One way to avoid the difficult task of clearing land was to raise livestock. Some farmers turned to sheep, figuring they might do well among all the low brush. In 1907, however, H. L. Russell, William Henry's successor at the College of Agriculture, warned farmers that sheep were susceptible to parasites and would not do well eating forest scrub. Farmers eventually found Russell was right. Dairying was the most obvious choice for livestock farmers. The College of Agriculture did not abandon farmers once on the land. It threw itself into experimenting with fertilizers that would enrich cutover soil. It established a branch station in Spooner, hosted a series of winter meetings with farmers in Bayfield and Ashland counties, and ran a demonstration farm near Superior. The school also assisted counties in organizing county agent programs. The ubiquitous problem of clearing stumps also prompted the College of Agriculture to experiment with stump-pulling devices to make the job easier. Russell even managed to buy up surplus military dynamite after World War I to blast stumps out of the ground.[168]

By 1911 the most northern parts of Burnett County were settled. In that year the Soo Line Railroad extended its trackage from Frederic to Lake Superior. The line crossed the St. Croix River near the Yellow River at a town called White City. Shortly thereafter the railroad commission began offering its allotted lands at public auction. One of their first buyers was Ed Peet, the enthusiastic promoter of the county. He proceeded to build the first hotel, which he called Danbury House, and re-christened the town with the same name.[169]

The yeoman farmer epitomized the cutover settler. They were family farmers, many of recent immigrant stock, who came with the expectation that hard work, family cooperation, and assisting neighbors would bring them an independent, if not a prosperous, life in rural America. While some immigrants came directly from Europe, many had first found jobs in factories, in mines, or on railroads. They were attracted to the cutover by the prospect of becoming an independent farmer and enjoying rural community life. Many settled in communities of "their own kind." Scandinavians and Germans continued to settle in the North Country but were now joined by Poles, Finns, and Latvians, as well as "Americans" from other regions of the country. Fishing and hunting opportunities also attracted many farmers from southern Wisconsin.[170]

Settlers to the cutover were lured to the region by confidence in the current scientific knowledge of the time and trust in the honesty of land dealers and future market opportunities. By 1920 many cutover farmers had made a good showing for themselves. The U.S. census for that year showed that land values in the cutover had increased to 47 percent of Wisconsin's average and that 33 percent of farm acreage here was improved. Burnett and Polk counties counted impressive gains in land brought under cultivation and improvements made.[171]

Despite these gains, the cutover was a comparatively poor region. "A land of plenty?" wrote one Burnett County pioneer. "For starters there were forest fires, poverty, horse flies, diphtheria, pregnancy, open air toilets, and 'blood, sweat, and tears.'" These smaller farms produced crops worth one-third less than the rest of the state. Part of the reason was that many farms were still in the frontier stage of development. Some settlers were part-time farmers either by choice or by economic necessity. These men worked periodically in the woods or in sawmills, leaving little time for farm expansion. Some farmers were simply not interested in participating in a commercial agricultural market. They instead preferred the enjoyment of the natural environment and the outdoor life of hunting and fishing, and they farmed to support this lifestyle. Families were able to supplement their income from harvesting cranberries and blueberries.[172] These pioneers, however, came with few resources, and since they did not enjoy the benefit of squatters' rights and homestead laws, they started off with more debt and smaller farms than farmers in the Lower St. Croix Valley. Farmers in the cutover would have fewer cushions to ward off blows from the failure of their soil, the weather, or the vagaries of the market.

When farms began to fail in the 1920s, farmers began to blame lumber companies for their financial woes for charging too much for their land and profiting unfairly. However, historian Robert Gough has argued, "Farmers coming to the cutover at the beginning of the twentieth century inherited this legacy of an unregulated, ill-informed, short-sighted, and loosely managed system of land distribution." Lumber companies did profit from the sale of cutover land, but no fortunes were made. The reasons for poverty and the failure to thrive in farming were more complex than this simple populist charge would warrant.[173]

When an agricultural depression of the 1920s followed in the wake of the high prices fetched during World War I years, many farmers faced the prospect of failure. In 1927 the Wisconsin Department of Agriculture began taking inventories of county lands. Their conclusion for the state's northern counties was less sanguine than reports earlier in the century. Bayfield County went from having 60 to 70 percent of its lands deemed suitable for agriculture in 1916 to 75 percent assessed as submarginal for agriculture. Recommendations were made that at least 20 percent of the land be reforested.[174]

In the 1920s, Prohibition also gave the cutover an unsavory reputation. During Prohibition many areas around the country, of course, ignored or even flaunted the Nineteenth Amendment. Northern Wisconsin residents were generally "wet" on the issue. Economic hard times also encouraged the making of moonshine from potatoes. Some people in Luck in Polk County reputedly made this a "sideline." The cutover's remoteness and frontier-like environment along with its position between the Twin Cities, Chicago, and Canada made it an ideal hideout for Chicago bootleggers. Al Capone allegedly spent $250,000 for a four-bedroom house equipped with machine-gun portals on Cranberry Lake in Sawyer County. In 1934 John Dillinger escaped to the cutover. His pursuit by the FBI and other law enforcement agents made national headlines and helped create the image of northern Wisconsin as a lawless place.

The eugenics movement of the period also colored outsiders' perceptions of cutover residents. Eugenicists feared that isolated settlements of single ethnic groups would result in physically degenerate "hillbilly" people. These attitudes mingled with the growing perception that much of the land in the cutover was submarginal agricultural land. The Jeffersonian yeoman farmer ideal also took a beating in the first two decades of the twentieth century. Rural Americans began to be viewed as "backward, immoral, and increasingly dangerous." State and federal experts were ready to step in to prevent northern Wisconsin from turning into another Appalachia.[175]

The cutover's reputation was not helped either by stories published in Milwaukee newspapers that focused primarily on forest fires, hunting accidents, bear attacks, and any violent crime. The presence of unassimilated Native Americans along with the North Woods' reputation from the logging era as a haven for houses of prostitution also gave it an exotic and sinful reputation in need of reform. However, the region unexpectedly got a reputation as an "upscale" vacation destination when Calvin Coolidge spent a ten-week summer vacation at Cedar Lake Lodge on the Brule River in Douglas County. Many elites became smitten by its "wilderness" ambience and thought it should become a resort location like New York's Adirondacks.[176]

For a variety of reasons, then, settlers stopped coming to the cutover in the 1920s. The number of farms increased by less than 3 percent, and acreage only increased by a little more than 4 percent. When the Great Depression hit in the 1930s, farming in the cutover reached crisis proportions. While farm products declined throughout the state and even the nation, cutover farm products fetched a proportionally lower price for their goods. The sandy soil in the Pine Barrens produced grasses lacking in nutrition. Potato growing was also limited since the soil was too light and lacked the necessary plant foods. Dry summers turned into droughts for this sandy-soiled region. Few farmers here practiced crop rotation or other means to improve pasturage. One observer described the cutover farms: "Cropped fields . . . where a few cattle . . . find scanty grazing. . . . These poorly fed and poorly cared for herds are the basis of a dairy industry. . . . The industry is, however, small-scale in character, with poor barns and equipment and meager returns." Farms were too scattered for creamery trucks to collect milk profitably. Farmers, however, began to find a summer market for their dairy products in the growing resort industry. To supplement their incomes, many farmers turned to fur farming of beaver, silver fox, and chinchilla rabbits, as well as gray and white rabbits for meat. Many also turned to commercial cranberry growing in the marshy, acidic soil that ran through the Pine Barrens. Local Indians harvested wild rice by hand to supplement the tables of sportsmen.[177]

When the remaining land in the cutover went unclaimed, the old stumps and brush acted as fire hazards. Tens of thousands of acres burned in the 1930s, convincing many experts in the state that farming had not "redeemed" the land in the cutover. To add insult to injury, cutover residents were also more likely to need relief than their counterparts elsewhere in the state, and relief rates there matched other depressed regions of the country like Appalachia and the Dust Bowl area of the Great Plains. This put tremendous pressure on county agencies that required more in taxes but faced declining property values and a declining tax base. Since the 1920s, tax delinquencies had become increasingly common in the cutover. This only increased during the Depression. Many farmers who could not pay their taxes simply gave up. "Almost as numerous as the occupied farms are the abandoned, tumbled-down farmhouses surrounded by fields going to waste," wrote Wisconsin geographer Raymond Murphy in 1931. "Sometimes only a few stones and a patch of quack grass remain to mark the site of a former home, and to give the impression of poor land and unsuccessful farming." The Pitted Sand Plain in northern Burnett, northwestern Washburn, and southeastern Douglas counties displayed characteristic features of farm abandonment. The first phase, Murphy noted, was that of "one little shack out in grassy barrens . . . occupied by an old man who formerly grew corn and a few other crops." Fire had destroyed much of the humus in the soil, and many farmers neglected to build it back up. "After a year or two of use the corn field 'got away' and now is a bare expanse of ripple-marked sand. Near the shack a few vegetables are still grown, but they hardly repay the effort, and the old, paralytic settler barely manages to exist." In the second stage, homes were abandoned, windows were broken, and cleared land was overgrown with quack grass. The third stage was marked by decayed, tumbled-down homes, and the growth of scrub oak and jack pine. Murphy found few orchards in the Pine Barrens since frost often struck any time of the year. He was also critical of local farming practices. "Instead of the use of scientific farm practices to combat handicaps of soil and climate, the common practice seems to consist of meeting declining yields by cutting down acreage until returns are not enough to pay the taxes on the land, and the county must take possession."[178]

What was particularly depressing about the Pine Barrens and the cutover in the early 1930s was the lack of young people and children. "The region is characteristically one of people past middle age—weatherworn old Scandinavians who came here with their wives and children many years ago," wrote a contemporary observer. "The children have grown up and gone. The Barrens does not hold its younger generation. No new settlers are moving in, and one gets the impression that when the present hardy survivors pass on there will be none to take their places."[179]

Northern Wisconsin, however, did not experience a net population loss in the 1930s. When the urban industrial economy collapsed, rumors circulated that subsistence farming was possible in the North Country. Many unemployed city workers joined the "back-to-the-land movement" that sprang up in the decade. Between 1935 and 1940, the population increased in the cutover, as did the number of farms. Most of these new settlers, however, "were not serious farmers. They saw themselves as temporarily eking out a semi-subsistence existence, squatting on or renting cheap land, or perhaps living on part of a relative's farm." Gough claimed that these "farmers" had "a negative effect on agricultural development in the cutover." They did not clear new land or raise crops for market. "The properties they left behind when they moved on contributed to the image of the region as filled with abandoned farms."[180]

Despite these problems, the industrious farmer managed to survive these economic hard times. By relying on family labor, the frugal household economy of the farm wife, off-farm work, and catering to the growing tourist industry, many farm families made it. Some even managed to redeem tax delinquent land before foreclosure. Gough found in his study that the majority of land that actually experienced foreclosure in 1930 belonged to individual speculators and land companies.[181]

However, New Deal agricultural programs were the last nails in the coffin for many family farmers in the cutover and across America. Large commercial farmers were able to take more advantage of these programs than subsistence and marginal farmers. The Agricultural Adjustment Act of 1933 aimed to raise farm prices through agreements to limit farm production. Farmers who had surplus acres and could still farm for the market as well as their families could participate in the program. The same held true for the Soil Conservation and Domestic Allotment Act of 1936. This act promoted conservation by encouraging farmers to remove land from cultivation. In sum, New Deal agricultural programs promoted the cash economy of commercial farming and undermined semi-subsistence farming. The Department of Agriculture also began to take the approach that lands that did not produce for the market were inefficient and unnecessary. These low-producing areas, it feared, would become another Appalachia if the surplus population not needed for commercial farming was not moved off the land. This view was reinforced by the migration of Kentuckians to northern Wisconsin. Through the 1930s, Wisconsin underwent a series of relocation programs sponsored by federal, state, and county governments.[182]

The Northern Wisconsin Settler Relocation Project, which began in 1934, specifically targeted the cutover. By 1940, when the project ended, $500,000 of federal funds had been used to purchase between four and five hundred farms in seven cutover counties, including Sawyer and Bayfield counties. Walter A. Rowlands and Dean Chris Christenson of the University of Wisconsin's College of Agriculture requested the money from the U.S. Department of Agriculture. L. G. Sorden, from the Agricultural Extension of the University of Wisconsin in Madison, headed the project. He was convinced this was the best course of action for both the farm families and for the future land use of the cutover. "All the settlers whose farms were purchased," he claimed, "were living on submarginal land, which was either too light and sandy, too stony and rough, or so isolated from markets that they were definitely uneconomic farm units." He felt the relocation was a great benefit since "as many as 80% of the families whose farms were purchased received public aid." They clearly were not prospering in the cutover. "On the average, $2000 was paid per farm," Sorden reported. "These 'farms' ranged from a tar paper shack in the woods to a few quite well-developed farms."[183]

Of the families who relocated, only 38 percent asked for new farms. They obtained financing from the Farm Security Administration. One-third decided to retire since they were too old to begin again elsewhere or take up a new line of work. The Wisconsin Rural Rehabilitation Corporation built "retirement homesteads" in northern Wisconsin for these people, complete with modern conveniences, a large garden, and even a small barn for milk cows. The county maintained these homes for a nominal fee. Others chose new occupations ranging from woodworking to general laborers to resort work to mercantile businesses. The title to the lands they vacated was transferred to federal, state, or county governments for forestry uses.[184]

Sorden defended the relocation project in 1979, citing the fact that nearly all the families who were approached chose relocation. "When the project was explained and when the families were given time to think it over and talk it over with other people in whom they had confidence," explained Sorden, "98% of these families were willing to sell and relocate." Counties also benefited because "this isolated settler relocation project immediately made possible a saving in school costs of more than $15,000 per annum by closing rural schools," Sorden noted. "In addition, several thousand dollars worth of school transportation costs were eliminated. Road costs were reduced by the elimination of maintenance and snow plowing. Relief costs were cut materially by placing many of these families in a position to make their own living." Sorden took great satisfaction in the role he played in this project, and he was confident that these people were given renewed hope "by their removal from isolated areas to established communities where they and their families [had] a chance to start over again with a more secure financial and social future."[185]

Robert Gough argues, however, that while farming did survive in the cutover after 1940, these farms either were worked part-time as a hobby or became much larger operations. In the 1930s, farmers with good land were encouraged by state experts to expand their holdings and turn to dairying. Others grazed horses for recreational riders, grew Christmas trees, pumpkins, or ginseng, or became orchard farmers. Most depended upon income from off the farm, especially the "farm" wife. "The new economic plan for the cutover which deemphasized farming and stressed reforestation and tourism," Gough argues, "did not attract new residents to northern Wisconsin or enrich the ones who already lived there." The unfortunate result was that even in 1990, the Wisconsin counties with the lowest per capita income were all in the cutover. They included Burnett and Sawyer counties, as well as Forest, Iron, and Rusk counties.[186]

The reduction in farms and the changed nature of those who remained affected the social fabric of the cutover. "To those of us who helped clear a stump-farm from the cut-over, there is nostalgia for the events of those times," recalled one old-timer, "for the feeling of pride when another acre of clover was added, for the excitement of a burning pile of stumps, or for the alarm when a wild-fire swept across the nearest hill." Gough reinforced that nostalgia. "No longer could the bonds of rural neighborhoods be fostered by school pageants and district business meetings in one-room schoolhouses," he lamented. James Kates also echoed these sentiments when he wrote, "With school consolidation, the daily rhythm of life now centered more on urban places with schools . . . encouraging the expansion of urban and commercial attitudes into the countryside once dominated by the values of yeoman farming. . . . For the people in the cutover committed to yeoman farming these were sad developments."[187]

As farming faded in the Upper St. Croix Valley, a new vision of how to order the landscape was gradually winning acceptance—the conservation movement and recreation. The myth of the yeoman farmer yielded to the myth of the North Woods.

4 | Up North

The Development of Recreation in the St. Croix Valley

In 1936 the twenty counties of northwest Wisconsin cooperated in a tourist brochure that promoted the region as "Indian Head Country." The name was derived from the shape of Wisconsin's St. Croix borderland that appeared to the imaginative as the silhouette of a human profile. Pierce County was the chin, St. Croix County the mouth, and Burnett County formed a prominent "Roman" nose. For the tourist boosters the choice of "Indian Head" was obvious. Not only did the large nose suggest the Indian profile on the "Buffalo" nickel then in circulation, but the Indian was the symbol of all that was uniquely American. The Indian was a symbol of wild, unrestrained nature. Never for a moment did the tourist promoters think of labeling the twenty-county area "Swedish Head Country" or "Polish Head Country." Such a label was, of course, ludicrous even if it did call to mind some of the people who had devoted their lives to the unsuccessful effort to bring agriculture to the cutover. That history was too recent, too painful, too prosaic. It would be as unspoiled nature—a romantic, even ridiculous, impossibility given the history of logging and farming—that the St. Croix region would be sold to the public. Although the river valley was not unspoiled, the robustness of its midwestern environment made its rebirth as a vacation haven possible. The river would be dredged of dead-head logs and silt, its banks riprapped, new forests grown, fish restocked, and wildlife returned. One could almost be convinced that they were seeing the St. Croix much like Little Crow did before the axe and plow came, but the river valley required the hand of man to return it to its former pristine state.

Ironically, as the St. Croix River emerged from wilderness to a developed and settled region, American attitudes toward nature and the wilderness were changing. During the colonial era, America was seen as a land of abundance and a refuge from Old World ills, but its primeval forests were also seen as a hostile wilderness filled with savage beasts and men. While it bestowed bounty on those able to meet its challenges, nature was a harsh taskmaster and extracted a heavy price from the unfit. The Industrial Revolution changed this relationship. Humans became the master of nature instead of the victim. Industrialization, however, also scarred and even destroyed nature's beauty and exposed its fragility. At the mercy of humans, nature was no longer to be feared but cherished.[1]

This appreciation of nature had its roots in the eighteenth-century Enlightenment when the natural world was held up as inspiration and a model for social organization. If human society followed the laws of nature, it could find peace and harmony. These beliefs found expression through political, economic, social, and artistic channels, which required nature to be experienced firsthand. This glorification of nature continued into the early nineteenth-century Romantic movement with its reaction against the ugliness of the Industrial Revolution. Nature was not only beautiful, sublime, and a guide to social order, but also a source of spiritual renewal for people severed from their rural roots in ugly urban cities. While the Industrial Revolution marred nature, it also ironically made nature more possible to enjoy. The invention of the steamboat and the railroad allowed people to experience natural wonders firsthand without foregoing many of the creature comforts of civilization. There they could experience spiritual renewal and regeneration from the more fast-paced and wearisome world of the city.

This cultural context shaped the way late eighteenth- and early nineteenth-century white explorers and settlers perceived the St. Croix River Valley. Its distinctive geographic formations, such as the "Old Man of the Dalles," provided explorers with navigation references, but also drew them into the unique splendors of the river valley. George Nelson, the Canadian fur trader who wintered in the valley in 1802–3, noted in his diary:

> Whenever this country becomes settled how delightfully will the inhabitants pass their time. There is no place perhaps on this globe where nature has displayed & diversified lands & water as here. I have always felt as if invited to settle down & admire the beautiful views with a sort of joyful thankfulness for having been led to them.[2]

Nelson's paean to the Upper St. Croix Country was most likely added to his diary many years after his winter in the valley. It reflects the power of a picturesque landscape to overcome the realities of Nelson's last days on the river: cold and wet spring weather, rapids, portages, mosquitoes swarming, and fear of a Dakota attack. Romanticism was necessary to transform a truly wild landscape into a picturesque retreat, and the mastery that came with technology and private property made possible the evolution of the Upper St. Croix from a battleground between the Ojibwe and Dakota to the white man's "Indian Head Country" vacation destination.

Although the exploitable natural resources enticed the first permanent settlers, tourists also began to venture into the St. Croix Valley early in the nineteenth century. One of the first steamboat tourists was an Italian named Giacomo Beltrami, who made the trip in 1823. He found the scenery and towering bluffs comparable to the beauty of the Rhine River.[3] Henry Schoolcraft's published account of his 1832 journey publicized the scenery to a national audience. "Its banks are high and afford a series of picturesque views," he wrote. In 1837 Joseph N. Nicollet, a French expatriate, followed Schoolcraft's path, exploring and mapping the Northwest Territories. He, too, was struck by Lake St. Croix's beauty. "The shores are rugged and steep, interrupted by lovely, sheltering coves," he related. "The shallows are plentiful. It is indeed a picturesque river."[4]

Steamboats and the "Fashionable Tour"

The first person to recommend the Upper Mississippi valley to tourists was George Catlin, a self-trained artist from Philadelphia. Catlin's ambition was to visually record North American Indians in their natural environment before they "vanished." In 1835 and 1836 he ventured to the North Country to paint the Sioux Indians. Catlin was so enamored by the country that he encouraged a "Fashionable Tour" by steamboat from St. Louis to the Falls of St. Anthony. He wrote:

> This Tour would comprehend but a small part of the great, "Far West;" but it will furnish to the traveler a fair sample, and being a part of it which is now made so easily accessible to the world, and the only part of it to which *ladies* can have access, I would recommend to all who have time and inclination to devote to the enjoyment of so splendid a Tour.[5]

Many adventuresome travelers responded to Catlin's recommendation and began to take steamboat tours of the Upper Mississippi River. In 1837 the widow of Alexander Hamilton, Elizabeth Schuyler Hamilton, toured to Fort Snelling and the Falls of St. Anthony. When she returned to the East, she encouraged others to take the "Fashionable Tour." It was, however, the artists of the period who painted the Upper Mississippi valley who enticed crowds to come here. In 1839 John Rowson Smith and John

Risley painted a panorama of the Upper Mississippi valley, with which they then toured the country. In the summer of 1848 Henry Lewis painted a panorama of the Mississippi between St. Louis and Fort Snelling that included scenes of the St. Croix. Lewis's paintings *Gorge of the St. Croix* and *Cheever's Mill* spanned hundreds of square feet of canvas. Within a decade at least eight to ten panoramas of the Upper Mississippi toured the country.[6]

Honeymooning couples, small parties, and even groups of a hundred or more soon made traveling the Upper Mississippi River a popular pastime. Chartered boats allowed the well-healed to avoid the immigrant throngs and freight stops common on the usual steamboat runs upriver. Tourists came from as far south as New Orleans, and when rail service reached the Mississippi, they came from Chicago, Pittsburgh, New York, Boston, and even Europe. Artists and writers found the region inspiring, and prominent politicians and journalists, such as Millard Fillmore and Thurlow Weed of New York, as well as other dignitaries made the Upper Mississippi an important stop on their travel itineraries. River towns made them feel like honored guests by welcoming them with gala receptions. "Indian Watching," done in complete "safety," added to the attraction of a trip up the Upper Mississippi and St. Croix rivers. A Dubuque newspaper advertised the trip as a "convenient and certain" way to watch Indians living in their native world.[7]

During the summer of 1849 the travel writer Ephraim S. Seymour of New York State made his "Fashionable Steamboat Tour" of the Upper Mississippi River. He started in St. Louis, made a stop in Galena, Illinois, and then traveled upriver to Fort Snelling and the Falls of St. Anthony. Unlike other tourists who stayed on the Mississippi, Seymour also ventured into the St. Croix Valley, collected information on Indians and lumbering, and detailed the scenery from the Willow River to St. Croix Falls. In 1850 he published *Sketches of Minnesota, the New England of the West,* which further promoted the scenic splendor of the St. Croix Valley to the American reading public. More importantly, Seymour was also the first to promote the healthful benefits of the climate from ills such as ague and consumption. In his book, Seymour related an encounter he had with an old friend from Galena whose health had been impaired by repeated attacks of cholera. The friend hoped a trip upriver to Minnesota might restore his health. "A few days spent in sporting and fishing among the brooks, rivers, and lakes of this bracing climate," Seymour proclaimed, "had rendered him quite robust and healthy." And he advised, "Such excursions might be recommended to many invalids, as far superior to quack medicines and expensive nostrums." Sportsmen, Seymour noted in his book, would also find the Lower St. Croix Valley a paradise. "Deer are killed here in great numbers. . . . The bear and the large gray wolf are often seen. Wild geese and ducks resort here in great numbers. . . . The best trout fishing in the northwest is said to be on the Rush River. They are caught in immense quantities, not only with hooks, but also with scoop-nets." The fields were "alive with [passenger] pigeons, which were constantly rising from the ground in large flocks."[8]

During the 1840s northerners began to lure southerners away from the lower latitude resorts that they had patronized, such as the Virginia Springs and the Harrodsburg Springs of Kentucky. Southerners had also sought cooler climes in the Hudson valley and Niagara Falls. In 1842 Daniel Drake wrote *The Northern Lakes: A Summer Resort for Invalids of the South,* which encouraged southerners to explore the Great Lakes region aboard ship. Drake claimed that by coming north of the 44-degree line of latitude one could escape "the region of miasms, musquitos [*sic*], congestive fevers, liver diseases, jaundice, cholera morbus, dyspepsia, blue devil and duns!" The gentle rolling of the boat and cool lake breezes, he claimed, could cure hysteria and even hypochondria. But before the era of widespread rail linkups, the Great Lakes were not easily accessible to those in the South, but the Upper Mississippi River's "Fashionable Tour" was an attractive alternative with all the same healthful benefits.[9]

In 1852 Edward Sullivan published *Rambles and Scrambles in North and South America,* which described his adventurous canoe trip down the Brule and St. Croix rivers. And in 1853 Elizabeth Fries Ellet traveled up the St. Croix in the comfort of a side-wheeler steamboat with thirty staterooms to "explore" the frontier. The result was the travelogue *Summer Rambles in the West,* which eloquently described the scenery of the Dalles and the Lower St. Croix Valley. She wrote of the Dalles:

> A scene presented itself which nothing on the Upper Mississippi can parallel. The stream enters a wild, narrow gorge, so deep and dark, that the declining sun is quite shut out; perpendicular walls of traprock, scarlet and chocolate-colored, and gray with the mass of centuries, rising from the water, are piled in savage grandeur on either side . . . some gigantic buttress uplifts itself in front of the cliffs, like a ruined tower of primeval days.[10]

The St. Croix, however, was primarily a working river. The only means of travel was by steam packets and freight boats carrying supplies, livestock, and pioneer settlers. The St. Croix logging boom north of Stillwater also hindered the free flow of river traffic, as did the seemingly endless stream of logs floating downriver. By summer's end the log run was finished, but the warmer, drier season lowered the water levels, exposing sandbars and narrower channels and making excursions more difficult. For example, in August 1859 an excursion steamer disembarked from Stillwater with thirty-five to forty citizens aboard. The *Kate* picked up more passengers at Marine and Osceola, bringing its number to nearly a hundred. For the occasion, the boat was decorated with banners and evergreens. Although it left early in the day, the steamer did not reach the Dalles until the following morning due to "unavoidable detentions on account of the low stage of the water and heavy freight" and was hung up on bars. Apparently the passengers were not very put out by the long trip as the delay was "amply atoned for, in the privilege of passing through the 'Dell' just as the sun was peeping over the mountains and dispelling the most beautiful mist and spray from that most beautiful and romantic spot."[11]

Despite the hazards of travel on the St. Croix, the towns along the water still enthusiastically planned for and promoted their attractions. Prescott, Hudson, Stillwater, Osceola, and Taylors Falls built hotel accommodations for both new settlers and tourists. In 1857 the four-story Sawyer House was completed in Stillwater. Its "spacious rooms for social events made it one of the outstanding hostelries in the development of Minnesota." Summer cottages were planned for the shores of Lake St. Croix. "The day is not far distant," claimed the *Stillwater Messenger,* "when nice cottages . . . will reflect their white and dancing shadows from the bosom of Lake St. Croix."[12]

Tourism in the St. Croix Valley got a boost from the Twin Cities when John P. Owens, the editor of the *St. Paul Minnesotan,* took an excursion on the steamboat *Humbolt* in 1853. "The little *Humbolt* is a great accommodation to the people of the St. Croix," he wrote. "She stops anywhere along the river to do any and all kinds of business that may offer, and will give passengers a longer ride, so far as *time* is concerned, for a dollar, than any other craft we ever traveled upon." The boat graciously stopped at Marine Mills to allow its hungry travelers to lunch at the Marine House. Owens also stopped in Taylors Falls and made an assessment of this town's accommodations. "This Chisago House, is *better furnished,* and as well kept— barring the inconvenience of having no meat and vegetable market at hand—as any house in St. Paul, St. Anthony, or Stillwater," he wrote. "We never hated to leave a place so much in our life, when absent from home."[13]

In June 1857 the *St. Paul Advertiser* identified Marine as the spot to visit, claiming, "To the invalid, the pleasure seeker, as well as the sportsman, no place affords more ample inducements for sojourn and recreation." In 1859

Stillwater welcomed regional visitors to its Fourth of July festivities. The steamer *Itasca* brought visitors from St. Paul and other stops along the Mississippi. The passengers enjoyed the annual parade, a German Singing Society, and tumblers from the Turner Society. After a cold supper in the armory, the visitors enjoyed a ball at the Sawyer House until the whistle from the boat summoned them for their late-night journey home.[14]

Minnesota, however, courted tourists more aggressively than did Wisconsin. As early as the 1850s, Minnesota was determined to create recreational retreats that could rival eastern resorts, such as Saratoga Springs in New York State. Many hoped it would become the playground of the wealthy. Minnesota historian Theodore C. Blegen has written that tourism in Minnesota began with the establishment of journalism in the territory. "Every newspaper was a tourist bureau," he claimed. James M. Goodhue, the editor of the *Minnesota Pioneer,* was a leading booster of the recreational attractions of the territory. He made appeals to residents all along the Mississippi to escape the epidemics of cholera and malaria that plague southern climes for the healthy air and cool breezes of the North Country. "'Hurry along through the valley of the Mississippi, its shores studded with towns . . . flying by islands, prairies, woodlands, bluffs—an ever varied scene of beauty, away up into the land of the wild Dakota, and of cascades and pine forests, and cooling breezes.'"[15]

John W. Bond, the premiere pamphlet promoter for Minnesota, wrote in 1853, "We have springs equal to any in the world." Rather than lure easterners, however, the ease of travel up the Mississippi made the target audience southerners. "Gentlemen residing in New Orleans can come here by a quick and delightful conveyance," Bond explained, "and bring all that is necessary to make them comfortable during the summer months, and at a trifling expense. For a small sum of money they can purchase a few acres of land on the river, and build summer-cottages." Bond intended to promote the Falls of St. Anthony, which he believed would "rank with Saratoga, Newport, and the White mountains" in New Hampshire. In 1854 Earl S. Goodrich, the editor of the *St. Paul Pioneer,* beckoned southerners to the cooler, more refreshing northern retreats with biblical allusions. "Miserable sun-burned denizens of the torrid zone," he wrote, "come to Minnesota all ye that are roasting and heavy laden and we will give you rest." In 1856 artist Edwin Whitefield, who had painted landscapes and residences in the Hudson River valley and the Mississippi River, arrived in St. Paul. As an artist and newly established land speculator, Whitefield captured the beauty of local lakes in his paintings to promote the area. His works inspired many tourists to the fashionable river tours and to explore Minnesota.[16]

The St. Croix Valley, however, remained somewhat off the beaten path for most tourists. It therefore fell to local artists to record the physical attractions of the valley. Robert Sweeny was a St. Paul pharmacist turned artist. In 1858 the Minnesota Historical Society commissioned him to paint flowers, plants, and Indian artifacts. He then turned his attention to the St. Croix and painted in a documentary-like fashion the lumber mill sloughs at St. Croix Falls, the wood arch bridge over the river at Taylors Falls, Indians coming ashore on Lake St. Croix, and the Dalles. His paintings and sketches depict the picturesque qualities of the wilderness. Augustus O. Moore followed Sweeny. His sketches aimed to show that in the St. Croix Valley man and nature could live harmoniously. Another artist of the St. Croix was Elijah E. Edwards, who was principal of the Chisago Seminary, as well as clergyman and writer. Many of his painting and sketches were of the Dalles with an eye for light and romantic views.[17] While these artists recorded and interpreted the St. Croix River, their influence only extended to the local region in attracting visitors.

At midcentury there was still an abundance of game in the St. Croix Valley to supply the sportsman with a wide variety of birds, animals, and fish. Local presses extolled the sporting opportunities of the valley. "The country surrounding our city is filled with game," boasted the *St. Croix Union* in 1854. "Not infrequently do we hear a sportsman relate the experience of deer shooting . . . or what sport they had in 'bagging' a drove of prairie chickens. Deer are so plentiful. . . . Our hunters have become so well acquainted with the habits of this animal and so adept in the use of the rifle that it is a matter of no common occurrence to find their tables well supplied with venison. . . . We have a great many streams filled with [trout], and it is fine sport for those who are disposed to engage in it."[18]

In the 1850s the area between Marine and Taylors Falls was called the "bear hunting ground." The innkeeper at the Marine Mills Hotel loved to serve this local delicacy to his visitors. From time to time a "General Bear Hunt" was organized out of Prescott as a two-day excursion for "all who desire to share in the sport." An amateur poet from Hudson enticed hunters with the following:

> Come on then, ye sportsmen with high boots, rifle and blanket, and I will
> shortly conduct you to the forests where my forefathers, as they chased the
> swift elk and the huge black bear, would proudly exclaim,
>> No pent-up willow huts contain our powers,
>> But the unbounded wilderness is ours.[19]

Since winters were cold and long, many hearty souls took to winter sports. In the winter of 1863 the *Stillwater Messenger* announced, "Members of the Skating Club and all others are invited to call and examine our stock of skates, skating caps, hoods . . . gloves, mitts, &c." A few months later the paper reported, "The warm days and cool nights we have had lately have made the skating good upon the lake, and large crowds are enjoying the sport during this pleasant weather."[20]

Sectional tensions put a damper on tourism in the North Country. While outright abolitionism was not a force in Wisconsin or Minnesota, "Free Soil, Free Labor, and Free Men" was. In the summer of 1860 a Mississippi slave owner vacationing at the popular Winslow House in St. Anthony brought along a slave woman named Eliza Winston. Winston had apparently been promised her freedom, and once on free soil she gained the support of an abolitionist and petitioned the Minnesota court for her freedom. The court sided with her and granted her request with no challenge from her master. Anti-abolitionist sentiment, however, had been aroused, whereby a mob proposed to send Winston back to her master and to tar-and-feather the abolitionist who aided her. The Underground Railroad whisked her to Canada, and the matter was legally ended. Hotel owners, however, feared the loss of southern tourists' patronage if those guests risked losing their personal servants. The *Stillwater Democrat* warned that the "'intermeddling propensities of Abolition fanatics' would keep nearly a hundred of wealthy Southerners and their Negro servants from spending the summer along the shores of Lake St. Croix." By the next spring, however, it was the Civil War that halted the southern tourists.[21]

Once the war was over, the St. Croix Valley resumed its place as an attractive destination. When Horace Greeley visited the St. Croix Valley in 1865, he was impressed not only by its wheat, but also by its healthy climate, and he recommended the area for those plagued by ague and chronic coughs. The *Stillwater Messenger* quickly echoed these sentiments. The paper even joined in the exaggerations that often accompanied the literature written about the health benefits of the area: "Pine emits an odor which is peculiarly healing and highly beneficial for invalids; hence it is no uncommon thing for small parties to take up their quarters in the wilderness, and spend the winter there with numerous gangs of lumbermen." Consumption sufferers, in particular, could find relief in the pineries of the upper St. Croix. A poem was even written about the health-giving pine trees:

> For health comes sparkling in the stream
> From Namekagon stealing;

There's iron in our northern winds,
Our pines are trees of healing.[22]

Health seekers from the South and East were also enticed back to the region after the war by handbooks such as *Tourists' and Invalids' Complete Guide and Epitome of Travel*. By the end of the 1870s, southerners began to come to the St. Croix again in noticeable numbers. "Capt. Jack Reaney came up on the steamer Knapp Tuesday," wrote the *Burnett County Sentinel*. "He informs us that the tourists from the south are coming up in large numbers, and many of them find their way to the St. Croix river."[23]

Between the 1860s and 1880s, making and selling photographs and stereograph images of the St. Croix became a profitable business. Creating landscape images of the river became the "bread and butter" for local commercial photographers. In 1875 John P. Doremus of Patterson, New Jersey, began photographing the river as part of a "floating gallery" on a boat that was "a little palace itself." "He started out from St. Anthony over a year ago," related the *Stillwater Lumberman*, "with the intention of taking views along the Mississippi and its tributaries down to New Orleans." The paper expressed appreciation for his carefully considered photos. "He takes it leisurely and does his work in fine shape, the views he has of the St. Croix being the best we have ever seen." Photographic documentation further stimulated tourists looking to escape the oppressive heat, humidity, and illness of the lower Mississippi. The St. Croix Valley's fame spread further when in 1885 Eastman's roll film was developed. In 1900 Kodak's Brownie camera made photography easier and cheaper for visitors to the St. Croix to share their experiences with friends back home.[24]

Railroads and the Growth of Tourism

During the 1870s and 1880s, as more travelers were moving by rail, steamboat tourism was gradually reduced from lengthy tours to day excursions on the river. With connections to St. Paul, Milwaukee, and Chicago, local steamboats enjoyed a steady stream of passengers. Excursions provided church, social, and work groups trips with a more relaxed form of social interchange. For example, in June 1875 firemen from Red Wing, "accompanied by their ladies," boarded the steamer *James Means* for a trip up to the Dalles. Not only did they enjoy the scenery, but they joined the Stillwater fire department in some social recreation. "In the evening the boat laid at the levee several hours," related the *Stillwater Lumberman,* "and allowed the firemen an opportunity to be entertained by the chief and members of the Stillwater department." Steamboat owners also made moonlight cruises available for the more romantically inclined. Excursionists who chose to stay in "the charmingly old-fashioned" town of Taylors Falls found comfortable accommodations at the local hotel. There they were "gladly welcomed and hospitably entertained." The president of the town council, L. K. Stannard, gave welcome addresses to St. Paul excursionists. Others found lodging in towns downriver before they began their return home by rail.[25]

Increasingly, St. Croix tourists came from the burgeoning Twin Cities, and guidebooks catered to this new class of travelers. They continued to stress many of the same themes of the earlier period. "Romantic beauty, historical incidents and legendary lore contributed towards making the Valley of the St. Croix River not only very interesting to the tourist," wrote guidebook author William Dunne, "but exceedingly valuable to students of either events or nature. Here within an hour's ride of the two leading cities of Minnesota, is a miniature Hudson, excelling, in some features, that famous river of the East. Along its shores fierce Indian battles have been fought, and its fertile, picturesque valley contains attractive cascades and waterfalls that rival the renowned 'Falls of Minnehaha.'" Throughout the guide, Dunne recounted the basic history of the St. Croix Valley of Indians, explorers, missionaries, fur traders, lumbermen, and settlers to stir the imagination and conjure ghosts of yesteryear. "We traveled past battle grounds and fishing nooks, past the old home of the deer and the moose, past where Poor Lo held full sway but a generation ago and we had enjoyed the day."[26]

The guidebook poetically described the landscape. "The sunlight stole through the embowering trees of the glen just enough to brighten into sparkling crystals the falling waters of Osceola Creek," wrote Dunne, calling the creek "a very beautiful gem of nature." Of the Dalles he wrote, "Shadows from jutting rocks and tall trees fall upon the water in strange contrast with the sun-brightened portions where the tree-topped rock walls of the Dalles are distinctly reflected in the seemingly quiet stream, yet, of quietness, 'tis but the semblance born far below the glassy surface. Between these two walls the river flows and eddies with depth and force." His description of Devil's Chair was especially evocative: "From a height of eighty feet his Satanic Majesty could view the whole extent of the beautiful landscape. Upon the footstool of his chair he could rest his weary feet or stand and address his kindred spirits of the Northwest during his councils with them." Dunne tempted the adventurous spirit of his readers with the suggestion that there were still the possibilities of new sights to discover. "In the attractive glens and curious ravines along the sides of the St. Croix there are yet to be found cascades and other scenic beauties that will, in the near future be noticed and highly appreciated."[27]

While trains brought visitors to the St. Croix Valley, they also made it possible for tourists to venture away from the river and to explore its inland lakes and towns. After 1868, trains ran to Center City, Lindstrom, Forest Lake, and Lake Elmo in Washington County, making them summer destinations for vacationers from the Twin Cities. The advantage of the lakes over the river was that they could be enjoyed throughout the summer, free from the sights and sounds of logs and crude lumbermen on the river. Although many Swedish immigrants farmed the area, the lakes themselves remained sparsely populated throughout the nineteenth century, even though the banks were largely prairie. Trains made frequent stops at Forest Lake to refuel, thus making it easy for visitors to attend to business in the city and return to the rejuvenating lake region with regularity. Railroad companies were the chief promoters and builders of the resort industry, and they encouraged people to buy a permanent house along the route. Initially, early lake visitors pitched tents for their stay. When a certain area proved its popularity, enterprising businessmen built resorts. This also made it possible for large parties of friends, church groups, social clubs, or businessmen to enjoy the great outdoors together.[28]

Lake Elmo emerged by the late 1870s as a premiere vacation spot. It was halfway between St. Paul and Stillwater and was promoted by the St. Paul & Sioux City Railroad. Many of St. Paul's fashionable class enjoyed boating, picnicking, and dancing under the stars. Dunne's guidebook described Lake Elmo: "A handsomely situated body of water, such a delightful place as we would expect to find in the undulating wooded district. . . . Its rustic seats and shaded walks, its neat pavilions, its boating and fishing make it a popular excursion resort for societies and schools." Elmo Lodge was equipped with "every modern accommodation and in the highest sense 'cares for' its guests." Its "up to the times" comforts attracted repeat guests.[29]

White Bear Lake, north of the Twin Cities and equidistant from Stillwater, attracted the cities' elites. Vacationers began coming to the lake as soon as a road was built in the early 1850s from St. Paul and could arrive by horse or carriage in just two hours. In 1857 an elegant Greek Revival hotel was built to accommodate the fashionable, and less pretentious lodges housed more modest clientele. By the Civil War, White Bear Lake was a popular resort welcoming holiday and weekend pleasure seekers and sportsmen. Once the railroad came after the war, the twenty-minute train ride turned White Bear Lake into a summer home retreat. It is "one of the brightest gems in the circle of lakes surrounding St. Paul and Minneapolis," wrote the St. Louis, Minneapolis & St. Paul Short Line promotional pamphlet. "White Bear is the oldest summer resort in the State, and consequently, is far advanced in many of the conveniences required by fashionable people who do not care to indulge in the wild and sometimes inconvenient modes of life found at our less developed watering places."[30]

Even Mark Twain wrote about White Bear Lake in his *Life on the Mississippi*. "There are a dozen minor summer resorts around about St. Paul and Minneapolis," Twain related, "but White Bear Lake *is* the *resort*." It possessed "the largest fleet of sail boats and yachts to be found in Minnesota," wrote Dunne. "On the evenings of the 'Regatta' and 'open air' concerts, White Bear Lake assumes the appearance of a gala night at Manhattan Beach, more than of what is generally expected at a suburban summer resort." By 1885 *Northwest Magazine* enthusiastically endorsed White Bear Lake as a resort area. "White Bear has pavilions, club houses and pleasure boats galore. But it has never become noisy and Coney-Islandised," the magazine noted. "It remains today a place for rest and pleasure rather than rioting and boisterous sports. It is fashionable without being fashion-ridden; popular and populous without being crowded."[31]

In 1884 the resort industry reached Lindstrom when Ida Van Horn Elstrom opened her Lake View House on the peninsula between the two Lindstrom lakes. Trains deposited guests at the nearby train station. When Elstrom married John W. Nelson, the newlyweds changed the name to the Lake View Hotel. When the hotel burned down in 1900, the couple opened the Villa Cape Horn resort on the lakefront west of their old establishment, complete with a dancing pavilion. The resort thrived well into the 1920s.[32]

After the Nelsons' success, other resorts began to appear on the local lakes. Besides the usual resort businesses, the Chisago Lakes area also attracted nonprofit camps. In 1906 the Minneapolis YMCA opened Camp Icaghowan on the eastern shore of Green Lake near Chisago City. The name of the camp derived from an Ojibwe expression that meant "growing in every way." It was an apt phrase for a camp dedicated to providing disadvantaged urban boys with a week or two of summer fun. The cost to families was a dollar a day. Charitable Minneapolis businessmen picked up the remaining cost. Camp Icaghowan won a special place in the hearts of the boys who summered there. The original camp lasted until World War II. After the war, the men who spent their youth there built a new Camp Icaghowan on the Wisconsin side of the St. Croix near Amery.[33]

By 1880 fishermen and campers had made the Chisago Lakes a well-established sportsmen's locale. "Camping out at the numerous lakes with which Stillwater is surrounded is a growing practice with our citizens," noted the *Stillwater Lumberman*. "The practice is a good one. It is not expensive, and as a means of promoting health none better will be found. . . . There are no fashionable calls to make or receive, no elaborate dressing for company. Everything is free and unrestrained. The male members of the family usually go out on the evening train or drive out . . . until the morning calls them back to business."[34]

The typical resort of this period in the North Woods catered to the patterns of social interaction of the elite. They operated on what was called the American Plan in a largely self-contained world. There was a main lodge, where all meals and social activities took place. Lodges were usually constructed of local materials, such as logs, with a large fieldstone fireplace as a centerpiece. Guests slept in simple cottages. Maintenance buildings, barns, an icehouse, a small farm, and a boathouse supplied all the basic necessities to run the resort. The single, family-owned cottage did not generally start to appear on the lakes until the turn of the century.[35]

Selling visitors on the St. Croix River itself was difficult because the logging industry dominated the river. The St. Louis, Minneapolis & St. Paul Short Line promotional pamphlet praised the beauty of the St. Croix but noted the fact that it was a working river and that visitors had to be aware of this. In 1869 a total of 270 steamboats plied the river between Prescott and Taylors Falls; by 1882 this number declined to only 77, in part because of logs congesting the river. "Ever present among the islands and along the low shores for several miles . . . are the evidences of the vast traffic in lumber that is carried on in this valley," the promoters warned. "The thousands of logs that lie 'hung up' on the shores, at which gangs of men are laboring, tugging and rolling, to get them afloat in the river: the miles of booms, the vast number of piles that are driven to prevent the logs from stranding . . . the dozens of steamers for town . . . and the numbers of men employed, all combine to form an array of business that is not seen in the ordinary routes of travel elsewhere in the west, and probably not in the world." Despite the obstacles in the river, however, St. Paul residents continued to patronize the St. Croix River and the Dalles during the second half of the nineteenth century because of its proximity. The St. Croix Valley was advertised to Twin City residents as simply a day trip. "In the brief period of a single day," wrote a railroad advertisement, "the appreciative 'sight seer' can here enjoy a variety of scenery, perhaps unequaled in America—if the world." The pamphlet promoted the Upper St. Croix more for sportsmen where outfitters and guides were ready to assist that type of traveler rather than cater to the fashionable.[36]

The St. Croix Valley also faced competition from the growth in recreation to the south. Between 1873 and 1893, southern Wisconsin also experienced its own tourist boom that attracted residents from Chicago and other southern climes. The Bethesda Mineral Spring Company promoted Waukesha as the "Saratoga of the West." The Chicago and Northwestern Railroad serviced the town, and soon Waukesha had thirty hotels and dozens of boarding houses that catered to summer visitors. Spas had become so popular during this period with the middle class that nearly every spring bubbling out of limestone substrata was promoted as a spa, such as in Madison, Beaver Dam, Sparta, Palmyra, Beloit, and Appleton.[37]

The St. Croix Valley did not sit idly by while watching as Minnesota and southern Wisconsin developed resorts. In 1873 Ebenezer Moore embarked on a plan to turn Osceola into another "Saratoga of the West" and invited the public to his St. Croix Mineral Springs two miles south of Osceola near Buttermilk Falls. Moore hoped both tourists and health seekers would flock to its "healing" waters. Before his vision was realized, however, Moore sold his interests in the springs to a partner from Eau Claire. In the spring of 1875 the new owners laid a foundation for a "mammoth hotel" aptly named the Riverside Hotel. "Messrs. Stephens, Williams & Fletcher, the proprietors of this property, are determined to make the springs a popular resort for both invalids and pleasure seekers," wrote the *Stillwater Lumberman*. "The location selected for the hotel is a delightful one, overlooking the river and affording a picturesque view of the surrounding country." The dining room seated two hundred guests, and the grounds were complete with a croquet course, a trout pound, a deer park, and a half-mile circular racetrack. A hydraulic pump brought spring water into all parts of the hotel. "It promises to become one of the most attractive summer resorts in the Northwest," boasted the *Lumberman*. This dream, however, never materialized. Although medical men endorsed the healthfulness of the waters, tourists never patronized the hotel. In 1873 a nationwide financial panic led to years of depression. Labor strife plagued railroads across the country. Higher rates and strikes tied up everyone's travel plans, and financial hardship reached every part of the country that relied on rail service. In 1885 the under-used hotel burned down.[38]

By the late 1880s railroads assumed most of the commercial transport into and out of the St. Croix Valley as residents and businesses came to appreciate their year-round efficiency, comfort, and dependability. Railroads eagerly assisted the tourist trade by coordinating their schedules with boat excursions. A friendly rivalry developed between the towns along the St. Croix over who attracted the most excursionists. The day-trip excursions offered from Minneapolis to Osceola made it the leading Soo Line city along the St. Croix. "Osceola largely leads the towns on the St. Croix," boasted the *Polk County Press* in October 1887. The Soo Line sold 335 excursion tickets out of Osceola that season. St. Croix Falls followed a distant second with 191 tickets sold. Marine was next with 130, and Dresser Junction sold a mere 56 tickets. With the growth of pleasure excursions, Osceola came into its own as a tourist town. It boasted that its waterfall was "unrivaled by any waterfall in the northwest." For its Fourth of July celebrations the town attracted one thousand people who enjoyed a parade, a baseball game between Osceola and St. Croix Falls clubs (Osceola won the pitcher's duel, 24 to 23), and a picnic on Eagle Point Bluff, "one of the finest groves in the valley." This event foreshadowed Osceola's popularity as a summer resort during the 1890s, when it was common for as many as 1,000 excursionists a day from the Twin Cities to visit the town's picnic groves. A notable excursion occurred in July 1891 when an African American Knights of Pythias lodge visited. Many people from Osceola joined them for dancing and baseball. Later they took a trip upriver together. "As many white people as colored attended the picnic," noted the *Polk County Press*. "All danced and rode together and a real nice time was enjoyed."[39]

Hunting, Fishing, and Tourism

One way to pass the time on the long, slow steamboat trips was to shoot geese from the boiler deck, which entertained both passengers and crew. Captain O. F. Knapp, of the steamer *Enterprise,* first introduced the practice in the mid-1860s. The sport caught on, and by the 1870s and 1880s groups chartered steamboats for this purpose. Southern tourists were especially enamored with this unique form of hunting. "Frequent notice has been made in these columns," wrote the *Stillwater Lumberman,* "of the rare sport furnished on the St. Croix by hunting geese with a steamboat. The time has now arrived for the full enjoyment of this sport and it is daily being indulged in." The paper provided a colorful description of how the sport was done: "As soon as the boat was headed down the lake a bulkhead was constructed around the forward guards of the lower deck so that the hunters could, if they choose, shoot from that place unobserved. Screens were constructed of blankets and placed around the railing in front of the boiler deck for the same purpose. All these precautions are rendered necessary as a boat cannot get within gun shot if any person's body or head is visible to them."[40]

In 1879 the *Burnett County Sentinel* noted, "Hunting and fishing parties are the order of the day in this vicinity." "A party of 5 passed through here from Marine en route for the upper Namekagon fishing and sporting," reported the *Sentinel* with interest and approval. Bounties were offered for certain animals that farmers considered pests. "The scalp of a lynx was worth . . . $3," reported the *Sentinel*. "There is a bounty of $6 on wolves, $3 on wild cats, and $2 on foxes." This made hunting a lucrative sport that aided the farmer and settler in dealing with these pesky animals. Indians, too, often redeemed these animals for their reward.[41]

By the 1870s popular sporting magazines, such as *American Sportsman* (1871), *Forest and Stream* (1873), *Field and Stream* (1874), and *American Angler* (1881), began publishing to encourage outdoor sports. This new breed of outdoorsmen approached wildlife in a practical, utilitarian manner. They were more interested in the sport and prize catches. Beginning in the 1870s many railroads organized hunting and fishing excursions. Some railroads owned their own resorts.[42]

In 1851, only three years after statehood, Wisconsin established defined hunting seasons in order to protect game, birds, and fish. Nonetheless, by the 1870s overhunting and the expansion of settlement made big game scarce along the Lower St. Croix. Conservationist ideas were slow to take root in this frontier region. When a rare moose was spotted near Rush City in the fall of 1877, the pursuit was on. The following excerpt from the *Stillwater Lumberman* provides an insight into the attitudes of residents toward the sport of hunting:

> A wild moose was foolish enough to call upon Frank La Suise, at that gentleman's residence . . . introducing himself to Frank's family by peering through the window of their residence. Frank not liking such familiarity, seized his gun and greeted the animal with a charge of buckshot, which caused the moose to take to the water, whence Frank followed in a canoe, blazing away at the "baste" as rapidly as he could load his gun. A broadside from Adam Dopp, who appeared on the scene, blinded the creature; so that Frank was enabled soon to dispatch it with a club. . . . It was the means of furnishing a very tender article of fresh meat for our citizen's dinner last Sunday.[43]

By the 1880s moose had even disappeared in the Upper St. Croix Valley. A killing of one was worthy of note. "A moose was killed near Clam Lake

Figure 15. The Devil's Chair in the Dalles was just one of the picturesque sites that ensured the establishment of the Interstate Park. *Outing Magazine*, March 1890.

Figure 16. Canoeists pass Angle Rock on the St. Croix. *Outing Magazine*, March 1890.

last week," the *Burnett County Sentinel* remarked with interest. "A very rare animal in these parts." An old-time settler reminisced in 1880 that the early days were his "happy days. Game was everywhere." In one fall season he had killed 130 deer, 16 elk, and 3 bears.[44]

If moose and other big game were no longer plentiful in the North Woods, fish, waterfowl, and deer still were. Unlike the moose, deer did not disappear from the St. Croix Valley with the retreat of the forest. Various kinds of berries flourished in the brush left in the loggers' path, on which deer feasted. Hunters in turn feasted on the deer. There seemed to be an endless supply of fish. In June 1877 a fishing party from Hudson set out for the Clam River. They returned "having caught seven hundred and fifty trout," recorded the *Lumberman*. In 1891 one hunter caught a sturgeon in the Namekagon River near Phipps weighing eighty-one pounds. The *Burnett County Sentinel* claimed, "This is said to be the largest fish ever taken out of a stream in this locality."[45] However, even fish populations were threatened as loggers dumped sawdust into the river, riverbanks eroded from the constant bombardment of logs headed downstream, and upstream deforestation led to soil erosion that in turn silted up the river and destroyed the natural habitat.

Unlike moose or other big game, however, fish stocks were easier to re-
plenish. In 1866 the Wisconsin State Legislature appointed a fish inspector.
This eventually led to fish stocking in the state's waterways. In 1880 over a
million brook trout were put into the streams of Wisconsin. In 1883 the
U.S. Fish Commission deposited 250,000 white fish and lake trout eggs
into Lake St. Croix. The federal government's initial interest in restocking
rivers and streams was to preserve commercial fishing. Sport fishing was
an indirect beneficiary of this program that kept the St. Croix and its tributaries stocked with fish. In 1895 the *Polk County Press* bragged, "There is no
county in Wisconsin, outside of the Lake Superior counties, where better,
or a greater variety of fishing can be found than in Polk county. Within a
circuit of ten miles there are fifteen lakes, and the St. Croix river, all well
stocked with pickerel, bass, pike and other fish, besides three fine trout
streams, well supplied with speckled and rainbow trout."[46]

By the 1890s, however, there was a growing awareness among Americans
that these natural resources were not in infinite supply. Middle-class urbanites, cooped up in gritty cities, increasingly developed an appreciation
for the outdoors. This changed attitudes about nature from one of simple
appreciation of the picturesque to the growing recognition that natural
treasures needed to be managed. Sportsmen were the first group to join the
conservation movement. By the turn of the century, "roughing it" came to
be seen as a critical part of individual character building as well as an opportunity to engage in a distinctive American cultural activity.

The Wisconsin Central Railroad enticed these more hardy travelers into
the far North Woods. The railroad company built a hotel in Ashland called
the Chequamegon that housed several trainloads of sportsmen and vacationers. In 1885 the Chicago, Milwaukee & St. Paul Railway produced a
pamphlet, *Gems of the Northwest*. In the brochure, outdoorsmen were de-
picted with the latest equipment in tents, fishing poles, and the like while
"roughing it" in the great outdoors. The formal attire in which sportsmen
were photographed, including ties, jackets, and vests, demonstrated the
elite nature of hunting during this time.[47]

Oddly enough, many men who made their fortunes exploiting nature
were among the first to build sanctuaries in wilderness areas. Factory owners and railroad men, as well as the doctors and lawyers who served them,
found the St. Croix one of the least spoiled havens in the Upper Midwest.
Like wealthy eastern robber barons who built rustic lodges in the Adirondacks, these midwestern business elites built lodges and estates along the
Upper St. Croix and Namekagon rivers. One example is that of the Velie
Estate. The John Deere Company originally owned the two-thousand-acre

site dating back to 1893. It was located in Douglas County, seven miles southwest of the Gordon Dam. In 1905 Velie built a twenty-four-room lodge to serve as a fishing club for company officials. It included a playhouse, fish hatchery, and stable. In the 1920s President Calvin Coolidge put the Brule River on the sportsman's map with his widely reported fishing trips to the region.[48]

Elite sportsmen lent support for early advocates for the preservation of wild places. Notable national conservationists, such as Wisconsin natives John Muir, Frederick Law Olmstead, Robert Underwood Johnson, and others, called for the protection of significant, monumental landscapes. In "A Voice for Wilderness," Muir wrote, "Thousands of tired, nerve-shaken, over-civilized people are beginning to find out that going to the mountains is going home; that wildness is a necessity; and that mountain parks and reservations are useful not only as fountains of timber and irrigating rivers, but as fountains of life." His efforts, and those of others, led to the eventual formation of the National Park Service in 1916.[49]

Interstate Park and the Last Stand of the Steamboat Men

The people of Minnesota and Wisconsin were far ahead of the nation in their embrace of conservation. Devastating forest fires and the overhunting of game produced a public clamor for statewide conservation programs. Minnesota led the way in 1891 with the establishment of its first state park at Itasca and the Pillsbury State Forest in 1899. As for Wisconsin, the receding forests in the St. Croix Valley alarmed many residents. In the 1890s J. Stannard Baker of the Baker Land and Title Company became interested in reforesting some of his own cutover lands. He planted trees on land he owned on Deer Lake six miles east of his home in St. Croix Falls, as well as in the village. In a period of eight to ten years, Baker planted thirty thousand trees. When asked why he bothered to do this when he would probably not live to see them mature, Baker said, "Some people in this world want big white monuments. I will take a green one." By 1950 many of the seedlings had grown thirty to forty feet with fourteen- and sixteen-inch trunks.[50]

The conservation movement arrived in the St. Croix Valley in full force, however, when George Hazzard organized a movement to create a state park in the Dalles area. In 1857 Hazzard had arrived in the St. Croix Valley by steamship just as evening set in. "Those who made the trip remember its impressions," Hazzard wrote years later. "To others it cannot be described." His love of the river grew as he served as a general agent for railroads and steamboat lines out of St. Paul. The Dalles of the St. Croix was always high on his list of recommended visits, and the grateful appreciation of tourists whom he had steered there convinced him that "in the Dalles there was great value to the States of Minnesota and Wisconsin." When the Dalles rocks were sought after for macadam roads, Hazzard conceived the idea of creating a park to preserve it for future generations.[51] His vision interested Oscar Roos of Taylors Falls, who deeded a considerable amount of acreage to the state of Minnesota, and state senator William S. Dedon of Taylors Falls. Other leading citizens from Taylors Falls, St. Croix Falls, St. Paul, and Madison, such as Harry D. Baker, son of J. Stannard Baker, worked together to get Minnesota and Wisconsin to pass the enabling legislation.

On February 25, 1895, the Minnesota Interstate Park was born out of the donated land. It took more lobbying in both state capitals to get appropriations needed for more land purchases. In March, park organizers brought a delegation of Minnesota legislators and more than a hundred "distinguished" guests to the Dalles to see for themselves the value of this scenic wonder. The *Taylors Falls Journal* followed up on this visit by printing a special pictorial issue extolling the beauty that should be saved. It was sent to both the Minnesota and Wisconsin state legislatures. On April 22, 1895, Minnesota passed the park bill by an overwhelming majority, although without appropriating all of the promised money, and Hazzard became its first commissioner.[52]

While the Minnesota side had received enthusiastic support in the nearby state capital city of St. Paul, the Wisconsin Legislature in Madison was much more reluctant to allocate funds for such a remote park. "Appropriations for the purchase of the necessary lands within the park limits were very difficult to obtain from Wisconsin legislators," recalled local resident and lobbyist Harry D. Baker, "because so few members of the legislature at that time knew anything about this part of the state." Unlike their neighboring state, Wisconsin had no state parks at that time. Even when members of the park committee brought photographs of the Dalles to Madison, most legislators still remained disinterested.[53]

Hazzard, however, was not a man to be put off easily. Even though a Minnesota resident, he lobbied Madison politicians. After petitioning and finally haranguing state senator John M. True of Baraboo for hours, Hazzard received an appropriation to purchase Dalles land. For several years Hazzard and others continued this effort, "getting only perhaps five or ten

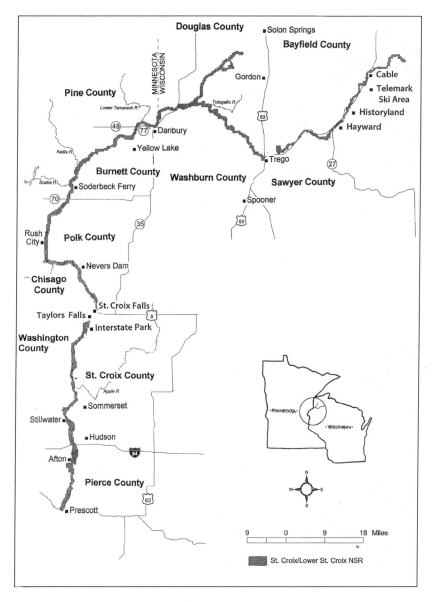

Map 8. Recreation era, St. Croix National Scenic Riverway (Source: National Park Service).

thousand dollars at each session of the legislature," recalled Baker, "at some sessions nothing at all." Some of the lands also had to be condemned by court action. By March 1899 the park promoters finally secured the support they needed in Madison for the Wisconsin side of the Interstate Park. It is "a monument to the energy and the enthusiasm and foresight not only of George H. Hazzard," declared Baker, "but to those men who had vision enough to see the possibilities of this picturesque and scenic area as the park that it has now become."[54]

Harry D. Baker also deserves much of the credit for the park's creation. Between 1901 and 1911, Baker wrote numerous letters to land owners in the area to acquire more land for the park. New land acquisitions, however, were neither easy nor popular. In a letter dated April 10, 1902, Baker browbeat a local landowner:

> The price you ask is at the rate of $25 per acre, which is over double what we would regard the land as worth. For any agricultural purpose it is certainly not valuable, as the natural meadows, which are a comparatively small proportion of the entire acreage comprise the only part of the land that is fit for anything but pasture. I have consulted with the other members of the commission, and we have decided to offer you $12.50 per acre for the entire tract. I doubt very much if any appraisers appointed in condemnation proceedings would value this land as high.[55]

He also authored entries for the Interstate Park in the *Wisconsin Blue Book*. In 1901 he wrote:

> Nowhere else are evidences of this power to rend and produce more magnificently portrayed than in the Dalles of the St. Croix, in Polk county, that matchless beauty spot just becoming known as the Inter-State Park, a veritable paradise of Nature's handiwork, owned jointly by Wisconsin and Minnesota.[56]

Despite the state of Wisconsin's initial reluctance to build the park, the Badger State eventually committed 1,734 acres compared to 292 from Minnesota. Much of this land in Wisconsin was available, oddly enough, because of the 1850s lawsuit between William Hungerford and Caleb Cushing that kept the falls of the St. Croix Falls from being developed as a manufacturing center."[57]

In 1898 the Interstate Park got a hearty endorsement from Warren Manning, a nationally renowned landscape architect and secretary of the American Park and Outdoor Art Association, when he visited the park.

We place upon the records of this American Park and Outdoor Art Association an expression of our appreciation of the work that has already been accomplished toward securing the Dalles of the St. Croix as forest reserve for the benefit of the citizens of Minnesota and Wisconsin, where native plants and animals that are fast being exterminated may be perpetuated and where they and the remarkably varied interesting geological conditions may be readily accessible to students.[58]

Manning's enthusiasm helped reinforce the St. Croix Valley's reputation as a mecca for artists, including Douglas Volk, director of the Minneapolis School of Art, and his father, Leonard W. Volk, founder and president of the Chicago Academy of Design.[59]

George Hazzard responded to public enthusiasm to visit Interstate Park by organizing the Twin Falls Association to develop tourism by coordinating railroad and steamboat excursions to the park. In November 1896 Captain John Kent asked the people of Osceola and Taylors Falls to help build a special excursion boat and enlist local support of towns upriver from Stillwater to keep the St. Croix navigable. It did not take much to convince the townspeople that the growing demand for excursions to the new park would mean an economic boon for them. The new Interstate Navigation Company raised $6,000 to build the *Gracie Kent*—"a neat little craft built for the State Park Business."[60]

The summer of 1897 began with great enthusiasm as the river towns saw hundreds of people pour into the valley to take the boat trip through the park. Hotels and restaurants were "crowded to their utmost to the satisfaction of all." The railroad grossed $5,418. By July, Kent installed bigger engines on the boat to make quicker trips. By August the *Gracie Kent* averaged excursion parties of five hundred per trip. However, by September low water levels caused major problems. In that month the *Gracie Kent* was supposed to have met five hundred passengers disembarking from the train at Osceola. The boat, however, ran aground on a sandbar at Cedar Bend and never reached the town. In another incident stranded passengers had to be ferried in skiffs to the train depots for their return trip. On another occasion passengers were forced to spend the night aboard the boat when it got hung up on a sandbar. They were less patient and understanding than their 1850s counterparts when their boat was snagged. The reputation of the Dalles cruises suffered as a result.[61]

The changing water levels were the result of the deliberate actions of the lumber barons from Stillwater. A clash between lumber's historic industrial uses of the river with the new tourist industry was inevitable. It was a battle between the past and the future. When the Interstate Park opened, the tourist industry was a promising venture for upriver towns whose lumbering and milling operations had recently collapsed. Only fleetingly did logging and tourism work in harmony. During the major log jam in 1886, thousands of curious folks came to Taylors Falls to gape. More often, however, fluctuating water levels made it difficult to coordinate train schedules with boats, so the railroads simply ended their excursions to the St. Croix, leaving upriver town residents irate. Captain Kent and the Interstate Navigation Company, however, still remained optimistic that the industry could make a go of it. He believed that the growing demand for recreation and excursions on the river would provide the leverage to file a lawsuit against the "Dam(n) Boom Company." Nothing, however, came of the lawsuit, and the Interstate Navigation Company struggled to make plans for the future. In February 1898 the Navigation Company sold the *Gracie Kent* to a New Orleans interest in order to purchase the bigger *Vernie Mac*. When the excursion season began in May, the new boat was not filled to capacity—a bad omen for the new business. Low water levels caused by Nevers Dam gave the *Vernie Mac* the same problems the *Gracie Kent* had—sandbars and delays. For the next three months no excursions made their way to the park by boat.[62]

When the budding tourism industry was ready for war, the Navigation Company contacted the War Department regarding the problems with open navigation on the St. Croix River. It found a sympathetic ear with Major Frederic V. Abbot, whose goal was to maintain open navigation for all parties on the river. Up to this time the U.S. Army Corps of Engineers had worked to facilitate logging. It removed sandbars and channeled the river to move logs and lumber south. Pressed by the tourist interests, Major Abbot asked the chief of engineer's permission to prosecute the loggers under the new River and Harbors Act of March 3, 1899. For a time it looked like the tourist industry would prevail. The corps initially supported maintaining a three-foot channel as far as St. Croix Falls, but heavy lobbying in Washington by the lumbermen frustrated the plan.[63]

In the meantime, the Navigation Company began to collect testimony for their own lawsuit against the St. Croix Boom Company regarding their "capricious" control over the water levels of the river. The Boom Company retaliated by purchasing an excursion steamer of their own for the sole purpose of giving the excursion business a bad name by finding every sandbar on the river with no thought or care for the comfort or convenience of those aboard. Boom Company owner William Sauntry then told the *Polk County Press* that his experiences demonstrated that the St. Croix River

Figure 17. The steamboat *Vernie Mac* at Taylors Falls, ca. 1900. Log drives made tourist excursions such as this one through the scenic Dalles problematic. Wisconsin Historical Society, WHi-6210.

above Stillwater was simply not navigable. This effort and Sauntry's ability to influence the secretary of war prompted the *Polk County Press* to wonder how much longer the people of the towns north of Stillwater would have to put up with the "insults inflicted by the company of which Sauntry is *boss lumberman.*"[64]

St. Croix residents heeded the call. In January 1900 Major Abbot held hearings in St. Paul and found himself confronted with sixty representatives from the communities affected. Frank B. Dorothy of St. Croix Falls presented a petition from one hundred residents asking for "free and unobstructed passage" of the river. The Northwest Ordinance, William H. C. Folsom of Taylors Falls argued, had promised all Americans free navigation on inland waterways, and he estimated that there were over 78,000 abandoned logs embedded in the riverbed that prevented the fulfillment of this promise. One of the most eloquent speakers was William Blanding of St. Croix Falls, who complained that the logging interests of Stillwater had driven small businesses from river towns. He pointed out that sawmills had to be shut down since they could not get any logs, power resources had been hindered, and commercial and pleasure boating were ruined. Lumbering, he argued, was a transient business whose main object was to exhaust the valley's timber supplies and move on. "When like wasting pestilence they have passed over the land and the coming fire has destroyed all traces of the footsteps and the overtaxed waters of the rivers and its tributaries once more flow free to all," Blanding opined, "then perhaps this dam company-ridden country may be allowed to make use of what natural resources these greedy tyrants have left in it."[65]

The railroads also expressed their vested interest in the condition of the St. Croix River. The excursion business was very lucrative. In 1897 the St. Paul & Duluth Railroad grossed $5,418 running trains between Taylors Falls and Stillwater, picking up and dropping off visitors to the Interstate Park. Low water levels, caused by lumbermen manipulating the river for log drives, the following season reduced their income to $1,340. If this were to continue, they would have to discontinue service.[66]

While the St. Croix Boom Company admitted they had inhibited navigation, they argued that they needed to control the river or they would be out of business. Their trump cards were the facts that they employed a large labor force; had a huge investment in land, transportation, machinery, and buildings; and supplied the nation with much-needed lumber. Downriver Mississippi towns would languish if logging were restricted. And they had every legal right under state charters to operate the boom and dam. The company also argued disingenuously that interfering with the dam would ruin smaller mills on the Upper St. Croix if they could not free-float logs on the river. What the company did not want to admit was that their control over the Nevers Dam had already ruined many a small mill owner upriver.[67]

Abbot tried to work out a compromise. He insisted that all parties share the river, with certain days allotted solely to the excursion business, such as Decoration Day, Independence Day, and the entire month of August when the winter log drives were generally over. The Army Corps also made an effort to clear the river approaches to Interstate Park of sandbars, but after several ineffectual seasons the effort was abandoned. Recreational use of the St. Croix River was effectively stymied until the logging industry's death rattle.[68]

That came in June 1914, when the last log was sent down the boom at Stillwater. Within six years that town's population dropped from twelve thousand to eight thousand residents. The implications for the future of other towns on the river were clear, and they desperately reached out to tourism. Marine Mills joined the Improvement Club of St. Croix Falls and the Commercial Club of Taylors Falls "to restore the most beautiful, scenic river in the world to its old time steamboat navigation."[69]

Many steamboat excursions continued to ply the waters of the St. Croix until the 1920s, but coordinated excursions between railroads, towns, and boats never fully recovered from the conflict with Nevers Dam and the Boom Company. Other forms of transportation, such as inter-urban street-cars, connected the St. Croix Valley with the Twin Cities and competed with railroads. In 1899 the Twin City Rapid Transit Company began to run cars between St. Paul and Stillwater. The streetcar made it much easier for those in the valley to go to the city for business and shopping. It also pro-vided even easier and cheaper access for Twin City residents to enjoy a day in the St. Croix Valley. A ride from Stillwater to St. Paul cost thirty cents and took an hour and ten minutes. By 1913 cars ran every thirty minutes. In the thirty-three years the inter-urban streetcars serviced the valley, thou-sands of tourists and day-trippers enjoyed the splendors of the St. Croix River. Many valley residents took the Sunday car to Wildwood Park on White Bear Lake, where they enjoyed the roller coasters, merry-go-rounds, and other amusements. However, in 1930 service was cut back to every two hours. And in the darkest days of the Great Depression the line was cut completely due to lack of money and competition with the automobile.[70]

Summer homes gradually became the more popular way to enjoy the St. Croix River. As early as July 1908, the *Stillwater Messenger* noted the grow-ing desirability of the St. Croix River for summer residency:

> Afton is becoming noted for its beautiful location as a summer resort and its next-to-nature charms is making it envied by all who see it and are influ-enced by "the call of the wild." . . . In fact, we often wonder why any place on Lake St. Croix is not the ideal place for summer homes.[71]

In 1910 the *Stillwater Gazette* predicted, "It will be but a few years before the banks of the St. Croix are dotted with summer homes." By the 1910s and 1920s most of the cottages along the Willow River belonged to resi-dents of Hudson. The land originally belonged to a local farmer who leased the land to local fishermen for $10 a year. Camping was free, and many families slept in tents. Residents of the Twin Cities were among the first to build summer homes north of Stillwater. These "dedicated fishermen and small-boat owners," wrote James Taylor Dunn, "all wanted to identify themselves with the river and become a part of its life."[72]

The onset of summer homes on the St. Croix was the most tangible sign of changing life on the river. A summer hiatus from the din and noise of

Figure 18. Three women near the St. Croix River at Hudson, Wisconsin, ca. 1870. Photograph by James Abajian. Wisconsin Historical Society, WHi-34514.

the cities became the desired goal for many Americans. In his noted book *Nature and the American,* Hans Huth documented the growing practice of ownership of country estates and summer homes for average Americans. "Most of the winter-weary townspeople, by going to a resort or to their own country homes, or even by visiting city parks and participating in some kind of summer sport," Huth writes, "could find respite from the city during the sultry months. For these summer pleasures the northern part of the country as far west as the Great Lakes was the favorite section." The St. Croix Valley became a prime destination for those seeking country delights.[73]

Dam the St. Croix!

Although a holdover from the logging era, Nevers Dam inadvertently contributed to the recreational attraction of the St. Croix. Built in 1890 by the Milwaukee Bridge & Iron Works, its fifteen gates created a fifteen-mile flowage. Fish and ducks found a home in its placid waters, as did boats with fishermen and hunters who were ready to make sport of them. The dam also served as a wagon bridge across the St. Croix. Besides the Sunrise Ferry, it was the only means to cross the upper river. Once tourists in their automobiles descended into the St. Croix River Valley, the dam bridge facilitated this new business. However, during the remainder of the twentieth century, dams on the St. Croix would be the most controversial issue dividing those who used and loved the river.[74]

As logging eclipsed, the growing demand for electricity sent power-generating companies into the St. Croix Valley in search of hydroelectric sites. In 1903 the Minneapolis General Electric Company began construction on a water-powered generating station at St. Croix Falls. The plant was completed by 1906 and was the first electrical power generated from outside the Twin Cities for the two urban areas. The success of this project encouraged the Minneapolis Electric Company and its successor, the Northern States Power Company (NSP), to purchase more sites on the swift-flowing river with the hope of lighting up the entire Twin Cities. By the 1920s the power company became the largest owner of frontage on the St. Croix and Namekagon rivers. To people like George Hazzard and those who lived along the St. Croix, the new power dam was one more example of the use of the river for the benefit of outside interests. The NSP was not as tyrannical as the lumbermen who controlled Nevers Dam, and water levels remained constant, but the river served their needs first and the needs of residents and tourists second.[75]

In 1911 George Hazzard helped organize the St. Croix River Association and served as its first president. Its membership reflected a variety of local interests; its goals were somewhat mixed and contradictory. Fishermen were angered by the power company's control over water levels that often left thousands of fish stranded in shallow pools. While sportsmen got the association to stock smallmouth bass in the river, they were not interested in Hazzard's goal of improving navigation in order to revive commercial and tourist steamboating on the river.[76]

Hazzard himself held contradictory ideas about the future of the river. He was no wilderness preservationist; rather, he was a classic Progressive Era proponent of the "wise use" of natural resources. So while he opposed the dams for the Minneapolis power companies, he favored even more intrusive obstructions along the river in the name of regional development. Initially he proposed to rebuild Nevers Dam with concrete in order to create deep, slack pools fit for excursion boats, rather than the swift-flowing river favored by sportsmen. Another of his pet projects was to revive a nineteenth-century proposal to dig a canal between Lake Superior and the Mississippi River via the Bois Brule River valley and the St. Croix. Hazzard had even gone so far as to try to recruit support for the canal from the Upper Mississippi Improvement Association. However, after a review, the Army Corps of Engineers found the canal idea "inadvisable, infeasible, impractical." Hazzard's ideas, however, had captivated the attention of Minnesota and Wisconsin governors, senators, and representatives who then created their own independent Superior and Mississippi Canal Commissions. The Army Corps, however, remained firm in its position, and Hazzard's dream died. This defeat and old age caused Hazzard to withdraw from public life, and the St. Croix River Association's first efforts to be a force in the valley came to a close. However, sportsmen and conservationists would continue to revive the association to deal with threats to the river.[77]

Early in the summer of 1923 Stillwater postman Ira King took over Hazzard's work. He and other residents formed Stillwater Council No. 347, United Commercial Travelers (UCT). King chaired its Committee on River Improvement. "We sincerely believe," King told the *Stillwater Gazette,* "that the St. Croix River is one of Stillwater's best bets and we are putting our best efforts forward to see what can be done to better present conditions." Their goal was to dig a deeper channel to accommodate pleasure boaters north of the town and commercial barge traffic south of it. The organization flooded the offices of senators and congressmen from Minnesota and Wisconsin with letters requesting $20,000 for channel improvements from Congress.[78]

The UCT's efforts bore fruit when the Army Corps of Engineers ordered surveys of the river. By 1925 the *Stillwater Gazette* optimistically reported that the "St. Croix Project" of the Corps of Engineers would be a great economic boon to the "entire Northwest." Citizens from Marine, Osceola, St. Croix Falls, and Taylors Falls also put in a request for the lifting of wartime regulations that restricted the flow of water over the NSP dam. Low water levels below the dam made navigation between Taylors Falls and Stillwater nearly impossible. The Northern States Power Company proposed a solution of building another dam near Prescott to improve water levels south of Stillwater. However, no dam was built on the St. Croix River. In 1928 the power company applied for a permit to build a dam on the St. Croix at Kettle Rapids. Sportsmen, however, complained that the dam would flood some of the finest smallmouth bass fishing areas on the river. The dam never materialized because engineers for NSP even questioned the feasibility of a large dam on the river. Innovations in coal-generated power plants prompted the NSP to let their permit elapse. However, the company owned 29,238 acres of land on the St. Croix River and could not help but be a factor in determining the future of the river.[79]

Frustrated by their lack of influence on the Army Corps of Engineers, the UCT decided to revive Hazzard's old organization and in the late 1920s renamed it the St. Croix River Improvement Association. Their goal changed from requesting a three-foot channel to requesting a six-foot one. Their chief target of animosity remained the Northern States Power Company. "Why a corporation is allowed to prostitute for its private gain a beautiful river like the St. Croix," reported John Dunn to the association in 1929, "I cannot conceive. I am positive from my long observation that if we had the natural flow of the river it would within two years make and keep a channel suitable for medium size boats. This opinion has been confirmed time and again by talks with men who have lived close to and on the river during and since steam boat times."[80]

While a three- or six-foot channel was not forthcoming, the St. Croix River Improvement Association did obtain funds from the Army Corps in the early 1930s for a snag boat and spring cleanups of the river to remove deadheads (partially submerged logs), overhanging trees on the river banks, and other debris that obstructed navigation. Another victory for the association came in the mid-1930s when commercial net fishing was ended and was replaced by the line and hook method preferred by sport fishermen. By 1935 the St. Croix River Improvement Association also got the Minnesota conservation commissioner to stock smallmouth bass in the river. However, the association was caught completely off guard when in 1931 the Minnesota Highway Department built a highway along the bluffs of the river south of Taylors Falls. In the process, the construction firm A. Guthrie & Co. blasted tons of rock and dirt into the St. Croix River, creating a large island that obstructed more than three-quarters of its channel. Despite protests and hearings with the Army Corps, the firm was never forced to clean up the debris.[81]

In the 1930s the Army Corps of Engineers became more responsive to the idea of economic revitalization of the towns on the Lower St. Croix. It revived a plan created during the Herbert Hoover years of dredging a nine-foot channel in the Mississippi. The plan required a series of locks and dams along the entire length of the river. The corps was permitted to use relief funds to hire workers and finance construction contracts. In 1936 a dam was built at Red Wing on the Mississippi that backed up water on the St. Croix. Although conservationists, such as the Izaak Walton League, complained that the creation of a slack-water pool would harm wildlife, the economic distress of the times caused such objections to fall on deaf ears. By 1940 the Army Corps of Engineers' feat of turning the Mississippi River into a giant canal made the federal agency appear to be critical for the economic well-being of many heartland river valleys and towns. Struggling farmers in Wisconsin and Minnesota appealed to the corps district headquarters in St. Paul to turn the St. Croix River into a "Little TVA of the North" with a dam at Kettle River. The Kettle River project united the interests of northern farmers, who wanted more electrical power for milking machines, with the Army Corps of Engineers, which wanted more dams to control flooding downriver. The project was not seriously considered, however, until after World War II.

Sportsmen on the Upper St. Croix

As the Lower St. Croix River adjusted to the decline in logging and the development of recreation, the Upper St. Croix and Namekagon rivers experienced their own growth in tourism centered more on the surrounding lakes and recovering forests. Tourism here focused primarily on family vacationers rather than elitist tours or excursions. Most of the first resort owners in the cutover had been loggers, guides, or farmers. Many farmers struggling in the cutover got into the resort business by simply allowing travelers to pitch a tent on their property or by taking in boarders who were primarily interested in hunting or fishing.[82]

Many early hunting and fishing hostels dating back as early as the 1880s and 1890s were hotels and boarding houses in towns and stopping places along the roads used by transient lumbermen and teamsters. Conditions were primitive. "Private rooms were available in the urban places," wrote local historian and resident Eldon Marple, "but bunks were the rule in the country, rudimentary lodging at best. . . . Few were the females who braved the rigors of resorting then." In 1885 two Hayward residents advertised in the *North Wisconsin News,* proclaiming, "Mr. J. N. Russell and Mr. Christie have for the past month been building a summer resort hotel out on Spider Lake." In 1894 Bill Cornick bought the hotel and called it a "Fisherman's Camp." By 1896 he built the first cottage on Spider Lake and one on Lost Land Lake. The Sawyer County plat book for 1897 ran advertisements for the Round Lake Park Place Summer Resort and Round Lake House Summer Resort; both were located seven to eight miles east of Hayward. An advertisement for the Cable House, in the town of Cable, proclaimed it was a "Sportsmen's Paradise." A Hayward livery stable operator eagerly catered to the tourist trade by claiming, "I keep first class conveyances for transporting people to various Lake Resorts, such as Round Lake, Spider Lake, Lost Lake, Sand Lake, and Lac-Court Oreilles Lakes."[83]

Early resorts were transitory. They might be successful for a few years, and then a fire or the death of an owner brought an end to the business. In 1888 F. D. Stone, the county sheriff, opened his Jericho resort on Grindstone Lake, which he nicknamed the "seaport on Grindstone." Guests were entertained with a steamboat ride through the Grindstone–Court Oreilles–Whitefish chain of lakes. Although the Jericho apparently did good business, the resort burned down in 1891. Another example of an early resort is Boulder Lodge, built on an old log-driving camp located on Ghost Creek. Jim Goodwins of Hayward started it as a fishing camp and stopping place and later turned it into a resort. William Cornick built cottages on Lost Lake in upper Sawyer County in May 1896. These lodges burned down in 1903. Cornick then built a lodge on Teal Lake about twenty miles east of Hayward in Sawyer County. In 1921 the Ross family took it over and operated it as Teal Lake Lodge. This lodge initially catered to anglers and hunters and later evolved into a family destination. Due to the persistent dedication of the Ross family, the lodge became a cornerstone institution in turning the Upper St. Croix region into a tourist destination.[84]

The growth in recreation, however, depended upon transportation. In 1902 the tireless promoter of agriculture in Burnett County, Ed Peet, also extolled the county's potential for recreation. "Burnett is a county filled with lakes . . . nearly all of the lakes are filled with fine fish and into many of the lakes run swift little streams in which the speckled trout is found," he wrote. "The finest lake in the county, if not the finest in the state, is Yellow Lake. . . . Should a railroad ever touch [it] there is no reason why it would not become one of the finest summer resorts in all the west."[85]

Railroads did come to recognize the potential of tourism in northern Wisconsin and Minnesota. In 1916 the Chicago & North Western Railway distributed a brochure entitled *Lakes and Resorts of the Northwest.*

"Hundreds of delightful lakes and resorts situated in Wisconsin," it beckoned, "are in a region where one may escape from the heat and dust of the city, where the nights are cool and restful, the days full of sunshine, and where there is an evenness in climatic conditions and a purity of atmosphere that cannot be surpassed." It added that there were "scores of fishing and hunting resorts, hidden away in the virgin forests of northern Wisconsin . . . where the lover of Nature may make camp amid innumerable lakes and streams, surrounded by forests, where the soft balsam of the pines pervades the air, where speckled trout are abundant in the streams, and black bass and muskellunge in the lakes."[86]

A night's ride in a Pullman car was all it took to deposit vacationers in the North Woods in time for breakfast. Included in the recommended towns with resorts were Hayward and Chisago City, Minnesota. The lakes of the Hayward area were "reached by beautiful drives through the heart of the pines." Round Lake was "an entrancing body of cold, clear water, fed by numerous streams." For the fishermen there were good trout streams, and for the hunter were grouse, partridge, duck, and deer during hunting season. Chisago City was described as having excellent fishing, good duck hunting, and very attractive scenery. The brochure provided a list of hotels, lodges, and boarding houses along with their rates and distance from the station.[87]

In 1912 fledgling resort operators combined to found the Fish and Game Protective Association of North Wisconsin in an attempt to merge conservation and tourism. They distributed a booklet advertising both sport and the beauty of the region. A series of resort owners associations followed in their footsteps. Together they promoted road development, including a route they dubbed "Big Fish Auto Route." By 1917 northern Wisconsin boasted four hundred resorts. In 1923 an association of more than two thousand business and resort operators established tourist bureaus in regional urban centers like Milwaukee and Chicago. Following the distribution of a blizzard of brochures and maps, the Wisconsin Land O'Lakes Association claimed that 700,000 visitors had been induced to head north for their vacation.[88]

More than railroads, it was highways that opened the Upper St. Croix to tourism. In 1912 Congress established the "10% fund" that diverted 10 percent of forest revenues to road construction. Four years later, the Federal Aid Road Act authorized $10 million for this purpose. A federal highway act was also passed in 1916 that required "states to establish highway departments in order that they might obtain, on a matching basis, federal subsidy for highway construction." As early as 1902, the American Automobile Association (AAA) was promoting recreational auto-tourism.[89]

The automobile democratized recreation and tourism in America. Families with more modest incomes from Milwaukee and Chicago were able to access the cooler breezes and refreshing lakes of northern Wisconsin. Civic boosters in both Minnesota and Wisconsin were quick to recognize the growing popularity of the automobile and the rising demand for family vacation destinations, and they cultivated auto-tourism for its economic benefits. Both states, as well as enterprising individuals, billed their local communities as the best destinations. The interwar years saw paved state roads reach into the Lower and Upper St. Croix Valley. Road building was actually quite easy in the sandy soils of the cutover, and by the early 1930s the Upper St. Croix and Namekagon River area was "welded together by a road system which is very elaborate for such a little used area." The roads, however, were not well built or well maintained, but this only added to the rustic charm of this wilder region.[90]

During these years more summer homes and humble sportsmen's lodges began to dot the lakes and rivers in the St. Croix Valley, including Lake St. Croix, which had been shunned by tourists during the logging era. Depressed land values in the cutover due to failed or struggling farms also added to the area's appeal for vacationers. "Prices are considerably lower than would be the case in a better agricultural section where agriculture would compete with resorts for the land," wrote University of Wisconsin geographer Raymond Murphy, and "summer visitors find relatively wild unsettled areas more attractive." Murphy, however, could not help but notice the contrast between the poor and abandoned farms of the cutover and the new summer homes being built. "Wooded shores of the larger lakes," he wrote, "are the sites of expensive summer homes and resorts which seem strangely out of place in this unfruitful country."[91]

The Brule River became a prime summer haven with beautiful seasonal homes. Its denizens enjoyed the trout in its waters and the game in its forests. In the cutover regions of Burnett, Washburn, and Sawyer counties whose poorer soil defeated many a farmer, vacationers discovered sandy-bottomed lakes with clear, sparkling water, unlike the "pea soup" waters of neighboring counties to the south with their richer, heavier soils. An example of a sportsman's lodge that developed in the 1920s was Kilkare Lodge located between Birch Island Lake and Fish Lake in Burnett County. Its board of governors, composed of businessmen from Chicago, "hand-picked" its members through mutual acquaintances, ensuring there were "carefully selected executives and professional men." Its pamphlet boasted, "When you join here, your associates are 'your kind of folks,' from every standpoint!" It offered a "fully appointed Club House, our own farm, complete commissary, a chancy golf course, three lakes . . . [for] swimming, boating, trail-riding, shooting and trout-fishing." By 1929 its dining room hosted two hundred patrons and was open year round.[92]

Through the years, Burnett County attracted many vacationers, and summer resorts sprouted along their shores. "Many lakes and river shore lands that we would not have taken as a gift thirty years ago," wrote one old-timer in 1976, "are beauty spots now and provide homes and retreats for many." One enterprising promoter was Iver Johnson. He began his career as a humble postmaster in Webb Lake, Wisconsin, with a side business of running a little Indian trading store. Any travelers who ventured into the store at mealtimes were always asked to dinner. One guest who visited in the mid-1920s changed Johnson's life. He was Gus Munch, a Chicago baseball pitcher and sports writer. Before Munch left, he asked the Johnsons to build a new cottage and promised to write an article in *Outdoor Life and Recreation.* Munch kept his promise and described Webb Lake as a "veritable paradise for the bass fisherman looking for virgin waters and as yet but little fished and less known. . . . 'Tis indeed a delightful country for the sportsman who wants to camp out in the wilderness and not be bothered by tourists. They don't get there. The roads are too bad, but the fishing! Well, go on up there, and try it yourself, you'll see!" Munch said Johnson could take care of two people interested in fishing in the area. Within a short time Johnson was receiving a dozen letters a day from sportsmen from all over the region wanting to be one of the two he could accommodate. Johnson immediately began to build a cabin, rented a vacated house on Fairy Lake, pitched a tent, and added on to his store. He then proceeded to book twenty to thirty fishermen at a time, charging $3 a day for room and board. He met the sportsmen at the train in Spooner and brought them over the rough country "to the land of beauty and good fishing." With his new business booming, Johnson built more cottages, a tavern, and a dancehall that booked well-known traveling bands. By 1933 Johnson bragged that he could "accommodate 50 . . . instead of being able to care for two." In a brochure he put together, Johnson assured his guests, "Every effort was utilized to make our cottages the best in this locality. They are all new and well screened." The resort offered both light housekeeping cottages and sleeping cabins. The light housekeeping ones were "completely furnished with good clean beds, bedding including linens, dressers, stoves, dishes, tables, chairs, rockers, rugs, and in fact everything that is needed except towels and tea towels." Guests who preferred not to cook could dine in the main building. Johnson also offered river trips, tube floats, and horseback riding. As tourism boomed, other residents catered to the overflow the Johnsons could not handle.[93]

As Johnson discovered, once tourist dollars started flowing, they were capable of watering many roots in a community. By 1931 grocery stores in Danbury, Gordon, and Solon Springs were doing a booming business catering to summer visitors. "Resort people from Eau Claire Lakes to the east and Bardon Lake to the southwest flock into Gordon daily during the summer season to get the mail and to shop," wrote an observer, "and this trade is the main support of the grocery stores, and is of importance to all the other business houses of the towns as well." Struggling farmers in the cutover found the summer season a prime opportunity to sell their milk and cream to resorts. And like the bygone days of logging, the recreation industry depended upon seasonal migration of labor. The local population could not supply the demand for caretakers, housekeepers, waiters, cooks, laundresses, guides, handymen, and gardeners. They had to be recruited from cities and places further south. Many university students from Madison found this seasonal employment a wonderful way to supplement their incomes.[94]

The increased recreation and automobile traffic was also a boon for the ferry business. During the 1920s and 1930s the Soderbeck family ferried automobiles across the St. Croix River just north of Grantsburg. The season opened with the spring ice breakup and ended with the winter freeze. Rather than drive down to St. Croix Falls or up to Danbury to use the bridges there, travelers chose the leisurely trip across the St. Croix that cost them fifty cents. The ferry passage was critical to the recreation business and the new breed of vacationers in their Model T's.[95]

The success of tourist recruitment to the North Woods, however, had its down side. The area was not prepared for the large numbers of tourists who began roaming the countryside in automobiles. Travelers who had difficulty finding or affording accommodations began a practice called "gypsying." Without a railroad to funnel them into specific locales at specific times, car travelers simply took to the roads, stopping and starting as their desires struck—and they camped wherever they pleased. While the gypsy tourists reveled in the new freedom from civilization, farmers found them a nuisance. Picnickers and campers often left their garbage behind and relieved themselves anywhere. Bolder ones helped themselves to local produce. Since the region needed the tourist dollars, local authorities were reluctant to harass the "gypsies." Communities had to find other ways to

Figure 19. Automobiles gave tourists of modest means an opportunity to explore the St. Croix Valley. Here several cars await transportation across the river via the Soderbeck ferry, ca. 1930. St. Croix National Scenic Riverway files.

house and police these wanderers. Public parks and campgrounds proved to be one answer, and local resort owners began catering to this more modest income group. After 1925, tourist and cabin camps gained popularity with those tired of car camping. Individual cabins offered outdoor, quasi-communal settings with family privacy in comfortable accommodations, sometimes including a central dining/recreation hall, central bathhouse, and gas station. Another complaint of locals was that these new tourists demanded accommodations that matched those in the cities or expected local residents to fit a stereotyped image of North Woods pioneers. But most residents swallowed their irritations and learned to please paying guests. By the 1940s cabin owners provided better furnishings, homey decorations, plumbing, bedding, and kitchen equipment. Travelers no longer had to carry makeshift households in their car, so travel became much easier, and both tourists and host communities were much happier.[96]

This democratization of lodging also led to a change in services. The American Plan, which had been the model for resorts throughout the North Woods in earlier years, gave way to what was called "housekeeping"

Figure 20. Walter and Virginia Ross were among the most successful pioneers of the "tourist frontier" in northern Wisconsin. They helped move the industry away from a reliance on male sportsmen toward a family resort experience. This photograph shows the main lodge at Teal Lake Lodge, ca. 1949. Wisconsin Historical Society, WHi-37958.

resorts. These were more individualistic and less expensive than those in the earlier American Plan. Housekeeping resorts were composed of a collection of small cottages where guests were expected to bring their own food and linens and fend for themselves. There was no main lodge as a focal point for social gatherings. Stays were much shorter as well. This was due to better roads and automobiles that allowed people to check out a variety of lodgings and restaurants in a single vacation.[97]

By the early 1930s Minnesota recognized how much money tourism generated and began a tourist bureau. In 1932 the Minnesota State Board of Health also began to distribute a booklet called *State Laws and Regulations Relating to Hotels, Restaurants, and Places of Refreshment, Lodging Houses and Boarding Houses.* This included resorts, summer camps, summer cottages, cabin camps, and tourist camps. By the end of the decade, Minnesota estimated it took in over a hundred million tourist dollars.[98]

The Great Depression of the 1930s slowed but did not stop this boom in tourism. Hard times found the North Woods with more cottages and resorts than people to fill them. "So rapidly have summer homes and hotels sprung up along the shores of better lakes," wrote an observer, "that today the number of commercial resorts is too great in proportion to the number of visitors for the business to be profitable." The Ross family, who operated Teal Lake Lodge, however, responded to the challenge by publishing *Teal Lake Tidings,* a resort newsletter intended to maintain connections and a sense of community among their varied guests, most of whom hailed from Chicago. The resort added modern cabins, bathrooms, and a tennis court, and owners encouraged off-season visits. During World War II, Ross, ever the booster, argued that it was a patriotic duty to take a vacation in order to restore one's energy for war work! Ross's efforts paid off with a regular clientele returning year after year. For less enterprising resort owners, however, these were hard times.[99]

Government Conservation and the Invention of the North Woods

In spite of problems in the private sector, the New Deal programs of Franklin Roosevelt ensured that the Depression years were critical in creating the image of the North Woods as a "land of sky blue waters." These years also saw the beginning of meaningful federal involvement in the fate of the river. With so many Americans unemployed and the economy severely shaken, Roosevelt sought ways to employ people using the resources of the federal government. Among his most noteworthy programs were the Civilian Conservation Corp (CCC) and the Works Progress Administration (WPA). Federal or state agencies that developed a conservation project could request a CCC company. Camps were formed in national and state forests or in soil conservation districts. The four-hundred-thousand-men-strong CCC, under the auspices of the U.S. Army, became renowned for its accomplishments in forest, soil, water, and wildlife conservation. These programs helped establish the trend for greater government involvement in preserving and protecting the St. Croix River, its tributaries, and its forests and in facilitating the development of recreation.

Since the 1880s, forest fires had been the most immediate and biggest problem for conservationists and residents in the cutover. Wisconsin quite early in its history had recognized problems associated with the depletion of its forests, and in 1867 the legislature formed a commission to assess the state's forest reserves. Increase A. Lapham, the commission chair, produced a study titled *Report of the Disastrous Effects of the Destruction of Forest Trees Now Going on So Rapidly in the State of Wisconsin.* The report provided a comprehensive view of the effects of logging and lumbering. It specifically noted that the elimination of the forest cover reduced stream flow and led to widespread flooding and erosion. Professors at the University of Wisconsin and their students contributed to the call for measures to stop the destruction of forest fires. These fires ravaged the North Country at the end of the nineteenth century, and in 1897 the Wisconsin Legislature finally initiated a program to monitor and preserve the state's natural resources. Members appointed a state forestry warden, who implemented a system of town wardens and volunteer firefighters. Some watchtowers were built, and two-way radios came into use. However, these efforts were not enough. In 1908, after a particularly bad season of fire damage, a citizens' committee formed to pressure the state legislature for better forest protection. Governor James O. Davidson appointed a State Conservation Commission to study the problem of forest fires and make recommendations to the legislature. In 1923 this eventually resulted in the creation of fire protection districts. In 1925 these districts were given $25,000.[100]

By the 1920s disenchantment from the indiscriminant promotion of agricultural settlement, especially in the cutover, encouraged Wisconsin conservationists to push for a constitutional amendment to allow for state-owned and state-managed forests. This was passed in 1924, thus overriding the Wisconsin Supreme Court's 1915 ruling that state land purchases to create forests were unconstitutional. In 1927 the state passed a forest crop law that encouraged counties and private owners to put land into forest plantation by implementing a tax plan that recognized the long-term nature of forestland use. The Wisconsin Conservation Department became an independent agency and was given responsibility for the protection of all forestlands. In 1929 a new zoning law made it possible for counties to zone lands to prohibit agricultural settlement. "It was a final recognition that an unsuccessful farmer, settled on unsuitable land in an isolated place," wrote historian Robert C. Nesbit, "was anything but a taxable asset to the county." A series of state and county forests were thus established in the cutover lands in the St. Croix Valley. These forests had few trees, but at least the land had been designated for forestry. The Roosevelt administration help to set the stage as well when in 1933 it created the Chequamegon National Forest, almost 900,000 acres in size, at the headwaters of the St. Croix. The Badger State was thus poised to take full advantage of New Deal conservation programs. The CCC built twelve camps in Wisconsin national forests, twelve in its state forests, and eight in its state parks. More than 92,000 "CCC boys" labored in Wisconsin forests during in the nine years the corps operated existed.[101]

Minnesota had also taken action to monitor, if not protect, its forests. In 1876 the Minnesota State Forestry Association was established, making it one of the earliest forestry organizations in the country. Members worked primarily on the grassroots level, through civic groups and fraternal organizations. After the Hinckley fire in 1894, Minnesota lawmakers took more direct action in forest management and preservation. Christopher Columbus Andrews, a leading conservationist, served as the first chief fire warden. Under his leadership Minnesota adopted a progressive philosophy of forestry management. In addition, the state established a School of Forestry at the University of Minnesota to train professionals and to establish a state nursery, state forest park preserves, and more state parks that emphasized natural resource conservation. The Conservation Commission, organized in 1925, served as an umbrella agency over forestry, fire prevention, game and fish, lands and timber, state parks, and state public campgrounds. By the mid-1920s the state of Minnesota offered a wide variety of well-managed recreational venues and facilities for visitors to the North Woods.[102]

The cutover region of Wisconsin and Minnesota was sorely in need of revitalization. The two state governments cooperated with county agencies to receive several CCC companies to help redevelop the St. Croix Valley. Most of the CCC workers were from urban areas, especially Chicago, although young men from local counties were also able to get work with the CCC. In June 1933 Company #647, composed of 198 young men, arrived in Burnett County from Fort Sheridan, Illinois. One of the camps, Camp Smith Lake, was located two miles east of Pacwawong Lake in the upper reaches of the Namekagon River in Burnett County. Others were Clam Lake Camp, Ghost Lake Camp, Sawyer Camp, and Camp Riverside nine miles northeast of Danbury. Wisconsin's Interstate Park hosted two camps, one from 1935 to 1940 and the second from 1938 to 1940. Pine County in Minnesota hosted a CCC camp, and on the Lower St. Croix River a camp was stationed in Bayport. This camp was entirely made up of veterans, primarily sailors and ex-marines.[103]

The main objective of CCC programs in the valley was, of course, reforestation and conservation. The young "CCC boys," as they were called, built fire roads and lookout towers, which finally helped end the devastating fires that plagued the region. Once the fires were brought under control, the forests had the chance to recover. The CCC then planted millions of trees from seedlings they grew in their own nurseries. Forest experts were enlisted to provide advice on location and type of soil in which to plant them.[104]

Figure 21. The Bayport Civilian Conservation Corps camp, ca. 1935. The site is now occupied by an Anderson Windows warehouse. St. Croix National Scenic Riverway files.

An example of charred cutover land that was returned to verdant forest in Burnett County was the handiwork of Camp Riverside. Shortly before the company arrived, a fire had swept through fifteen hundred acres in the county. The inexperienced young CCC men had a big job ahead of them. For the next six years they cleared debris and planted nearly 2.5 million jack pine, Norway pine, white pine, and spruce trees. They built 75 miles of fire roads and laid 107 miles of fireproof telephone lines to the Burnett and Washburn county fire protection districts, as well as connecting lines to forestry units elsewhere in the state. The CCC also erected two fire towers—the Sterling Tower in Polk County and the McKenzie Tower in Burnett County. Firebreaks were cut across areas that posed high fire hazards, such as in the Jack Pine barrens. Water table surveys were made in order to install fire-fighting wells and pumps.

The Riverside CCC also constructed a 130-foot, two-span, timber bridge across the St. Croix River along the historic St. Croix Trail that marked the overland route from the Twin Cities to Bayfield, Wisconsin. This overland portage had been abandoned when the railroads came to the North Woods; now it was revived for recreational use. The camp boys also built an earthen dam on Loon Creek, which enters the Yellow River just before it joins with the St. Croix River, to raise the water level of the Loon Creek chain of lakes. They added four more parks and equipped them with campsites that increased recreational opportunities in Burnett County. Three thousand pheasants raised from chicks were released into the second-growth forests for sportsmen.[105]

As Burnett County reclaimed its forests and its farmland diminished, the county government found its tax rolls reduced. One way it raised money was to cater to hunters. Cabins and house trailers on government property were leased to enhance recreational usage of land in the St. Croix Valley. These small, crude shelters dotted the St. Croix and Namekagon rivers in Burnett County until they were taken over by the National Park Service in the 1970s.[106]

In addition to forestry programs, fish propagation and river and stream improvements were among the more significant programs of the CCC. One of the legacies left by the logging era was the problem that dozens of streams and rivers along the St. Croix, including the St. Croix itself, silted when forest was removed and nothing was left to hold the soil in place. The banks of rivers had also been severely eroded by increased rainwater runoff and log drives. Many lakes had been so silted that they became more like swamps and muskegs. Alder took root where once there had been blue water. In the nineteenth century, many hunters and trappers had reconciled themselves to the inevitability of the disappearance of game. Fishermen, however, had fully expected their sport to continue unabated after the forests had been logged over and lands turned to farms. When fish numbers began to decline in the nineteenth century throughout Wisconsin, there was an outcry to do something about it. Unfortunately, early stocking practices were not carefully considered. In 1881 carp were introduced into rivers and streams in southern Wisconsin because they were able to survive in warm and semi-stagnant water. The carp, however, bred quickly and made any future efforts in promoting more desirable species difficult. By 1935 the state began to hire men to clean out the carp and expanded native fish hatcheries. Inexperienced volunteers, however, improperly dumped untold numbers of fry into streams where most died. With the assistance of the CCC, the Wisconsin Conservation Commission began the practice of allowing fry to mature and releasing them under more careful supervision.[107]

The Civilian Conservation Corps along with the Wisconsin Conservation Department built a fish hatchery near Hayward on the Namekagon River. The CCC also planted a grove of conifers on the grounds of the hatchery and along the road leading up to it. Streams were cleared of debris, and the CCC recruits built diversion dams where temperatures and water speeds could be carefully monitored. The lakes in Burnett County were mapped for depth and type of bottom, and "fish refuges" were built and sunk to nurture fish life. Along barren riverbanks, the CCC planted trees, which prevented erosion and silting and provided shade and cooler waters for fish habitat. The agency also placed V-shaped log dams in streams to create shallow spawning beds.[108]

Although fishing recovered in the valley and attracted sport fishermen, tension arose between visitors anxious to amass a string of trophy fish and locals who during the Depression years saw fish as food for today and a future source of money. Fred Etcherson, for example, had lived in the St. Croix Valley nearly all his life, and waste of its bounty irritated him. "When I go there to fish, I get two . . . if I got someone who is going to go home with me to help eat them trout, I might take four," he related. "I want one left for tomorrow." But the visitors "catch more than they can take care of, so they throw them away." Federal authorities worked with state conservation departments to reform fishing laws, regulating bag limits, fish size, and seasons.[109]

Among the variety of work done by the CCC was combating pests and disease in the forest. Planting efforts included a program to limit the spread of white pine blister rust and insect control to stop the spread of grasshoppers and moths. Grasshoppers were poisoned. Slashing and burning the timber in infected areas curbed the jack pine tussock moth. Wildlife was counted and surveyed and their habitats enhanced. Roadside picnic areas and state parks were developed or enlarged. Facilities such as visitor centers were built and equipped with rest room facilities. The CCC landscaped the grounds. Nearby timber stands were cleaned up, as well as springs and ponds, and nature trails and access roads were laid out.

Since it was located primarily in farm country, the main mission of the Bayport CCC Camp was soil conservation. With supervision from the Federal Soil Conservation Corps, the CCC recruited ninety-one Washington County farmers, who had eleven hundred acres of land, for a five-year soil conservation program. The men spent considerable time fluming and riprapping gullies and constructing drainage ditches along roads. They planted at least 250,000 trees on farms to prevent erosion and assisted in rebuilding fences.[110]

In the Wisconsin Interstate Park, the CCC developed, under the direction of the technical staff of the National Park Service (NPS), a survey of the park and a master plan for trails and rustic shelters. All buildings were built from stone that came from a quarry in the park, following the guidelines of Albert Good, an architect noted for his rustic designs. The CCC boys cleared a trail along the St. Croix River by removing rocks using a variety of methods, such as crowbars, blocks and tackles, and even the "Indian method" of "fire and water." In this method, rocks were heated by a fire for several hours or even days. Once heated through, cold water was poured on top to crack and shatter them. In 1937 one camp razed the old bathhouses in the park, brought in new sand for a beach on the river, and built a trailside shelter and a log restroom. A second CCC camp completed a new bathhouse, a shelter along Horizon Rock trail, a park office, a picnic shelter by the beach, as well as picnicking and parking areas.[111]

The over-used Minnesota side of the Interstate Park also needed a lot of work. In 1937 it welcomed 327,496 visitors. On some Sundays 10,000 people crowded into the park. CCC workers built a variety of structures ranging from a refectory, stone curbing, stone retaining walls, and a stone drinking fountain. When the CCC was disbanded, the Works Progress Administration built a twenty-two-acre campground and a shelter-refectory combination building housing restrooms, laundry, kitchen, and utility facilities.[112]

The WPA also constructed overlooks along the St. Croix River. The first one it built sits atop a ridge located approximately one to two miles south of Stillwater and offers a long, southward view of the St. Croix River. The site is still in use. WPA workers also constructed a wayside rest area near the St. Croix Boom site. It functions as a scenic overlook, historic site, picnic spot, and rest stop. With assistance from the National Youth Administration, the WPA constructed low stone retaining walls encircling the cliff bank high above the river and placed stone fire rings. Curvilinear sidewalks meander through the site, and a steep stairway leads visitors down to the banks of the St. Croix River.[113]

Another project of the WPA was building dams in the region. One of the most notable projects was the dam on the North Fork of the Chief River in Sawyer County along the Tiger Cat Flowage and Round Lake. The area had been plagued by a drought that dried up old springs, which then in turn dried up lakes, forests, and over-cultivated land. Round Lake "dropped so low that some lakeshore residents had to take a long walk to go swimming." The North Fork of the Chief River, however, had a vast drainage network with good water reserves. The Tiger Cat dam, built in 1937 with WPA funding, held back the flow of Chief River tributaries and raised the lake levels and water table in the popular resort area. The project was in some respects too successful. Residents of Round Lake complained there was sometimes too much water, which washed away their beaches, piers, and shorelines.[114]

The Works Progress Administration also hired artists, musicians, and writers to promote the arts and culture in America. Many writers were employed to write state guidebooks. These books essentially provided a brief history of the particular state, its resources, and unique characteristics. *Wisconsin: A Guide to the Badger State* (1938) and *Minnesota: A State Guide* (1938) each included the St. Croix Valley in its tours. The publication of the guides was evidence of a growing interest in the history and culture of the states. In 1938 the state of Minnesota completed a statewide survey of historical and archaeological sites that allowed these resources to be included in recreational planning. In Wisconsin, the prolific historian Louise Phelps Kellogg stimulated public interest in the early history of the St. Croix and other sites involved in the fur trade–Indian era. Her work was included in the WPA guides and was likely, if inadvertently, responsible for tourist promoters dubbing the Upper St. Croix Valley "Indian Head Country." One detail the guidebooks did not include was the lingering poverty of the cutover region. Federal officials, however, compared the region to Appalachia and they hoped water development projects and recreational enhancements would enhance the Upper St. Croix Valley.[115]

The St. Croix River counties of Washburn in Wisconsin and Pine in Minnesota were among the areas in greatest need. The sandy soil was not responsive to the best efforts of farmers, and most eventually gave up. Pine County led its state in tax delinquencies. This prompted the Minnesota Natural Resources Board to recommend reforestation. Beginning in 1935 federal and state officials worked together to create a forest reserve near the Kettle River. Its potential for recreational use made the site ideal for the New Deal Program. This tract was owned by Northern States Power Company, which leased it to the Department of the Interior. In the latter part of 1935 and into 1936 the CCC established the St. Croix Recreational Demonstration Area (RDA). It comprised 30,000 acres, which made the St. Croix RDA the largest in the country. The large but poor area included scenic views along the high banks of the Kettle and St. Croix rivers. Existing farmhouses and outbuildings were demolished to make way for five developed areas, including facilities for a park administration and visitor center, three group camps, and one public campground and day-use area.

Between 1936 and 1942, the CCC and the WPA constructed buildings and scenic overlooks, cut roads and trails, erected dams, installed fish-rearing ponds, and landscaped portions of the RDA. They also planted pine, spruce, and hardwood trees.[116]

In 1943 the federal government donated the newly restored forest to the state of Minnesota. While most newly created state and county forests aimed to serve the dual interests of commercial logging and sporting activities, the unique scenic appeal of the rugged high banks of the St. Croix at Kettle Falls led the state of Minnesota in 1943 to preserve the historic RDA as the St. Croix River State Park.[117] The entire St. Croix Recreational Demonstration Area historic district was designated a National Historic Landmark (NHL) in September 1997 because of its landscape architecture, recreation and culture, transportation, and domestic use.

As unemployment dropped in the late 1930s from its high of 25 percent to approximately 14 percent, it became more difficult to continue to get money and men for the regional development program. When the threat of war for the United States seemed imminent in the spring of 1941, many camps were turned into noncombatant training centers. In December, after Pearl Harbor, the CCC was disbanded completely. Work left uncompleted was finished by day laborers from the WPA.

The CCC and the WPA had a lasting influence on the St. Croix Valley in reforestation; professional management of state forests, lakes, and streams; and control of soil erosion. Once these resources were preserved and used more wisely, wildlife and fish flourished. Through New Deal programs the National Park Service and numerous other federal agencies were brought into cooperation with Wisconsin's and Minnesota's conservation departments. Left in the wake of the Depression were a core of professional staff, established planning procedures, and, most important of all, a commitment in the region to activist government engagement with environmental development and protection. State and federal forests, together with CCC/WPA improvements, changed the face of the North Country. After World War II, when large numbers of tourists returned to what had been the Upper Midwest cutover, they found the forests, fish, and game in flourishing condition. To be sure, the old-growth forests were gone for good, but to the urban visitor the "weed trees" and well-ordered ranks of pine plantations that replaced them looked enough like nature to be an illusion of wilderness and welcome relief to the city. No longer would images of a battle-scarred landscape apply to the Upper St. Croix. Indeed, one ad agency could now promote the former cutover to millions of television viewers as "the land of sky blue waters."[118]

Return of the Tourist

In the wake of war and home front rationing, a prosperous America returned, anxious for a slice of the "good life" and ready to "see the USA in their Chevrolet." North-bound autos brought families who had not only the means to travel but also the desire and ability to buy their own place on a lake. The obtainable goal for many a returning GI was a seasonal home "Up North." The demand for summer cottages created a market for vacation homebuilders. How-to books and manuals on building a vacation home had been around since the early part of the century. In 1934 the first edition of *Popular Science Monthly*'s *How to Build Cabins, Lodges, and Bungalows: Complete Manual of Constructing, Decorating, and Furnishing Homes for Recreation or Profit* was published. It competed with Ralph P. Dillon's very popular *Sunset Cabin Plan Book,* which had reached its sixth edition in 1938. The National Plan Service of Chicago published a promotional booklet of models and floor plans of forty-eight cottages, encouraging its readers to "invest in health and happiness." "Take your family away from the grind, routine, and rush of the city," they recommend, "to the beautiful lakes, the cool shaded forests, and the invigorating air of

Figure 22. This gas station in Spooner, Wisconsin, appealed to auto tourists by using the "Indian Head" logo devised to promote northwest Wisconsin and by providing fast food and restrooms. Photograph by the L.L. Cook Company, ca. 1941. Wisconsin Historical Society, WHi-40468.

the country." Those interested could purchase a blueprint of any of the models offered, or have the National Plan Service design one. Models ranged from modest, rustic-looking one-room cottages to five-room retreats, equipped with baths, eat-in nooks, and porches. Some were intended for family use; others were designed for resort operators.[119]

The boom in waterfront cottage construction was both a bane and a blessing to the resort owners who had helped to develop a tourism industry in the Upper St. Croix Valley. The influx of new cottagers revealed the resurgent demand of urbanites for a piece of the North Woods, yet directing a portion of that traffic toward their resorts required renewed imagination and vision. The town of Hayward, Wisconsin, on the Namekagon River had its share of both in the years after World War II, and it provides a case study of how St. Croix Valley towns responded to the challenges of prosperity. More than anyone else, Cal Johnson was responsible for Hayward's ability to secure prosperity for its resort owners. One rainy morning in July 1949, Johnson hooked a mammoth muskellunge, and after a suitably epic battle, he landed the behemoth. He took it to a local taxidermist, and the fish was measured at 67.5 pounds and 60 inches—a new world record. The fact that Johnson was an outdoor writer who published his account of the record catch in a feature article in the national magazine *Outdoors* made the event a national story. Better yet, a controversy erupted over who deserved the record when Hayward citizens championed the 70-foot-long musky that Louie Spray, a local fisherman, had earlier caught. Either way, town boosters proclaimed, Hayward was the "Musky Capital of the World." Johnson's record fish toured sports shows across the Midwest, helping to link forever the name *Hayward* with the image of a giant musky. Capitalizing on this boon, resort owners organized the National Musky Festival Association. The first festival drew five thousand visitors. State legislators cooperated by making the musky the state fish. By the 1970s the town boasted a giant walk-in musky that houses the National Freshwater Fishing Hall of Fame. In the wake of Johnson's great catch every American fisherman knows Hayward is the place to catch the nation's largest freshwater game fish. A mixture of taxidermy, public relations, and a genuinely big fish served to guarantee the survival of the town's resort businesses.[120]

Another visionary for the recreational development of the North Woods was Hayward native Anthony Wise. When stationed in the Bavarian Alps during World War II, Wise conceived of a project that would invigorate the depressed climate of northern Wisconsin—an alpine skiing resort. As a 1947 Harvard Business School graduate, Wise was keenly aware of the fact that while tourist dollars came to the area with the warm weather, they left like departing flocks of geese for the south before winter chills set in. Alpine skiing was largely foreign to Wisconsinites, although Scandinavian immigrants had introduced *skikjoring,* or skiing, to the region in the mid-nineteenth century. By the 1880s ski-jumping clubs had been organized in Minneapolis and Red Wing, Minnesota. And in 1883 the Skiing Association of the Northwest was organized. By the 1930s Minnesotans enjoyed the "Winter Sports Week," which included a variety of activities, including hockey, skiing, skating, ice boating, curling, tobogganing, bobsled and sleighing parties, snow sculpture, and costume parades. Wisconsinites, however, were slower to embrace the sport, and establishing skiing there would require salesmanship.[121]

Fortunately, Wise proved to be not only a dreamer but also a shrewd businessman. In 1947 he wrote a report to the Hayward Chamber of Commerce complete with facts and figures concerning the area's potential for a ski resort. His analysis was based on a survey conducted by the Charles W. Hoyt Company of New York in 1944 and 1945 of recreational habits and desires of residents in states north of the Ohio River and east of the Mississippi. The survey probed the current interest in winter sports vacations. He also studied winter sporting activities in New England for comparison. What Wise found was that the greatest interest was in skiing. Tobogganing and skating were just side attractions for most people. The study helped to convince local resorts and hotels to experiment with winter operations, if Wise developed a ski resort. Wise and a partner went to work and carved out a rustic (read *crude*) ski facility. Desperate to make the facility pay, Wise contacted the Wisconsin Tourist Board in Chicago. "The Telemark Company has just finished construction of the finest downhill ski area in the Midwest," he wrote in December 1947. "Naturally we are very desirous of tapping the big Chicago market." Wise explained to them that he and his partner, H. B. Hewett of Minneapolis, were ex-GIs and had sunk their entire savings as well as risking loans to the tune of $10,000 to buy a 370-foot-high mound of Ice Age rock and gravel with a vertical drop of 2,400 feet three miles east of Cable, Wisconsin. They had also built a rustic one-room day lodge at the base, cleared a path for two downhill runs, and installed a towrope. They, however, had little money left for advertising and appealed to the tourist board for help. If the board sent vacationers to their ski resort, they would return to them a 10 percent commission. Fortunately a handful of skiers from Chicago found their way north, and Wise contracted with a ski instructor from Norway to conduct a school for novice flatlanders.[122]

Success came slowly, although within eight years Wise's Telemark resort was attracting 17,000 skiers a season. During that time he added Ski Inn, Norway Lodge, and Ski Lark Motel. At the end of two decades Telemark attracted 100,000 midwesterners per season and became known as the most innovative ski resort in the Midwest. It was equipped with T-bars and ski lifts, and it boasted a certified ski instruction school. In 1961, after a particularly light season for snow, Wise introduced snowmaking machines to northern Wisconsin. Federal aid from the Small Business Association made this possible by supplementing loans from local banks. Wise was also the first to use snowcats to groom the trails. These "pioneering efforts in snow grooming, with the help of creative blacksmiths from local lumber camps, provided good skiing conditions throughout the long season in spite of the intemperate climate." Reminiscent of an era when lumbermen prowled these woods, outdoor employees were dressed in "lumberjack" uniforms with red shirts, stag pants, high leather boots, and a high-crowned Scotch hat. In the spring, lift ticket-holders received a free roast beef dinner at the Blue Ox Feast, and skiers enjoyed fresh maple syrup during the Maple Sugar Feast.[123]

In 1972 the four-season Telemark Lodge opened for vacationers and conventioneers. The multi-million-dollar resort, surrounded by nearly one million acres of state and national forests, included the Telemark Recreational Community of single-family homes, townhouses, and condominiums complete with a sixty-three-kilometer cross-country ski complex, a golf course, tennis courts, riding stables, trout ponds, canoe rentals, and bicycle and hiking trails. Not content to rest solely on his success in the ski resort business, Wise also proved to be a major booster of his hometown of Hayward. Beginning in 1960, Wise built upon the "Lumberjack Saturday" during the annual musky festival. His interest in local history encouraged him to preserve elements of the region's colorful past. By inviting lumberjacks from all over the country to compete in an annual Lumberjack World Championships, skills once needed for survival in the lumber camps and woods were on display. Teams came from Australia, Germany, Japan, British Columbia, and heavily forested areas in the eastern and western parts of the United States. For three days contestants vied to see who was best at chopping, speed climbing, tree topping, and log rolling. When the contests concluded, visitors could step back in time at Wise's Historyland—a living history village of a church, hotel, bunkhouse, blacksmith shop, and wannigan. In addition, Historyland hosted other festivals, such as Voyageur Wild River Days and Indian powwows at the Ojibwe Indian Village. Looking back on what made this endeavor such a success,

Tony Wise concluded that it was a $150,000 loan from the Small Business Administration. Reflecting later on the role of the federal government in his success, Wise remarked, "As far as we're concerned . . . the United States government stepped in when nature failed."[124]

Telemark was not the only ski resort to open in the wake of World War II. In the fall of 1950 Lee Rogers and Walter J. Peterson opened Trollhaugen ski hill just east of Osceola on land purchased from Paul Neilsen. It, too, had a modest beginning with one towrope and three slopes. By 1956, however, Trollhaugen provided skiers with five towropes and six slopes. A rustic chalet was on hand to warm skiers and provide some hospitality. During these early years an average of 5,000 skiers traversed the slopes annually. Trollhaugen suffered from light snow in the early 1960s, and not to be outdone by Telemark, it installed a snowmaking system. Throughout the 1960s, T-bar lifts were added, a chair lift was installed, and more slopes were opened. By the end of the decade 80,000 skiers swooshed their way down the hills in one season. Trollhaugen also boasted that it had one of the best national ski patrols and ski schools in the Midwest.[125]

Dam the St. Croix, Again

The successful transition of the St. Croix Valley to recreation and tourism, however, did not end the concern regarding the viability of the river for future generations. With so many tourists and sportsmen enjoying the St. Croix Valley together with valley residents and a growing number of Twin Cities migrants, issues of control over the river and its proper usage once again emerged. The major catalysts to public outcry once again concerned Nevers Dam.

In 1945 the River and Harbors Act authorized a study of the St. Croix River basin. Among the proposals was rebuilding Nevers Dam to a height of forty-five feet. Damming the river would create a thirty-mile lake between Nevers Dam and the Kettle River and would create a reservoir from there to Danbury. The proposed project would affect seventy-five miles of the St. Croix River.[126] This time the opposition came from the Northern States Power Company and conservationists. The motive of conservationists was obvious—to preserve wildlife habitat and fishing streams. Nor was public opinion in the towns on the lower river fully behind the project, not after so much work and money had already been spent on promoting tourism. The Northern States Power Company, however, was the real foe of the dam project. In the past the company had cooperated with the federal government in conservation efforts, and now it felt it was being edged out by the Rural Electrification Administration (REA) program, a creation of the New Deal. The REA had done much to bring remote rural areas into the twentieth century by creating the means for farmers to purchase electric power cheaply. But if the REA were behind the Kettle River dam project, the NSP would have a new competitor—the federal government. On the side supporting the dam were the Army Corps and farmers. They, however, failed to create a united effort. The sole aim of farmers was to create hydroelectric power, which was somewhat at odds with the goal of their allies, the Army Corps, whose aim was flood control and improved navigation. In 1947, after a long, bitter struggle, dam permits were denied.

Because the proponents of the "Little TVA of the North" did not put forth a unified and concerted effort, the St. Croix River was spared during the 1940s and 1950s—the age of unprecedented dam building in America. Again in 1953 the Federal Power Commission refused to grant the Wisconsin Hydroelectric Company a permit to build a twenty-five-foot dam on the Namekagon River. The decision was based on the "unique" recreational features of the river. "The canoeist has the illusion of being in a forest primeval, far from civilization," wrote the presiding judge. The Namekagon case was a milestone in recognizing that energy development and the deep-rooted need for unspoiled recreation should be balanced. However, even the judge in the Wisconsin Hydroelectric case was forced to admit that in the case of the North Woods river, wilderness was an "illusion." By 1953 there were twenty-three dams and hydroelectric plants in the St. Croix Basin. The upper Namekagon had five small dams including electric generating stations at Trego and Hayward.[127]

In 1947, reenergized by the fight against the Kettle Dam project, the St. Croix River Improvement Association returned to its original name, the St. Croix River Association. While water flow and channel depths remained on their agenda, the association expanded its concerns to include sewage disposal and river pollution; parks, roads, and bridges along the river; wildlife and fish propagation; and pleasure boating. The organization that had once been among the biggest supporters of damning the St. Croix now emerged as a voice for protecting what was left of its natural areas.[128]

The valley's natural areas were largely preserved in a series of state parks and wildlife refuges that were created in the St. Croix River Valley. These were encouraged by the National Conference on State Parks that formed in 1921. National Park Service director Stephen Mather had become concerned about the overuse of gems of the NPS, such as Yellowstone, Yosemite, and the Grand Canyon national parks. He felt that state parks could fill the need for recreation and scenic enjoyment and take pressure off national parks. The new organization's slogan became "A State Park Every Hundred Miles." Wisconsin and Minnesota had already proven by then to be on the cutting edge of the state park movement with six each. Many other states followed and established state park boards or commissions with the authority to manage scenic and recreation areas that would come to include state monuments, beaches, lakeshores, parkways, waysides, and historical markers. Partnerships were formed between state park superintendents, departments of conservation, state comptrollers, and local boosters in a concerted effort to preserve natural settings, market them for public consumption, and glean a bit of pride in sharing their part of the American scene.[129]

On both the Wisconsin and Minnesota sides of the river, partnership efforts spurred conservation initiatives. In 1945 Alice M. O'Brien donated 180 acres of riverfront land north of Marine that had belonged to her father, William O'Brien, to the state of Minnesota. Within two years the Minnesota State Legislature officially established it as a state park. In 1951 Alice O'Brien donated an additional 15 acres, and in 1958 S. David Greenberg donated a 66-acre island. Over the years the William O'Brien State Park has grown through personal gifts and through the efforts of the Minnesota Parks Foundation to 1,343 acres. At the same time, the state of Wisconsin developed the Crex Meadows Wildlife Area in Burnett County. The property contains 25,000 acres of restored wetlands made possible through the construction of miles of dikes and a series of water-control structures. The marsh is home to several species of birds and animals and is used by hunters and naturalists. The William O'Brien State Park and the Crex Meadows Wildlife Area have contributed yet another dimension to the development of leisure activities by offering bird watching, wildflower appreciation, naturalists' talks, and walking tours. Through the establishment of the St. Croix River State Park and the Carlos Avery Wildlife Area, the state of Minnesota further expanded these opportunities.[130]

The combination of state stewardship and private commitment to conservation was demonstrated in 1947 when the state of Wisconsin added to its St. Croix forestlands through the acquisition of the Soren Jensen Uhrenholdt farm. Uhrenholdt had jealously protected a grove of virgin pine on his farm. As he explained to his son-in-law, the famed nature writer Sigurd F. Olson, the great pines were a symbol of nature's persistence, and he felt that as long as the trees survived so too would his family's connection with the land. When Uhrenholdt passed away, his children sold the farm to the state but donated the ninety-seven acres of tall timber as a memorial. On August 15, 1947, a few months after his death, the Uhren-holdt Memorial Forest was dedicated in the town of Seeley "among the lofty pines he preserved," wrote his neighbor Eldon Marple, as "a fit tribute to a man of vision who had the courage and wisdom to carry out practices he knew to be right."[131]

Another example of local preservation efforts in the St. Croix–Namekagon River Valley was the revival of the old Portage Trail along the Namekagon River near Hayward, once used by Indians, fur traders, and explorers, as a hiking route. In 1965 a local society under the leadership of Lyman Williamson set its goal to preserve what was left of the trail. Thanks to an 1855 government land survey and some help from old-time hunters and farmers, the group was able to locate sections of the old trail. William-son laid out the route and obtained permission from property owners to make improvements and allow hikers to use it. On May 22, 1966, the his-toric trail was reopened with the help of more than two hundred Boy Scouts of the Chippewa Valley Council and their leaders, who cleared out the logs, brush, and debris strewn on the path.[132]

Despite all these efforts at preserving areas along the St. Croix River, it was not enough to protect the unique qualities of the river. By the mid-twentieth century the St. Croix River Valley began to take on a suburban quality, in part from the dramatic growth of communities on the river and its close proximity to Minneapolis and St. Paul. The commuting time and the ease of travel between the Twin Cities and the St. Croix were drastically changed thanks to the automobile. Before long, towns and resorts along the St. Croix, particularly Stillwater, functioned as virtual suburbs of the metropolitan area. "With the end of the 1940s more and more city dwell-ers, having first come into the valley as summer renters," wrote James Tay-lor Dunn, "were establishing themselves in Marine as permanent residents, commuting to their jobs in Stillwater, St. Paul, and Minneapolis." The various parks and recreation areas attracted these newcomers, but at the same time a surge in residential population threatened to drastically alter the ambience of the area.[133]

Saving the St. Croix

In 1957 Theodore A. Norelius, editor of the *Chisago County Press,* was the first to voice the opinion that the St. Croix's scenic and recreational features deserved national attention. The Northern States Power Company still owned extensive tracts of land along the St. Croix

River. For years it had graciously allowed sportsmen access to the game and fish on its land. Norelius, however, feared that in an era when nuclear power seemed to be the future of electrical energy, the utility might be tempted to sell its lands to real estate developers who would subdivide the land for cottages. "If Mr. Public has a place or places to play in the future," he asserted, "now is the time to consolidate all efforts here in the upper Midwest and ask for a gigantic St. Croix Federal Park, perhaps named the 'River of Pioneers National Park.'" Only in this way, he argued, could sportsmen avoid signs stating "Private" and "No Trespassing." "This is your land," Norelius wrote. "Protect it and preserve it for all posterity."[134] Utilizing the federal and state parks lines of communication, U. E. Hella, director of the Minnesota State Park System, lobbied Howard Baker, director of the National Park Service's Midwest Region. The proposal also won the support of Minnesota senator Hubert H. Humphrey, but the idea was stymied in the House of Representatives, where it withered in committee.

While the "River of the Pioneers National Park" idea faded, the national mood was slowly shifting toward greater environmental activism. One of the most articulate voices calling for change was a writer who had spent much of his youth in the St. Croix region, Sigurd F. Olson. In his influential books, *The Singing Wilderness* (1956) and *The Listening Point* (1958), Olson drew attention to the threats against the environment. He argued that the unique values and beauty of the North Country were very fragile and threatened by development. Adding much force to his warnings was Rachel Carson's *Silent Spring,* published in 1962. Her seminal book created a national sensation over the devastating harm caused by pesticides on wildlife. As the 1960s progressed, the environmental movement enjoyed a revival of public interest and support not seen since the Progressive Era's conservation movement. Along with these noteworthy books was a 1959 National Park Service report, *Our Fourth Shore: Great Lakes Shoreline Recreation Area Survey.* This work helped bring the National Park Service into the Great Lakes region. While it did not specifically target the St. Croix River Valley, the river would never have received the attention it later did without this report. Thanks to it, three new national parks were proposed on the Upper Great Lakes, including the Apostle Islands on Lake Superior.[135]

On September 3, 1964, the U.S. Congress passed the Wilderness Act, which marked another significant shift in American attitudes toward nature and the outdoors. This act required federal agencies to identify and preserve areas as part of a national wilderness preservation system including regions unspoiled and uninhabited by humans. The aesthetic and spiritual qualities of nature were once again valued for their own sakes, but this time they were perceived as benefiting the public good. Emphasis turned from parceling natural resources as commodities to a more holistic approach based on ecological values that ultimately would benefit everyone.[136]

In 1964 a conflict erupted that brought into collision the new idea of the St. Croix River as an environmental amenity with the old notion of it as a resource for exploitation. The Northern States Power Company announced it intended to build a new coal-fired power plant on the river just south of Stillwater. The project included a giant generating station, a 785-foot smoke stack, and coal yards that would stretch for a quarter mile along the river bank and reach a height of 35 feet. Outrage was greatest downwind, on the Wisconsin side of the river, where towns feared pollution and the destruction of their beautiful river views. On the Minnesota side of the river, the project was seen as an economic boon for local taxing bodies and a needed source of employment. The power company pointed to the growing need for electricity in the Twin Cities, which was the nation's fifteenth largest metropolitan area and was projected to grow by more than 350,000 new residents by the end of the decade. The giant coal smokestack, it seemed, was part of the metro area's growing shadow over the river. Divided between project boosters and opponents, the St. Croix River Association was unable to offer effective opposition to the plant. An ad hoc citizens group, "Save the St. Croix," fought a dogged rearguard action, but they could not stop the power plant project. Their opposition, however, was not without effect. By putting Northern States Power Company and local politicians through a media wringer, they helped to pave the way for a greater public commitment to river protection.[137]

In the summer of 1964, just as the power plant fight was coming to a head, the U.S. Department of the Interior issued a report calling for the preservation of the St. Croix as a wild and scenic river of national significance. The man behind the proposal was Gaylord Nelson, U.S. senator from Wisconsin. Nelson, who would go on to become the "father of Earth Day," had grown up in the St. Croix Valley. As a Boy Scout and as an adult, he had frequently canoed and fished along the river. He had long wanted to take action to preserve the river, and he was politician enough to know that the dispute over the Northern States Power plant created the opportunity for action. The Northern States Power Company owned 37 percent of the lands along the Upper St. Croix, and in the midst of the power plant fight, they badly needed to demonstrate that their proposal was not the beginning of industrializing the river. Nelson knew the company's executives, and he had previously discussed with them plans for river protection. The senator would get his river park, and NSP would get its power plant. Since Nelson had little legal chance of stopping the project anyway, it was an excellent bargain.[138]

Eventually, the proposal for a St. Croix River national park was folded into a broad-based bill, the Wild and Scenic Rivers Act, designed to preserve a handful of America's finest river environments. The movement for a national approach to river protection had been building through the 1950s, an era of rapid dam and infrastructure development. President Lyndon B. Johnson made a wild river bill part of his "Great Society" program, and in his 1965 State of the Union address he argued, "The time has come to identify and preserve free-flowing stretches of our great rivers before growth and development have made the beauty of the unspoiled waterway a memory." Yet, while Johnson's support eased winning congressional support for the St. Croix, Nelson was aware that for any preservation plan to work on the river, it would have to include the diverse interests of the people in the valley itself. Here he was fortunate to be able to draw upon the legacy of federal-state cooperation and intrastate cooperation that had been forged during the first half of the twentieth century. The specifics of the St. Croix portion of the Wild and Scenic Rivers Act were hammered out over several years and drew upon the work of a Joint State-Federal Task Force created to study the Northern States Power plant proposal as well as an interagency task force that called for the federal government to purchase a half-mile corridor along the upper river. With the exception of private property owners along the river, the idea of a St. Croix River park had strong support in Wisconsin and Minnesota. Nelson, however, encountered unexpected opposition from the National Park Service, which objected to including the suburbanized Lower St. Croix in the bill. Nelson had proposed to protect the stretch of river south of Taylors Falls through easements and zoning, something the National Park Service felt was "beyond the capacity of the agency." The upper river was, according to the park service, "a classic northwoods stream," but the lower river was best left to local government management.[139]

Eventually Nelson was forced to accept half a loaf. The Wild and Scenic Rivers Act only included the St. Croix and Namekagon rivers as far south as Taylors Falls. Yet another Army Corps of Engineers' plan to put a dam on the Upper St. Croix forced preservationists to act while they could. Among the unique features of the bill that was signed by President Johnson in October 1968 were the provisions for cooperation between the public and private interests and between state and federal governments. At the heart of these provisions was the Northern States Power Company's commitment

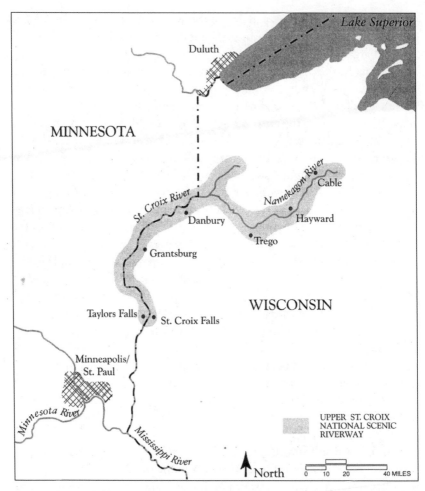

Map 9. The original St. Croix National Scenic Riverway, 1968 (Source: National Park Service).

to enter into a cooperative agreement with the National Park Service and the states of Wisconsin and Minnesota. The National Park Service received 7,000 acres of corporate riverfront lands. Minnesota received 13,000 acres and Wisconsin 5,000 acres for state parks along the river. The privatization of the St. Croix, a process that had begun with the first land sales in the 1830s, came to an end in 1968, and public ownership became the new trend. The real losers in the plan were several hundred owners of riverfront vacation homes, people who loved the river and knew it intimately. They were forced to sell their properties in order to create a 400-foot use area for

Figure 23. From left: Senator Gaylord Nelson (Wisconsin), Wisconsin governor Patrick Lucey, and Senator Walter F. Mondale (Minnesota) inspect new construction on the river at the time of the creation of the Lower St. Croix National Scenic Riverway, October 1972. Wisconsin Historical Society, WHi-56794.

canoeists, fishermen, and campers. Once and for all, plans for dams and industry were halted on the Upper St. Croix. The banks of the river became a national park, reserved for outdoor recreation.[140]

Only at the last minute and with great reluctance did Senator Nelson agree to drop the Lower St. Croix from inclusion in the Wild and Scenic Rivers Act. He did, however, ensure that the act directed the Department of Interior to study the lower river for future inclusion in the wild and scenic river system. When that study was concluded, Nelson and his Senate co-sponsor, Walter F. Mondale of Minnesota, had a proposal that called for managing the high bluffs and rocky outcrops of the river between Taylors Falls and Stillwater as a "scenic" zone, while the river below would be designated a "recreational" zone. Unlike the canoe-friendly waters of the upper river, the Lower St. Croix was dominated by power boat enthusiasts. Local townspeople and thousands of Twin City residents flocked to the river each summer weekend for skiing and swimming. For them, their boats were like a summer home, and the river was their playground. Under the Nelson-Mondale bill the recreational use of the river would be protected and managed by the National Park Service "in cooperation with the states of Minnesota and Wisconsin and local units of government." These same agencies would cooperate to manage "commercial development." Protection was to be achieved by limited land acquisition in fee simple and a broad program of zoning control and scenic easements.[141]

Urgency spurred this legislative initiative. In October 1970 the Calder Corporation of St. Paul announced plans to build a series of ten-story high-rise condominiums along the high bluff of the river at Hudson, Wisconsin. The project would double the size of the town and constitute a major scenic intrusion. Many people rightly saw the proposal as merely a sign of things to come. Real estate in the Lower St. Croix was red hot as scores of new bedroom communities took root. For successful executives and their doctors and lawyers, the St. Croix River was an attractive residential option offering recreation, scenery, and proximity to the city. The Calder project in Hudson was a warning that, as one activist put it, "this place was up for sale and was going to the highest bidder." Such a realization motivated Nelson and Mondale, yet while it created an opportunity for action, the monied property interests involved necessitated caution as well.[142]

Senator Mondale was particularly sensitive to the concerns of St. Croix frontage owners. His wife was from the river town of Afton, Minnesota, and he had many friends who lived and boated along the river. At a 1971 hearing on a lower river protection bill, he caught an earful from them because his bill called for "taking too much fee property." Gamely, Mondale tried to rally support for land protection. "I love the St. Croix," he said. "I want to protect it for the people who live there and for the people who want to use it responsibly so as not to destroy that area of the river . . . but the crush is on . . . and unless we can protect it some way, it's going to go." Saving his best argument for the end, he reminded his audience, "Then everyone's property values will be destroyed."[143]

Property rights objections became even more of a concern when James Watt became the head of the Bureau of Outdoor Recreation, the federal agency that had taken the lead in planning wild river protection. Watt, who years later as Ronald Regan's secretary of the interior would become an enemy of environmentalists and a champion of property holders, believed that the Lower St. Croix plan drafted by his department called for too much federal involvement. He justified attacking his own planners on the grounds of President Richard M. Nixon's "New Federalism" program, which called for returning power from Washington, D.C., to state and local governments. Watt may have succeeded in sinking the St. Croix initiative all together had not a little old-fashioned politics trumped his ideological stance. The Lower St. Croix proposal affected the districts of two young Republican congressmen, each of whom faced tough reelection races. Congressmen Vernon Thomson and Albert Quie could not afford James Watt's ideological purity; they wanted a St. Croix bill for their constituents. They went over Watt's head to Secretary of the Interior Rogers B. Morton, a man who knew the value of holding congressional seats. Together they redrafted the St. Croix bill so it looked like the New Federalism. The Republican compromise divided the lower river between the federal "scenic zone" and the "recreational" zone, from Stillwater to the Mississippi, which would be jointly managed by Wisconsin and Minnesota.[144]

In 1972, with Minnesota and Wisconsin politicians, both Republicans and Democrats, all in favor of "saving" the Lower St. Croix, the Lower St. Croix National Scenic Riverway Act was passed. It was in many ways a pioneering piece of legislation since it bore little resemblance to a traditional national park, where the federal government owned and managed most of the land. Rather, it was a system of cooperation between government entities. On one hand, the bill's line of division between federal and state management zones further subdivided the valley, continuing a process that had begun with the creation of the first county boundaries in the 1830s. On the other hand, the act created a mechanism by which private and public concerns and local, state, and federal concerns about the river could be negotiated. In that sense the St. Croix National Scenic Riverway restored to the valley a unity that was natural.

One thing the creation of the scenic riverway did not achieve in 1972, however, was an end to controversy about the river. In fact, public management of the river ensured that public controversy would be the rule; such is the nature of democratic government. Within weeks of the passage of the Lower St. Croix bill, there were congressional efforts to amend it. Most of the legislative problems flowed from the Bureau of Outdoor Recreation's unrealistic cost projections for scenic easements, which turned out to be off by more than 100 percent. Once purchased, easements proved to be an administrative nightmare for the National Park Service, which found itself reviewing suburban homeowners' improvement plans, as well as marina construction proposals, bridge construction, and boating congestion issues.[145]

The Lower St. Croix National Scenic Riverway did help to create a forum by which suburban sprawl could be checked along the river. Still, it was up to each municipality along the river to embrace or reject the partnership opportunities the new park offered. Emblematic of the way many river towns responded was the experience of Stillwater. Rather than be swallowed up in the Twin Cities' commuter world, Stillwater sought to retain its historic charm and identity, as well as cater to tourists. By the mid-1970s the town began an ambitious restoration program. The buildings and homes from the era of the lumber barons and lumberjacks were painstakingly restored and renovated into shops, restaurants, and museums. "This combination of natural and man-made attractions, plus some carefully orchestrated festivals," wrote the *Pioneer Press,* "have created in Stillwater an increase in tourism unlike most other towns in the state." Picnickers now enjoyed the spot where the old Stillwater boom collected its logs. The town's annual "Lumberjack Days" and revived steamboat excursions celebrated the river's heritage. Reflecting on the impact of the scenic riverway, journalists gushed, "The 'birth place of Minnesota' now is experiencing a robust rebirth of its own."[146]

Scenic river status, however, also brought a loss of control to Stillwater and other river towns. In 1989, for example, a proposal to hold an air show along the Stillwater riverfront was opposed by the National Park Service as against the spirit of the Wild and Scenic Rivers Act. A chance to attract an estimated 250,000 tourists to the town was lost when federal opposition led to the withdrawal of the Air Force Thunderbolts, a cornerstone of the entertainment. The controversy weakened and deeply divided the Minnesota–Wisconsin Boundary Area Commission, a key partner in managing the lower river. At the same time, the National Park Service was at loggerheads with the Minnesota Department of Transportation, which wanted to replace a historic steel truss lift bridge in Stillwater. Some local people wanted a new bridge that would reduce traffic in the town's historic downtown. Others preferred to keep the old bridge as a key component in that historic district. The dispute dragged on for more than a decade. Park Service involvement in the issue underscored the degree to which towns like Stillwater had to face national scrutiny when making what had once been local decisions.[147]

One of the National Park Service's management partners on the Upper St. Croix has been the St. Croix Band of the Lake Superior Ojibwe. This tribe controls a 137-acre tract along the river that has been the subject of an on-again, off-again cooperative agreement with the Park Service. More important to the development of tourism in the valley, however, has been the tribe's establishment of gaming enterprises at Danbury and Turtle Lake. While this type of recreation is not based upon an enjoyment of the region's protected natural resources, it has added diversification to the valley's tourist economy. Long pitied or belittled by white neighbors, the St. Croix Ojibwe have become major players in the local economy. In Burnett County, Wisconsin, they have become the largest private employers, with more than 2,500 people, Indian and non-Indian, working for various tribal enterprises.[148]

By 1993 the well-entrenched tourist industry along the upper river prompted the Department of Natural Resources in Wisconsin to begin the Northern Initiatives Project for northern Wisconsin. This project was the result of an internal review of the DNR that revealed that the agency played a greater role in the economic well-being of the northern half of the state than in the south. Decisions made by the DNR relating to deer permits, fishing bag limits, and the like had a profound impact on which areas tourists chose for their vacation destination. The DNR invited the residents of northern Wisconsin to share in the decision making of how to manage the state's natural resources. The initiative concluded that northern Wisconsin had successfully transformed itself into:

> a unique and distinguishable regional entity . . . with its reputation for clean air, water, healthy forests and abundant public opportunities. It is a place where sound science shapes environmental policy, guides sustainable management and ensures expenditures that yield commensurate environmental benefits.[149]

In July 1997 *National Geographic* commented on the changing face of the valley. "Today the St. Croix is an asset for its own sake, no longer for its wild rice or furs or timber, nor for other purposes of man's pocketbook, but now for his silence and his soul."[150]

The effort to preserve the St. Croix Valley continues through the efforts of a group of volunteers who formed the Friends of the St. Croix Headwaters (FOTSCH) in the fall of 2004. Its stated purpose is to protect the headwaters of the St. Croix River basin, left out of the national riverway legislation. In November 2006 the Totogatic River achieved the status of Outstanding and Exceptional Resource Waters for its excellent water quality, high recreational and aesthetic value, and high quality of fishing. FOTSCH's next goal is to have the Totogatic added to the state's wild and scenic river designation. The group also recognizes the economic importance of tourism to the region and seeks to find a balance between protecting the environment and local culture on one hand and generating income and employment on the other.[151]

At the dawn of another century and millennium, the recreational experience along the St. Croix National Scenic Riverway is perhaps more structured than at any time in its history. Local, state, and federal park units have imprinted the landscape with standardized signage, trails, structures, and campgrounds. Private concession operators supplement these public resources with additional recreational facilities and tourist amenities. Developed areas have been "returned" to their "natural" states in an attempt to enhance the riverway's "wild" and "scenic" character.[152]

These artificially created parks, preserves, and resorts became valued for their "natural" characteristics. When Americans took to the woods, they took along many of the comforts of home and generally expected to find modern conveniences. Throughout the country, entrepreneurs began to oblige their patrons with transportation to resort areas, developed campgrounds, well-equipped housekeeping cabins, and provisions. Scenic vistas, monumental landscapes, and the experiences they evoked became commodities for sale. Tourism was also not a natural product, as Aaron Shapiro has noted, but "is developed, managed, and packaged by people and organizations with often competing interests." It has been dependent upon "paid vacations, better roads, improved lodging, promotion, and people's desires to escape the heat and toil of cities during the summer." Rather than a remote, pristine retreat, the North Woods continues to be part of the economic web connecting city and country, people and commodities.[153]

The St. Croix is not a wild river, but Congress has designated it a scenic river, and its scenic qualities constitute its principle economic asset. That asset is today managed with a rigor that would have been unimaginable to the American Fur Company and that would astonish even an organizational genius like Frederick Weyerhaeuser. But it is useful to recall that the valley has long been managed by its human population, from the family hunting zones established by the Ojibwe to the 160-acre farms that sprang from federal homestead law and immigrant dreams. Our management reflects our desires and our values. Over time, from furs to lumber to wheat to dairy to scenery, we have varied how we manage the valley based upon our economic needs. The Bible says the "Earth abides," and history teaches us that people change. The landscape of the beautiful St. Croix River is a product of abiding, persistent nature and the restless, changeable human societies that have called the valley home. The St. Croix is what we have made it, and it will be what we dream it to become.

Nearly a century later, Ray Stannard Baker's vision of a modern life balanced between urban amenities and a natural landscape has come to fruition with the formation of the St. Croix National Scenic Riverway. The St. Croix has, in many ways, been the quintessential Upper Midwest river—abundant in resources, used, abused, yet valued, loved, restored, and ever argued over. Her story flows from a contradictory nation's heartland, swirling in the economic, political, or cultural mainstream of a flawed, stolid, ever changing people's quest to build a new world.

Notes

Introduction

1. Ray Stannard Baker [David Grayson, pseud.], *Adventures in Contentment* (New York: Doubleday, 1907), 1.

2. James Taylor Dunn, *The St. Croix: Midwest Border River* (1965; reprint, St. Paul: Minnesota Historical Society Press, 1979); W. H. C. Folsom, *Fifty Years in the Northwest*, ed. E. E. Edwards (1888; reprint, Taylors Falls, Minn.: Taylors Falls Historical Society, 1999), 34.

3. Dunn, *St. Croix: Midwest Border River*, 5.

4. Frederick Jackson Turner, *The Frontier in American History* (New York: Henry Holt & Co., 1920), 1–38; C. Vann Woodward, *The Burden of Southern History* (Baton Rouge: Louisiana State University Press, 1993), 3–26.

5. Patricia Nelson Limerick, "What on Earth Is the New Western History?" in *Trails: Toward a New Western History*, ed. Patricia Nelson Limerick, Clyde A. Milner II, and Charles E. Rankin (Lawrence: University Press of Kansas, 1991), 81.

6. Folsom, *Fifty Years in the Northwest*, 34.

7. James Grey, quoted in Walter A. Rowlands, "The Great Lakes Cutover Region," in *Regionalism in America*, ed. Merrill Jensen (Madison: University of Wisconsin Press, 1965), 334.

Chapter 1. Valley of Plenty, River of Conflict

1. Throughout the book we use the terms *Dakota* and *Ojibwe* to describe the two principal Indian groups who lived in the St. Croix Valley. The Dakota (a name they used to describe themselves that roughly translates as "friends") are perhaps better known by the name *Sioux*, a disparaging term derived from the Algonquin Indian word *na-towe-ssiwa*, spelled by the French as *Naudoweissious*, which means "people of an alien tribe" (Guy Gibbon, *The Sioux: The Dakota and Lakota Nations* [Malden, Mass.: Blackwell, 2003], 2–3). The Sioux were composed of several groups united by language, tradition, and geography, including the Lakota in the Missouri River area to the west and the Dakota in the Minnesota and Mississippi river valleys to the east. The Dakota group most consistently involved

in the history of the St. Croix was the *Mdewakanton.* During the nineteenth century they were the only Dakota group reported living or hunting along the river. Their Algonquin enemies are here referred to as the *Ojibwe,* a word that means "puckered up," in reference to a distinctive style of moccasin. The name *Chippewa,* which has dominated official and anthropological literature, has a similar meaning, as does *Ojibwa,* which is more commonly used to describe related groups of people in Canada. The Ojibwe usually refer to themselves as *Anishinabe,* or "original people" (for a history of the Anishinabe/Ojibwe, see Barry M. Pritzker, *A Native American Encyclopedia: History, Culture, and Peoples* [New York: Oxford University Press, 2000], 406–9).

2. Henry Rowe Schoolcraft, *Narrative Journal of Travels through the Great Chain of American Lakes to the Sources of the Mississippi River in the Year 1820* (Albany: E. & F. Hosford, 1821), 309–12 (quote on 309); Zebulon M. Pike, *An Account of Expeditions to the Sources of the Mississippi and Through Parts of Louisiana, to the Sources of the Arkansaw, Kans, La Platte, Pierre Jaun, Rivers* (Philadelphia: C. & A. Conrad, 1810), 94–95.

3. Nicholas Perrot, "Memoir on the Manners, Customs, and Religion of the Savages of North America," in *The Indian Tribes of the Upper Mississippi Valley and the Region of the Great Lakes,* ed. Emma Helen Blair (Lincoln: University of Nebraska Press, 1996), 119.

4. Elliott West, *The Contested Plains: Indians, Goldseekers, and the Rush to Colorado* (Lawrence: University of Kansas Press, 1998), 17; Ruth Landes, *The Mystic Lake Sioux: Sociology of the Mdewakantonwan Santee* (Madison: University of Wisconsin Press, 1968), 28.

5. Perrot, "Memoir on the Manners," 130–31.

6. Ibid.; Gary Clayton Anderson, *Kinsmen of Another Kind: Dakota-White Relations in the Upper Mississippi Valley, 1650–1862* (St. Paul: Minnesota Historical Society, 1997), 6–7; Louis Hennepin, *A Description of Louisiana,* ed. John Gilmary Shea (New York: John G. Shea, 1880), 227–39; Jonathan Carver, *Travels through the Interior Parts of North America in the Years 1766, 1767, and 1768* (London: J. Walter and S. Crowder, 1778), 264.

7. Landes, *Mystic Lake Sioux,* 195–97.

8. Daniel S. Wovcha, Barbara C. Delaney, and Gerda E. Nordquist, *Minnesota's St. Croix River Valley and Anoka Sandplain* (Minneapolis: University of Minnesota Press, 1995), 19; William W. Warren, *History of the Ojibway People* (St. Paul: Minnesota Historical Society Press, 1984), 97; Hennepin, *Description of Louisiana,* 322–23; Anderson, *Kinsmen of Another Kind,* 7.

9. Landes, *Mystic Lake Sioux,* 28–29; Perrot, "Memoir on the Manners," 120–21; Hennepin, *Description of Louisiana,* 322–23; Paul Kane, quoted in Shepard Krech III, *The Ecological Indian: Myth and Reality* (New York: Norton, 1999), 131.

10. Landes, *Mystic Lake Sioux,* 162–70; Hennepin, *Description of Louisiana,* 323.

11. Hennepin, *Description of Louisiana,* 323, 244–45.

12. Perrot, "Memoirs on the Manners," 103; Henry Rowe Schoolcraft, *Schoolcraft's Expedition to Lake Itasca: The Discovery of the Source of the Mississippi,* ed. Philip P. Mason (East Lansing: Michigan State University Press, 1958), 90; Joseph N. Nicollet, *Journals of Joseph N. Nicollet: A Scientist on the Mississippi Headwaters, with Notes on Indian Life, 1836–37,* ed. Martha Coleman Bray (St. Paul: Minnesota Historical Society, 1970), 146; Samuel W. Pond, "The Dakotas or Sioux of Minnesota as They Were in 1834," *Minnesota Historical Society Collections* 12 (1889): 345.

13. S. W. Pond, "Dakotas or Sioux of Minnesota," 345.

14. Anderson, *Kinsmen of Another Kind,* 2–3, 13; Pierre de Charlevoix, *Journal of a Voyage to North America* (London: R. and J. Dodsley, 1761), 1:280.

15. Pierre Esprit Radisson, *Voyages of Peter Esprit Radisson,* ed. Gideon Scull (New York: Peter Smith, 1943), 201–9.

16. Perrot, "Memoir on the Manners, 159–65; R. David Edmunds and Joseph L. Peyser, *The Fox Wars: The Mesquakie Challenge to New France* (Norman: University of Oklahoma Press, 1993), 16–17; William W. Warren, *History of the Ojibway People* (St. Paul: Minnesota Historical Society, 1984), 129–30.

17. Warren, *History of the Ojibway People,* 120; David L. Fritz, "Historic Resource Study: St. Croix National Scenic Riverway," unpublished draft (Omaha, Neb.: National Park Service, July 1989), 14.

18. "Geographical Names in Wisconsin," *Wisconsin Historical Collections* 1 (1849): 113; James Taylor Dunn, *The St. Croix: Midwest Border River* (1965; reprint, St. Paul: Minnesota Historical Society Press, 1979), 28–29; *Stillwater Messenger,* 25 July 1917; *Stillwater Messenger* clipping in the Hjalmar Otto Peterson Papers, Minnesota Historical Society, St. Paul; Jean Baptiste Bénard de La Harpe, "Le Suer's Voyage up the Mississippi," *Wisconsin Historical Collections* 16 (1902): 185–86.

19. Meyer, *History of the Santee Sioux,* 11–13; Anderson, *Kinsmen of Another Kind,* 36–37.

20. Anderson, *Kinsmen of Another Kind,* 36.

21. Meyer, *History of the Santee Sioux,* 10–11; Hennepin, *Description of Louisiana,* 202–3; Richard White, "The Winning of the West: The Expansion of the Western Sioux in the Eighteenth and Nineteenth Centuries," *Journal of American History* 65, no. 12 (1978): 319–43.

22. Anderson, *Kinsmen of Another Kind,* 27.

23. Ibid., 22–23.

24. Warren, *History of the Ojibway People,* 80–86.

25. Ibid., 155–62; Anderson, *Kinsmen of Another Kind,* 46–49.

26. Anderson, *Kinsmen of Another Kind,* 44–49.

27. Ibid., 51–54; Kenneth P. Bailey, ed. and trans., *Journal of Joseph Marin, French Colonial Explorer and Military Commander in the Wisconsin Country, August 7, 1753–June 20, 1754* (n.p.: privately printed, 1975), 93–94.

28. Bailey, *Journal of Joseph Marin, Junior, 1753 and 1754,* 72–75, 93–97.

29. Arthur S. Morton, *A History of the Canadian West to 1870–71* (Toronto: University of Toronto Press, 1973), 231–35; Grace Lee Nute, "Marin versus La Verendrye," *Minnesota History* 32, no. 4 (December 1951): 226–38; Francis Parkman, *Montcalm and Wolfe* (Boston: Little, Brown, 1908), 85–86.

30. Anderson, *Kinsmen of Another Kind,* 59; Alexander Henry, *Travels and Adventures in Canada and the Indian Territories between the Years 1760 and 1776* (Edmonton, Alberta: M. G. Hurtig, 1969), 187–88, 194–200.

31. Warren, *History of the Ojibway People,* 243–44.

32. Ibid., 244–49.

33. Ibid., 10, 220, 279–97, 337; A. Henry, *Travels and Adventures,* 153.

34. Melissa Connor et al., *Archeological Investigations along the St. Croix National Scenic Riverway, 1983* (Lincoln, Neb.: National Park Service, 1985), 320–40.

35. Carver, *Travels through the Interior Parts of North America in the Years 1766, 1767, and 1768,* 104–6, 476–77; James Stanley Goddard, "Journal of a Voyage, 1766–67," ed. Carolyn Gilman, in *The Journals of Jonathan Carver and Related Documents,* ed. John Parker (St. Paul: Minnesota Historical Society Press, 1976), 190.

36. Carver, *Travels through the Interior Parts of North America,* 100, 104–6.

37. Ibid., 476–77, 533; William E. Lass, *Minnesota: A History* (New York: Norton, 1998), 67–69; Stephen H. Long, *The Northern Expeditions of Stephen H. Long: The Journals of 1817 and 1823 and Related Documents,* ed. Lucile M. Kane, June Holmquist, and Carolyn Gilman (St. Paul: Minnesota Historical Society Press, 1978), 50, 96.

38. Peter Pond, "Narrative of Peter Pond," in *Five Fur Traders of the Northwest,* ed. Charles M. Gates (St. Paul: Minnesota Historical Society, 1965), 39–66; Charles Gautier de Verville, "Gautier's Journal of a Visit to the Mississippi, 1777–1778," *Wisconsin Historical Collections* 11 (1888): 102–5.

39. Augustin Grignon, "Seventy-two Years' Recollections of Wisconsin," *Wisconsin Historical Collections* 3 (1857): 244–47, 288; Harold Hickerson, *Ethnohistory of the Chippewa in Central Minnesota* (New York: Garland Publishing, 1974), 153–56.

40. George Nelson, *A Winter in the St. Croix Valley: George Nelson's Reminiscences, 1802–03,* ed. Richard Bardon and Grace Lee Nute (St. Paul: Minnesota Historical Society, 1948); Michel Curot, "A Wisconsin Fur-Trader's Journal, 1803–04," *Collections of the State Historical Society of Wisconsin* 20 (1911): 396–471; Thomas Connor [pseud. of John Sayer], "The Diary of Thomas Connor," in *Five Fur Traders of the Northwest,* ed. Charles M. Gates (St. Paul: Minnesota Historical Society, 1965), 243–78. The alleged Thomas Connor journal is unsigned and has since been generally accepted to be the work of John Sayer, a veteran partner in the North West Company. For more details see Douglas A. Birk, *John Sayer's Snake River Journal, 1804–1805: A Fur Trade Diary from East Central Minnesota* (St. Paul: Institute for Minnesota Archaeology, 1989).

41. Curot, "Wisconsin Fur-Trader's Journal," 465; Birk, *John Sayer's Snake River Journal,* 7; Nelson, *Winter in the St. Croix Valley,* 43; Bruce M. White, "The

Regional Context of the Removal Order of 1850," in *Fish in the Lakes, Wild Rice, and Game in Abundance: Testimony on Behalf of Mille Lacs Ojibwe Hunting and Fishing Rights,* ed. James M. McClurken (East Lansing: Michigan State University Press, 2000), 203.

42. Nelson, *Winter in the St. Croix Valley,* 21; Connor [Sayer], "Diary of Thomas Connor," 266–67.

43. Curot, "Wisconsin Fur-Trader's Journal," 416, 420, 438; Nelson, *Winter in the St. Croix Valley,* 20.

44. Curot, "Wisconsin Fur-Trader's Journal," 439; Nelson, *Winter in the St. Croix Valley,* 21.

45. Nelson, *Winter in the St. Croix Valley,* 44, 46.

46. Curot, "Wisconsin Fur-Trader's Journal," 451; Connor [Sayer], "Diary of Thomas Connor," 255; Birk, *John Sayer's Snake River Journal,* 32.

47. Francois Victor Malhiot, "A Wisconsin Fur-Trader's Journal, 1804–05," *Wisconsin Historical Collections* 19 (1910): 195–96; Nelson, *Winter in the St. Croix Valley,* 46; Curot, "Wisconsin Fur-Trader's Journal," 433, 464.

48. Curot, "Wisconsin Fur-Trader's Journal," 423.

49. George Nelson, *My First Years in the Fur Trade: The Journals of 1802–1804,* ed. Laura Peers and Theresa M. Schenck (St. Paul: Minnesota Historical Society Press, 2002), 91.

50. Sylvia Van Kirk, *Many Tender Ties: Women in Fur-Trade Society, 1670–1870* (Norman: University of Oklahoma Press, 1980), 91–92; Nelson, quoted in ibid., 91; Connor [Sayer], "Diary of Thomas Connor," 270.

51. Birk, *John Sayer's Snake River Journal,* 29; Curot, "Wisconsin Fur-Trader's Journal," 412, 418, 421–22.

52. Jacqueline Peterson and Jennifer S. H. Brown, *The New Peoples: Being and Becoming Métis in North America* (Lincoln: University of Nebraska Press, 1985), 63; Curot, "Wisconsin Fur-Trader's Journal," 447; Birk, *John Sayer's Snake River Journal,* 24–25, 31.

53. Johann Georg Kohl, *Kitchi-Gami, Life Among the Lake Superior Ojibway* (St. Paul: Minnesota Historical Society Press, 1985), 111; Ruth Landes, *The Ojibway Women* (New York: Columbia University Press, 1938), 66; Warren, *History of the Ojibway People,* 299.

54. Robert E. Ritzenthaler, "Southwestern Chippewa," in *Handbook of North American Indians,* vol. 15, *Northeast,* ed. Bruce G. Trigger (Washington, D.C.: Smithsonian Institution, 1978), 747–49.

55. Ibid., 754–55; Warren, *History of the Ojibway People,* 100–101; Harold Hickerson, *The Chippewa and Their Neighbors: A Study in Ethnohistory* (New York: Holt, Rhinehart, 1970), 52–63.

56. Jeanne Kay, "The Land of La Baye: The Ecological Impact of the Green Bay Fur Trade, 1634–1836" (Ph.D. diss., University of Wisconsin–Madison, 1977), 31, 264; Wovcha, Delaney, and Nordquist, *Minnesota's St. Croix River Valley,* 19–20.

57. Alice Outwater, *Water: A Natural History* (New York: Basic Books, 1996), 19–31; C. A. Johnson, J. Pastor, and R. J. Naiman, "Effects of Beaver and Moose on Boreal Forest Landscapes," in *Landscape Ecology and Geographic Information Systems,* ed. Roy Haines-Young, David R. Green, and Steven Cousins (London: Taylor & Francis, 1994), 237–39.

58. Johnson, Pastor, and Naiman, "Effects of Beaver and Moose," 246–48.

59. Anderson, *Kinsmen of Another Kind,* 109.

60. W. H. C. Folsom, *Fifty Years in the Northwest,* ed. E. E. Edwards (1888; reprint, Taylors Falls, Minn.: Taylors Falls Historical Society, 1999), 233.

61. Harold A. Innis, *The Fur Trade in Canada* (Toronto: University of Toronto Press, 1956), 186–88.

62. Warren, *History of the Ojibway People,* 9–10; Bruce White, *The Fur Trade in Minnesota: An Introductory Guide to Manuscript Sources* (St. Paul: Minnesota Historical Society Press, 1977), 36, 41, 57.

63. Letter, Louis Grignon to John Lawe, 10 January 1820, *Wisconsin Historical Collections* 20 (1911): 146–47.

64. Hickerson, *Chippewa and Their Neighbors,* 74–75; Hickerson, *Ethnohistory of the Chippewa,* 201–2, 209; Birk, *John Sayer's Snake River Journal,* 11.

65. Frances Densmore, *Chippewa Customs* (Washington, D.C.: Smithsonian Institution, 1929), 132–35; Warren, *History of the Ojibway People,* 26, 129, 139, 355; Bailey, *Journal of Joseph Marin,* 95; Nelson, *Winter in the St. Croix Valley,* 45.

66. Warren, *History of the Ojibway People,* 163–74.

67. Ibid., 174; Hickerson, *Ethnohistory of the Chippewa,* 101–7; Nelson, *Winter in the St. Croix Valley,* 33–34.

68. Harold Hickerson, *Mdewakanton Band of Sioux Indians* (New York: Garland Publishing, 1974), 169–71.

69. Thomas Forsyth, "Journal of a Voyage from St. Louis to the Falls of St. Anthony, in 1819," *Wisconsin Historical Collections* 6 (1908): 213.

70. Anderson, *Kinsmen of Another Kind,* 82–83; Zebulon Montgomery Pike, *Sources of the Mississippi and the Western Louisiana Territory* (Baltimore: Fielding Lucas, 1810), 24–26.

71. Henry Rowe Schoolcraft, *Travels through the Northwestern Regions of the United States* (Albany: E & F. Hosford, 1821), 304–7.

72. "Treaty with the Sioux, etc., 1825," in *Indian Treaties, 1778–1883,* ed. Charles J. Kappler (New York: Interland Publishing, 1972), 250–55.

73. Anderson, *Kinsmen of Another Kind,* 135–37; Schoolcraft, *Schoolcraft's Expedition to Lake Itasca,* 84; Charles E. Cleland, "Preliminary Report of the Ethnohistorical Basis of the Hunting, Fishing, and Gathering Rights of the Mille Lacs Chippewa," in *Fish in the Lakes, Wild Rice, and Game in Abundance,* ed. James M. McClurken (East Lansing: Michigan State University Press, 2000), 24.

74. Nancy Goodman and Robert Goodman, *Joseph R. Brown: Adventurer on the Minnesota Frontier, 1820–1849* ([Rochester, Minn.]: Lone Oak Press, 1996), 128–30.

75. Ibid., 105–6; Schoolcraft, *Schoolcraft's Expedition to Lake Itasca,* 84; Anderson, *Kinsmen of Another Kind,* 138.

76. Schoolcraft, *Schoolcraft's Expedition to Lake Itasca,* 84; Goodman and Goodman, *Joseph R. Brown,* 109.

77. William Johnson to Jane Schoolcraft, 23 October 1833, *Michigan Pioneer and Historical Collections* 37 (1909): 198–99.

78. Anderson, *Kinsmen of Another Kind,* 107–8; Schoolcraft, *Travels through the Northwestern Regions,* 311–13, 323; Hickerson, *Mdewakanton Band of Sioux Indians,* 172–73.

79. Schoolcraft, *Schoolcraft's Expedition to Lake Itasca,* 51, 88; Goodman and Goodman, *Joseph R. Brown,* 107; Edmund Danziger, *The Chippewas of Lake Superior* (Norman: University of Oklahoma Press, 1979), 71.

80. "Treaty with the Chippewa, 1837," in Kappler, *Indian Treaties,* 491–92; Cleland, "Preliminary Report of the Ethnohistorical Basis," 30.

81. Henry Rowe Schoolcraft, *Personal Memoirs of a Residence of Thirty Years with the Indian Tribes on the American Frontiers* (Philadelphia: Lippincot, Grambo, 1851; repr., New York: Arno Press, 1975), 29–30.

82. Anderson, *Kinsmen of Another Kind,* 146, 148, 153–55; "Treaty with the Sioux, 1837," in Kappler, *Indian Treaties,* 493–94.

83. "Treaty with the Chippewa, 1837," 492; Cleland, "Preliminary Report of the Ethnohistorical Basis," 31.

84. Folsom, *Fifty Years in the Northwest,* 235–37, 263–66.

85. Ibid., 270–71; J. N. Davidson, *In Unnamed Wisconsin: Studies in the History of the Region Between Lake Michigan and the Mississippi* (Milwaukee: Silas Chapman, 1895), 160–61.

86. Folsom, *Fifty Years in the Northwest,* 268; Davidson, *In Unnamed Wisconsin,* 161; Warren, *History of the Ojibway People,* 11–12.

87. Philander Prescott, *The Recollections of Philander Prescott: Frontiersman of the Old Northwest, 1819–1862,* ed. Donald Dean Parker (Lincoln: University of Nebraska Press, 1866), 171–72; Cleland, "Preliminary Report of the Ethnohistorical Basis," 23–24; Anderson, *Kinsmen of Another Kind,* 172–73; Samuel W. Pond, "Indian Warfare in Minnesota," *Minnesota Historical Society Collections* 3 (1880): 131–33; George Copway, *The Traditional History and Characteristic Sketches of the Ojibway Nation* (London: Charles Gilpin, 1850), 58–59.

88. Folsom, *Fifty Years in the Northwest,* 265–67; E. D. Neill, "Battle of Lake Pokegama," *Minnesota Historical Society Collections* 1 (1872): 141–44.

89. Davidson, *In Unnamed Wisconsin,* 162; Cleland, "Preliminary Report of the Ethnohistorical Basis," 43.

90. Cleland, "Preliminary Report of the Ethnohistorical Basis," 43.

91. O. M. Thatcher, "The Mission in Folle Avoine," Yellow Lake Pamphlet File, Burnett County Historical Society, Danbury, Wis., 3; *Stillwater Messenger,* 26 January 1877.

92. Folsom, *Fifty Years in the Northwest,* 87–88.

93. Ibid., 47.

94. Cleland, "Preliminary Report of the Ethnohistorical Basis," 44; Folwell, *History of Minnesota,* 1:209.

95. Folsom, *Fifty Years in the Northwest,* 98.

96. B. M. White, "Regional Context of the Removal Order," 165, 188, 214, 269.

97. Bruce M. White, "Indian Visit Stereotypes of Minnesota's Native People," *Minnesota History* 53 (Fall 1992): 103–5; Folsom, *Fifty Years in the Northwest,* 99.

98. B. M. White, "Regional Context of the Removal Order," 171, 202–4, 302–3.

99. *St. Croix Union,* 16 February 1855.

100. Anderson, *Kinsmen of Another Kind,* 177; *St. Croix Union,* 3 March 1855, 31 January 1856, 18 February 1856, 9 January 1857.

101. *St. Croix Union,* 2 February 1855, 19 December 1856.

102. Copway, *Traditional History and Characteristic Sketches,* 65–67.

103. Resolution 2264, in *St. Croix Indians of Wisconsin: Hearings before the U.S. House Committee on Indian Affairs, 23 July 1919* (Washington, D.C.: Government Printing Office, 1919), 3–8.

104. Photograph of the Clam River drive crew, 1902, Album 20.34a, Archives and Manuscripts Division, Wisconsin Historical Society, Madison.

105. Resolution 263, in *Condition of Indian Affairs in Wisconsin: Hearings before the Committee on Indian Affairs, United States Senate* (Washington, D.C.: Government Printing Office, 1910), 31.

106. James Fitting et al., *An Archeological Survey of the St. Croix National Scenic Riverway: Phase I Report* (Jackson, Mich.: Gilbert/Commonwealth, 1977), 68–69.

107. Ronald N. Satz, *Chippewa Treaty Rights: The Reserved Rights of Wisconsin's Chippewa Indians in Historical Perspective* (Madison: Wisconsin Academy of Sciences, 1991), 72–73.

108. Resolution 263, 16–27.

109. Paula Delfeld, *The Indian Priest: Father Philip B. Gordon, 1885–1948* (Chicago: Franciscan Herald Press, 1977).

110. Fitting et al., *Archeological Survey of the St. Croix National Scenic Riverway,* 35, 65–66.

111. Satz, *Chippewa Treaty Rights,* 98–100.

112. Proceedings of 1837 Chippewa Treaty, in Satz, *Chippewa Treaty Rights,* 142.

Chapter 2. River of Pine

1. George Nelson, *A Winter in the St. Croix Valley: George Nelson's Reminiscences, 1802–03,* ed. Richard Bardon and Grace Lee Nute (St. Paul: Minnesota Historical Society Press, 1948), 41; James Allen, "Journal and Letters of Lieutenant James Allen," in Henry Rowe Schoolcraft, *Schoolcraft's Expedition to Lake Itasca: The Discovery of the Sources of the Mississippi,* ed. Philip P. Mason (East Lansing: Michigan State University Press, 1958), 221–22.

2. John N. Vogel, *Great Lakes Lumber on the Great Plains: The Laird, Norton Lumber Company in South Dakota* (Iowa City: University of Iowa Press, 1992), vii–xiii.

3. William G. Rector, *Log Transportation in the Lake States Lumber Industry, 1840–1918* (Glendale, Calf.: Arthur H. Clarke Co., 1953), 90–94.

4. W. H. C. Folsom, *Fifty Years in the Northwest,* ed. E. E. Edwards (1888; reprint, Taylors Falls, Minn.: Taylors Falls Historical Society, 1999), 262–63; William W. Warren, *History of the Ojibway People* (St. Paul: Minnesota Historical Society Press, 1984), 18–20.

5. Folsom, *Fifty Years in the Northwest,* 81–82, 92–93, 303; Nancy Goodman and Robert Goodman, *Joseph R. Brown: Adventurer on the Minnesota Frontier, 1820–1849* ([Rochester, Minn.]: Lone Oak Press, 1996), 140–41.

6. Goodman and Goodman, *Joseph R. Brown,* 142, 140.

7. Folsom, *Fifty Years in the Northwest,* 99.

8. Ibid., 82–83; Edward W. Durant, "Lumbering and Steamboating on the St. Croix River," *Minnesota Historical Collections* 10, no. 2 (1905): 648–49; James Taylor Dunn, *The St. Croix: Midwest Border River* (1965; reprint, St. Paul: Minnesota Historical Society Press, 1979), 90–94.

9. Durant, "Lumbering and Steamboating," 649.

10. Goodman and Goodman, *Joseph R. Brown,* 142; Agnes M. Larson, *History of the White Pine Industry in Minnesota* (Minneapolis: University of Minnesota Press, 1949), 53–54.

11. A. M. Larson, *History of the White Pine Industry,* 71–78; Folsom, *Fifty Years in the Northwest,* 82–83, 92–93; James Fitting et al., *Archeological Survey of the St. Croix National Scenic Riverway: Phase 1 Report* (Jackson, Mich.: Gilbert/Commonwealth, 1977) 69; *St. Croix Union,* 6 March 1855.

12. *St. Croix Union,* 6 March 1855; A. M. Larson, *History of the White Pine Industry,* 79.

13. *St. Croix Union,* 6 March 1855.

14. *St. Croix Union,* 24 April 1855; Stephen B. Hanks, "Memoir of Captain S. B. Hanks," vol. 3, Minnesota Historical Society, St. Paul; A. M. Larson, *History of the White Pine Industry,* 72–73.

15. A. M. Larson, *History of the White Pine Industry,* 195; *St. Croix Union,* 6 March 1855.

16. A. M. Larson, *History of the White Pine Industry,* 127–28; James Bracklin, "A Tragedy of the Wisconsin Pinery," *Wisconsin Magazine of History* 3, no. 1 (September, 1919): 43–51.

17. Bracklin, "Tragedy of the Wisconsin Pinery," 43–51; Bruce M. White, "Regional Context of the Removal Order of 1850," in *Fish in the Lakes, Wild Rice, and Game in Abundance: Testimony on Behalf of Mille Lacs Ojibwe Hunting and Fishing Rights,* ed. James M. McClurken (East Lansing: Michigan State University Press, 2000), 204, 220, 284–87.

18. George B. Engberg, "Labor in the Lake States Lumber Industry, 1830–1930" (Ph.D. diss., University of Minnesota, 1949), 64–69; newspaper clipping dated 2 January 1930, Hjalmar Otto Peterson Papers, Minnesota Historical Society; A. M. Larson, *History of the White Pine Industry,* 74–75; William M. Blanding, "The Upper St. Croix Valley and Its Early Settlers," unpublished manuscript, ca. 1890, William H. Blanding Papers, Wisconsin Historical Society, River Falls Area Research Center.

19. *St. Croix Union,* 3 March 1855; Nils P. Haugen, "Pioneer and Political Reminiscences," *Wisconsin Magazine of History* 11, no. 2 (December 1927): 148–49; James Johnston, "Memoir," Wisconsin Historical Society, River Falls Area Research Center, 61, 72–73.

20. Robert F. Fries, *Empire in Pine: The Story of Lumbering in Wisconsin, 1830–1900* (Madison: State Historical Society of Wisconsin, 1951), 21.

21. A. M. Larson, *History of the White Pine Industry,* 19–23, 55–56, 63–64.

22. Ibid., 24; Folsom, *Fifty Years in the Northwest,* 43, 56–60.

23. Folsom, *Fifty Years in the Northwest,* 374–75; A. M. Larson, *History of the White Pine Industry,* 14–15.

24. A. M. Larson, *History of the White Pine Industry,* 88–89; Malcolm Rosholt, *The Wisconsin Logging Book, 1839–1939* (Rosholt, Wis.: Rosholt House, 1980), 190–95.

25. Rector, *Log Transportation in the Lake States,* 18.

26. A. M. Larson, *History of the White Pine Industry,* 88–91.

27. Ralph Clement Bryant, *Logging: The Principles and General Methods of Operation in the United States* (New York: John Wiley, 1914), 347–49.

28. Rector, *Log Transportation in the Lake States,* 40–41, 69, 110.

29. *St. Croix Union,* 2 December 1854; Rector, *Log Transportation in the Lake States,* 116–17.

30. Rector, *Log Transportation in the Lake States,* 121–25; *St. Croix Union,* 10 March 1855.

31. Rector, *Log Transportation in the Lake States,* 134–35.

32. Ibid., 115–16; William G. Rector, "The Birth of the St. Croix Octopus," *Wisconsin Magazine of History* 40, no. 3 (Spring 1957): 171–78.

33. Rector, "Birth of the St. Croix Octopus," 177.

34. *Taylors Falls Reporter,* 29 April 1865.

35. Fred C. Burke, *Logs on the Menominee: The History of the Menominee River Boom Company* (Marinette, Wis.: privately printed, 1946), 21; Rector, *Log Transportation in the Lake States,* 251–53; *Stillwater Messenger,* 2, 16 February 1872.

36. Rector, *Log Transportation in the Lake States,* 100; J. William Trygg, *Composite Map of United States Land Surveyor's Original Plats and Field Notes,* Minnesota Series (Ely, Minn.: J. Wm. Trygg, 1966).

37. *Stillwater Lumberman,* 18 June 1875.

38. *Polk County Press,* 4 March 1888; Rector, *Log Transportation in the Lake States,* 104–6; Namekagon River Improvement Company Articles of Organization,

15 November 1884, St. Croix County Register of Deeds Collection, Wisconsin Historical Society, River Falls Area Research Center; *Taylors Falls Reporter,* 30 October 1869; Bryant, *Logging,* 350.

39. *Taylors Falls Reporter,* 1 January 1885; *Burnett County Sentinel,* 12 March 1880, 29 April 1881.

40. W. H. C. Folsom, "Lumbering in the St. Croix Valley," *Minnesota Historical Collections* 10 (1905): 645–75; *Stillwater Lumberman,* 25 June 1875.

41. James Johnston, "Memoir," 61–63; Haugen, "Pioneer and Political Reminiscences," 150.

42. Haugen, "Pioneer and Political Reminiscences," 150; James Johnston, "Memoir," 60–64.

43. Haugen, "Pioneer and Political Reminiscences," 151.

44. James Johnston, "Memoir," 64; *Stillwater Lumberman,* 14 May 1875.

45. *Stillwater Lumberman,* 4 May 1877, 30 April 1875.

46. *Stillwater Lumberman,* 29 June 1877, 9 July 1875.

47. "What do you know about the *Lady Bernice,* the logging queen," Logging Clipping File, n.d., Burnett County Historical Society, Danbury, Wis.; *Stillwater Lumberman,* 14 May 1875.

48. *Northwestern Lumberman,* 8 May 1875; *Stillwater Lumberman,* 14 May 1875, 26 May 1877.

49. *Stillwater Lumberman,* 8 December 1877; Folsom, *Fifty Years in the Northwest,* 706–7.

50. Folsom, *Fifty Years in the Northwest,* 706–7.

51. Ibid.; *Burnett County Sentinel,* 21 June 1878.

52. *Taylors Falls Journal,* 10, 17, 24 June 1886; *Minneapolis Tribune,* 20 June 1886; *Polk County Press,* 26 June 1886.

53. *Polk County Press,* 10 July 1886; Rector, *Log Transportation in the Lake States,* 259–60; *Taylors Falls Journal,* 8, 15 July 1886.

54. Rosholt, *Wisconsin Logging Book,* 41; J. C. Ryan, *Early Loggers in Minnesota* (Duluth: Minnesota Timber Producers Association, 1975), 13.

55. Ryan, *Early Loggers in Minnesota,* 22–23, 27–30.

56. Ibid., 24–26; Rosholt, *Wisconsin Logging Book,* 95–102.

57. Ryan, *Early Loggers in Minnesota,* 3:41–43; Rosholt, *Wisconsin Logging Book,* 62–66.

58. Robert W. Wells, *"Daylight in the Swamp!"* (Garden City, N.Y.: Doubleday, 1978), 222–23.

59. Rector, *Log Transportation in the Lake States,* 164–5; *Taylors Falls Journal,* 19 January 1877, 30 November 1877, 17 August 1882.

60. Rosholt, *Wisconsin Logging Book,* 79–80.

61. *Stillwater Lumberman,* 3 December 1875, 18 May 1877; Fitting et al., *Archeological Survey of the St. Croix National Scenic Riverway,* 90.

62. *Mississippi Valley Lumberman,* 13 April 1877; Rector, *Log Transportation in the Lake States,* 204, 232.

63. *Mississippi Valley Lumberman,* 25 March 1887, 1 February 1895; A. M. Larson, *History of the White Pine Industry,* 153.

64. Folsom, "Lumbering in the St. Croix Valley," 645–75.

65. Folsom, *Fifty Years in the Northwest,* 413–14.

66. William G. Rector, "Lumber Barons in Revolt," *Minnesota History* 31, no. 1 (March 1950): 33–39.

67. Mark Wyman, *The Wisconsin Frontier* (Bloomington: Indiana University Press, 1998), 268–69; Fries, *Empire in Pine,* 148–55.

68. Eldon M. Marple, *The Visitor Writes Again* (Hayward, Wis.: The County Print Shop, 1984), 69–73, 87–95.

69. Ralph W. Hidy, Frank Ernest Hill, and Alan Nevins, *Timber and Men: The Weyerhaeuser Story* (New York: Macmillan, 1963), 96; Fred W. Kohlmeyer, *Timber Roots: The Laird, Norton Story, 1855–1905* (Winona, Minn.: Winona County Historical Society, 1972), 128.

70. Hidy, Hill, and Nevins, *Timber and Men,* 96–97; Kohlmeyer, *Timber Roots,* 130.

71. Kohlmeyer, *Timber Roots,* 127–28.

72. Rosholt, *Wisconsin Logging Book,* 152–53; Rector, *Log Transportation in the Lake States,* 107; Wyman, *Wisconsin Frontier,* 264.

73. Folsom, *Fifty Years in the Northwest,* 60–61; Augustus B. Easton, *History of the Saint Croix Valley* (Chicago: H. C. Cooper, Jr., 1909), 2:23–24; Rector, *Log Transportation in the Lake States,* 118, 124, 134, 144, 270; Kohlmeyer, *Timber Roots,* 130; Hidy, Hill, and Nevins, *Timber and Men,* 97.

74. Hidy, Hill, and Nevins, *Timber and Men,* 96–97, 157; *Polk County Press,* 26 February 1887, 2 March 1889; Rosemarie Vezina, *Nevers Dam: The Lumberman's Dam* (St. Croix Falls, Wis.: Standard-Press, 1965), 18–22.

75. In 1914 logging expert Ralph Clement described a bear-trap gate as composed of "two rectangular leaves each of which has a length equal to the width of the sluice. They are fastened to the bottom of the sluice by hinges on which they turn. The upstream leaf overlaps the downstream one when the leaves are down and the gate open." When the gate is open, it lies flat on the bed of the dam with the water rushing over it. A hand-operated wheel raised or lowered the gate. The advantage of this design is that one person working the gate is able to release a large amount of water very quickly. For more see Ralph Clement, *Logging: The Principles and General Methods of Operation in the United States* (New York: John Wiley, 1914), 355–57.

76. Ibid., 12–13; Rector, *Log Transportation in the Lake States,* 108; *Burnett County Sentinel,* 27 September 1889, 4 October 1889.

77. Vezina, *Nevers Dam,* 17–18; *Polk County Press,* 21 September 1889; Nevers Dam, National Register of Historic Places Registration Form, n.d., State Historic Preservation Office, Wisconsin Historical Society, Madison.

78. Vezina, *Nevers Dam,* 35–37; *Polk County Press,* 23 June 1883, 16 February 1884, 16 November 1889; Frederic Abbot to George Seymour, 18 December 1899,

William H. C. Folsom Papers, Wisconsin Historical Society, River Falls Area Research Center; miscellaneous newspaper clippings, Folsom Papers.

79. Kohlmeyer, *Timber Roots,* 175.

80. *Mississippi Valley Lumberman,* 5 June 1903, 11 December 1903.

81. James Willard Hurst, *Law and Economic Growth: The Legal History of the Lumber Industry in Wisconsin, 1836–1915* (Cambridge, Mass.: Harvard University Press, 1964), 382, 462; James Willard Hurst, "The Institutional Environment of the Logging Era in Wisconsin," in *The Great Lakes Forest: An Environmental and Social History,* ed. Susan L. Flader (Minneapolis: University of Minnesota Press, 1983), 151–55.

82. *Stillwater Messenger,* 16 March 1858, 12 July 1859, 30 August 1859; Hurst, *Law and Economic Growth,* 130; *Stillwater Lumberman,* 3 August 1877.

83. *Stillwater Lumberman,* 11 May 1877, 3 August 1877, 28 September 1877, 5 October 1877; Hidy, Hill, and Nevins, *Timber and Men,* 126.

84. Hidy, Hill, and Nevins, *Timber and Men,* 127; Hurst, *Law and Economic Growth,* 130–31; A. M. Larson, *History of the White Pine Industry,* 296–97; *Stillwater Lumberman,* 5 October 1877.

85. Hurst, *Law and Economic Growth,* 94–95; Kohlmeyer, *Timber Roots,* 141; Sharon Tarr, *Spooner: A History to 1930* (Spooner, Wis.: privately printed, 1976), 2.

86. Ryan, *Early Loggers in Minnesota,* 6–7; Albert C. Stuntz Diaries, 1858, 1863–1865, 1867–1869, 1882, Wisconsin Historical Society, River Falls Area Research Center; Kohlmeyer, *Timber Roots,* 140–42; Vezina, *Nevers Dam,* 13.

87. Stewart Holbrook, *Holy Old Mackinaw: A Natural History of the American Lumberjack* (Sausalito, Calf.: Comstock, 1938), 139; Robert W. Wells, *Fire at Peshtigo* (New York: Prentice Hall, 1968), 227–36.

88. Wells, *Daylight in the Swamp!* 133–34.

89. Easton, *History of the Saint Croix Valley,* 2:1247; Holbrook, *Holy Old Mackinaw,* 140–44.

90. Eldon M. Marple, *A History of the Hayward Lakes Region: Through the Eyes of the Visitor Who Came and Stayed* (Hayward, Wis.: Chicago Bay Grafix, 1976), 25–26; Kohlmeyer, *Timber Roots,* 142–43.

91. *Stillwater Lumberman,* 20 July 1877, 17 August 1877; *Burnett County Sentinel,* 9 May 1879.

92. *Stillwater Lumberman,* 20 July 1877.

93. Kohlmeyer, *Timber Roots,* 143; Hidy, Hill, and Nevins, *Timber and Men,* 148.

94. Lass, *Minnesota: A History* (New York: Norton, 1998), 244–45; Kohlmeyer, *Timber Roots,* 143.

95. James Johnston, "Memoir," 60; Hidy, Hill, and Nevins, *Timber and Men,* 150–51; John Ilmari Kolehmainen, "In Praise of the Finnish Backwoods Farmer," *Agricultural History* 24 (January 1950): 2; Engberg, "Labor in the Lake States Lumber Industry," 208–9.

96. Carl Kuhnly, "Homesteading in Wisconsin, Up Hill—Down Hill All The Way," pamphlet, 1985, Burnett County Historical Society, Danbury, Wis.; Vince

Pleska, "Lumberjack's Wife Axes Glory Myth," *St. Paul Sunday Pioneer Press,* 30 November 1980; "Local History of Joint District 5 Siren and Daniels," pamphlet, n.d., Burnett County Historical Society, Danbury, Wis.

97. Engberg, "Labor in the Lake States Lumber Industry," 67–87; Ryan, *Early Loggers in Minnesota,* 3:4–7.

98. *St. Croix Union,* 28 March 1856; Sherman E. Johnson, *From the St. Croix to the Potomac—Reflections of a Bureaucrat* (Bozeman: Montana State University, 1974), 28; Dunn, *St. Croix: Midwest Border River,* 105–6.

99. *Burnett County Sentinel,* 13 August 1880, 13 April 1888; Lewis C. Reimann, *Hurley—Still No Angel* (Ann Arbor, Mich.: Northwoods Publishers, 1954), 1–5; Eldon Marple, *The Visitor Who Came to Stay: Legacy of the Hayward Area* (Hayward, Wis.: Country Print Shop, 1971), 78–80; Fred Etcherson, oral history interview by David J. Olson, 30 July 1970, Wisconsin Historical Society, River Falls Area Research Center.

100. S. E. Johnson, *From the St. Croix to the Potomac,* 28.

101. *Burnett County Sentinel,* 19 January 1883; Kohlmeyer, *Timber Roots,* 273; Etcherson interview.

102. Etcherson interview.

103. *Stillwater Messenger,* 4 July 1930; Vogel, *Great Lakes Lumber on the Great Plains,* 13.

104. Clifford E. Ahlgren and Isabel F. Ahlgren, "The Human Impact on the Northern Forest Ecosystems," in *The Great Lakes Forest: An Environmental and Social History,* ed. Susan L. Flader (Minneapolis: University of Minnesota Press, 1983), 33–39.

105. Randall Rohe, "Lumbering, Wisconsin's Northern Urban Frontier," in *Wisconsin Land and Life,* ed. Robert C. Ostergren and Thomas R. Vale (Madison: University of Wisconsin Press, 1997), 121–30.

106. *Pine County Courier,* 21 March 1900; *Polk County Press,* 20 February 1892; Dunn, *St. Croix: Midwest Border River,* 103.

107. *Stillwater Gazette,* n.d., clipping file, Folsom Papers.

108. Kohlmeyer, *Timber Roots,* 177–78.

109. Hjalmar Otto Peterson, "Highlights from Stillwater History," 4 July 1930, clipping file, Hjalmar Otto Peterson Collection, Minnesota Historical Society, St. Paul.

110. Proceedings of a Council, 20 July 1837, National Archives and Records Service, RG 75, T494, Documents Relating to the Negotiation of Ratified and Unratified Treaties, Reel 3, 548–68.

Chapter 3. "The New Land"

1. Brian Page and Richard Walker, "From Settlement to Fordism: The Agro-Industrial Revolution in the American Midwest," *Economic Geography* (October 1991): 281–84. See this article for a discussion of the Northwest Land Ordinance and settlement patterns.

2. Hildegard Binder Johnson, "Towards a National Landscape," in *The Making of the American Landscape,* ed. Michael P. Conzen (Boston: Unwin Hyman, 1990).

3. History Network of Washington County, *Minnesota Beginnings: Records of Saint Croix County, Wisconsin Territory, 1840–1849* (Stillwater, Minn.: Washington County Historical Society, 1999), 2.

4. Nancy Goodman and Robert Goodman, *Joseph R. Brown: Adventurer on the Minnesota Frontier, 1820–1849* ([Rochester, Minn.]: Lone Oak Press, 1996), 152–56.

5. Ibid., 160–67; History Network, *Minnesota Beginnings,* 2; William E. Lass, *Minnesota: A History* (New York: Norton, 1998), 99.

6. Goodman and Goodman, *Joseph R. Brown,* 168–69.

7. Ibid.

8. Ibid., 174; History Network, *Minnesota Beginnings,* 5; Alice E. Smith, *From Exploration to Statehood* (Madison: State Historical Society of Wisconsin, 1973), 508.

9. Goodman and Goodman, *Joseph R. Brown,* 208.

10. History Network, *Minnesota Beginnings,* 6; W. H. C. Folsom, *Fifty Years in the Northwest,* ed. E. E. Edwards (1888; reprint, Taylors Falls, Minn.: Taylors Falls Historical Society, 1999), 52; David L. Fritz, "Historic Resource Study: St. Croix National Scenic Riverway," unpublished draft (Omaha, Neb.: National Park Service, July 1989), 170.

11. Dorothy Eaton Ahlgren and Mary Cotter Beeler, *A History of Prescott, Wisconsin: A River City and Farming Community on the St. Croix and Mississippi* ([Prescott, Wis.]: Prescott Area Historical Society, 1996), 14; History Network, *Minnesota Beginnings,* 6, 9.

12. History Network, *Minnesota Beginnings,* 9.

13. Ibid., 6, 9.

14. Ibid., 9; Mark Wyman, *Wisconsin Frontier* (Bloomington: Indiana University Press, 1998), 178.

15. Lass, *Minnesota,* 101; William E. Lass, "Minnesota's Separation from Wisconsin: Boundary Making on the Upper Mississippi Frontier," *Minnesota History* 50, no. 8 (Winter 1987): 310.

16. Lass, *Minnesota,* 102.

17. Folsom, *Fifty Years in the Northwest,* 34; Lass, *Minnesota,* 102.

18. A. E. Smith, *From Exploration to Statehood,* 663; Lass, *Minnesota,* 102–3; Lass, "Minnesota's Separation from Wisconsin," 310–16 (quotes on 312, 314).

19. Lass, *Minnesota,* 102–3; A. E. Smith, *From Exploration to Statehood,* 663; Lass, "Minnesota's Separation from Wisconsin," 310–16.

20. Lass, *Minnesota,* 103–4; Lass, "Minnesota's Separation from Wisconsin," 317–18; A. E. Smith, *From Exploration to Statehood,* 672–73; Folsom, *Fifty Years in the Northwest,* 104–5.

21. Lass, *Minnesota,* 103–4; A. E. Smith, *From Exploration to Statehood,* 673–79.

22. Folsom, *Fifty Years in the Northwest,* 144–45; History Network, *Minnesota Beginnings,* 11.

23. Willard E. Rosenfelt, *Washington: A History of the Minnesota County* (Stillwater, Minn.: The Croixside Press, 1977), 15, 93–97; History Network, *Minnesota Beginnings,* 11; Lass, *Minnesota,* 106–7.

24. Lass, *Minnesota,* 107; Wyman, *Wisconsin Frontier,* 175–76; James Taylor Dunn, *The St. Croix: Midwest Border River* (1965; reprint, St. Paul: Minnesota Historical Society Press, 1979), 50, 176–78.

25. Wyman, *Wisconsin Frontier,* 175–76; Dunn, *St. Croix: Midwest Border River,* 50, 176–78.

26. Richard N. Current, *The Civil War Era, 1848–1873* (Madison: State Historical Society of Wisconsin, 1976), 59.

27. Harold Weatherhead, *Westward to the St. Croix: The Story of St. Croix County, Wisconsin* (Hudson, Wis.: St. Croix County Historical Society, 1978), 27; *Stillwater Messenger,* 12 June 1860.

28. E. S. Seymour, *Sketches of Minnesota, the New England of the West* (New York: Harper & Bros., 1850), 193–94.

29. *St. Croix Union,* 2 June 1855, 27 June 1856, 15 September 1855.

30. *Stillwater Messenger,* 31 May 1859; *St. Croix Union,* 2 June 1855, 24 November 1855.

31. *St. Croix Union,* 1 December 1855, 23 January 1857; *Stillwater Messenger,* 6 October 1857.

32. *St. Croix Union,* 2 June 1855, 11 August 1855.

33. Folsom, *Fifty Years in the Northwest,* 563; *St. Croix Union,* 26 May 1855, 20 November 1854.

34. Dunn, *St. Croix: Midwest Border River,* 55; *St. Croix Union,* 12 May 1855; *Stillwater Messenger,* 31 May 1859; *St. Croix Union,* 2 June 1855, 24 November 1855; Seymour, *Sketches of Minnesota,* 179.

35. Current, *Civil War Era,* 45.

36. *St. Croix Union,* 23 January 1855.

37. *St. Croix Union,* 27 March 1855, 20 March 1855.

38. Paul W. Gates, "Frontier Land Business in Wisconsin," *Wisconsin Magazine of History* 52, no. 4 (Summer 1969): 322.

39. Duane Griffin, "Wisconsin's Vegetation History and the Balancing of Nature," in *Wisconsin Land and Life,* ed. Robert C. Ostergren and Thomas R. Vale (Madison: University of Wisconsin Press, 1997), 98–100.

40. Oliver Gibbs and C. E. Young, "Sketch of Prescott, and Pierce County," *Wisconsin Historical Collections* 3 (1857): 453–65; Ahlgren and Beeler, *History of Prescott, Wisconsin,* 5, 40; Folsom, *Fifty Years in the Northwest,* 212.

41. J. William Trygg, *Composite Map of United States Land Surveyor's Original Plats and Field Notes* (Ely, Minn.: J. Wm. Trygg, 1966), sheet 7; Genevieve Cline Day, *Hudson in the Early Days* (Hudson, Wis.: Star-Observer, 1963), 19.

42. Day, *Hudson in the Early Days,* 47, 57; Folsom, *Fifty Years in the Northwest,* 154; *St. Croix Union,* 18 November 1854.

43. Weatherhead, *Westward to the St. Croix,* 36–38.

44. Ibid., 40.

45. Ibid., 40–42; letter from H. H. Montman to his parents, 1858, Wisconsin Historical Society, River Falls Area Research Center.

46. Weatherhead, *Westward to the St. Croix,* 42.

47. Ibid., 47.

48. Trygg, *Composite Map of United States Land,* sheet 8; *St. Croix Union,* 24 April 1855; Folsom, *Fifty Years in the Northwest,* 87, 131–32.

49. Current, *Civil War Era,* 90–91; as quoted in John Giffin Thompson, "The Rise and Decline of the Wheat Growing Industry in Wisconsin: The Use and Abuse of America's Natural Resources" (Ph.D. diss., University of Wisconsin-Madison, 1909; repr., New York: Arno Press, 1972), 24–25.

50. Horace Greeley, quoted in Weatherhead, *Westward to the St. Croix,* 66–67.

51. Seymour, *Sketches of Minnesota,* 182–83; *St. Croix Union,* 17 November 1855; *Stillwater Messenger,* 25 January 1859, 30 August 1859.

52. John G. Rice, "The Swedes," in *They Chose Minnesota: A Survey of the State's Ethnic Groups,* ed. June Drenning Holmquist (St. Paul: Minnesota Historical Society Press, 1981), 250.

53. Ibid.

54. Anna Engquist, *Scandia—Then and Now* (Stillwater, Minn.: The Croixside Press, 1974), 9–18; Folsom, *Fifty Years in the Northwest,* 298, 310–11; Emeroy Johnson, "Early History of Chisago Lake Reexamined," Minnesota Historical Society Collection, n.d., 215–16; Emeroy Johnson, *An Early Look at Chisago County* (North Branch, Minn.: Chisago County Bicentennial Committee, 1976), 17, 19; Robert C. Ostergren, *A Community Transplanted: The Trans-Atlantic Experience of a Swedish Immigrant Settlement in the Upper Middle West, 1835–1915* (Madison: University of Wisconsin Press, 1988), 161.

55. Engquist, *Scandia—Then and Now,* 31, 5, 31–33.

56. *St. Croix Union,* 6 March 1855.

57. *St. Croix Union,* 12 September 1856.

58. *St. Croix Union,* 12 May 1855; Rosenfelt, *Washington,* 11; Folsom, *Fifty Years in the Northwest,* 356, 334.

59. Folsom, *Fifty Years in the Northwest,* 306–7.

60. Rosenfelt, *Washington,* 101; *St. Croix Union,* 10 November 1855.

61. Jim Cordes, *Reflections of Amador* (North Branch, Minn.: Review Corporation, 1976), 5; Rosenfelt, *Washington,* 12; *St. Croix Union,* 12 September 1856.

62. Dunn, *St. Croix: Midwest Border River,* 66–98; Folsom, *Fifty Years in the Northwest,* 104–5, 141–42.

63. Current, *Civil War Era,* 45, 47.

64. *St. Croix Union,* 12 June 1857.

65. *Stillwater Messenger,* 5 June 1860.

66. *St. Croix Union,* 13 November 1857.

67. Current, *Civil War Era,* 43–44.

68. Ibid., 236–37.

69. Folsom, *Fifty Years in the Northwest,* 146, 155–56, 334; Ahlgren and Beeler, *History of Prescott, Wisconsin,* 31; Rosenfelt, *Washington,* 12; Current, *Civil War Era,* 238–39; *St. Croix Union* and *Stillwater Messenger,* n.d.; *Stillwater Messenger,* 3 November 1857; Lass, *Minnesota,* 154.

70. *St. Croix Union,* 8 May 1857, 15 December 1855, 8 May 1857; *Stillwater Messenger,* 22 September 1857; Folsom, *Fifty Years in the Northwest,* 329.

71. Rosenfelt, *Washington,* 12–13; *Stillwater Messenger,* 22 September 1857.

72. Day, *Hudson in the Early Days,* 47, 57; Weatherhead, *Westward to the St. Croix,* 39–40; Current, *Civil War Era,* 79, 554.

73. Current, *Civil War Era,* 240–41; *Stillwater Messenger,* 1 March 1859.

74. *Stillwater Messenger,* 1 March 1859.

75. *Stillwater Messenger,* 24 May 1859, 19 February 1861.

76. Thompson, "Rise and Decline of the Wheat Growing Industry," 27–29.

77. Current, *Civil War Era,* 92–93.

78. *St. Croix Union,* 3 November 1855; Folsom, *Fifty Years in the Northwest,* 148, 213, 410; *Stillwater Messenger,* 8 December 1857; *St. Croix Union,* 22 September 1855, 19 May 1855, 14 July 1855.

79. Current, *Civil War Era,* 197–236.

80. Ibid., 296–335.

81. *Stillwater Messenger,* 12 December 1865.

82. Current, *Civil War Era,* 374–81.

83. Thompson, "Rise and Decline of the Wheat Growing Industry," 33, 68–70.

84. *Stillwater Messenger,* 13 August 1861, 27 February 1866, 30 August 1864; Theodore C. Blegen, *Minnesota: A History of the State* (Minneapolis: University of Minnesota Press, 1963), 342.

85. Current, *Civil War Era,* 430; Alice E. Smith, "Caleb Cushing's Investments in the St. Croix Valley," *Wisconsin Magazine of History* 28, no. 1 (September 1944): 15–17; Harry D. Baker, Baker Land and Title Company, 1879–1958, transcript of interview by W. H. Glover, 17 January 1950, Harry D. Baker Papers, Wisconsin Historical Society, River Falls Area Research Center.

86. Current, *Civil War Era,* 434.

87. A. E. Smith, "Caleb Cushing's Investments in the St. Croix Valley," 18.

88. Ibid.

89. Baker interview; Nelson Lawson, town clerk, "Luck Township," in *Recollections of 1876: Polk County's First Written Histories* (Osceola, Wis.: Polk County Press, 1876–78).

90. Current, *Civil War Era,* 415–16.

91. Ibid., 416–23.

92. Ibid., 417–18.

93. Blegen, *Minnesota,* 304–5; *Rush City Post,* June 1879.

94. Robert C. Nesbit, *Urbanization and Industrialization* (Madison: State Historical Society of Wisconsin, 1985), 3:87–89.

95. Current, *Civil War Era,* 430–31.

96. Ibid., 431–32; Blegen, *Minnesota,* 296–97.

97. Blegen, *Minnesota,* 296–99; Rosenfelt, *Washington,* 184–85; *Stillwater Messenger,* 9 June 1871.

98. Current, *Civil War Era,* 440–43; Nesbit, *Urbanization and Industrialization,* 111; James Willard Hurst, *Law and Economic Growth: The Legal History of the Lumber Industry in Wisconsin, 1836–1915* (Cambridge, Mass.: Belknap Press of Harvard University Press, 1964), 272. See chapter 11, "The Battle of the Piles," in Dunn, *St. Croix: Midwest Border River,* for a complete account of the rivalry between Hudson and Stillwater for railroad access.

99. Hurst, *Law and Economic Growth,* 272–73; Folsom, *Fifty Years in the Northwest,* 231, 253.

100. Hurst, *Law and Economic Growth,* 273.

101. *Polk County Press,* 13 October 1883.

102. *Polk County Press,* 31 May 1884.

103. *Polk County Press,* 15 September 1883.

104. *Polk County Press,* 15 September 1883, 29 September 1883, 16 May 1885, 12 February 1887.

105. *Polk County Press,* 26 February 1887, 27 August 1887, 29 October 1887.

106. Nesbit, *Urbanization and Industrialization,* 147.

107. Thompson, "Rise and Decline of the Wheat Growing Industry," 357–59, 70–74.

108. Nesbit, *Urbanization and Industrialization,* 10–12.

109. Ibid., 15.

110. Eric. E. Lampard, *The Rise of the Dairy Industry in Wisconsin: A Study in Agricultural Change, 1820–1920* (Madison: State Historical Society of Wisconsin, 1963), 57–96.

111. Ibid., 59, 97–101.

112. Joseph Schafer, *A History of Agriculture in Wisconsin* (Madison: State Historical Society of Wisconsin, 1922), 159–61.

113. Lass, *Minnesota,* 166–67; Blegen, *Minnesota,* 393.

114. *Polk County Press,* 14 November 1885; Thompson, "Rise and Decline of the Wheat Growing Industry," 82–85.

115. *Polk County Press,* 20 October 1883.

116. Nesbit, *Urbanization and Industrialization,* 14.

117. *Polk County Press,* 31 January 1885, 7 February 1885.

118. *Polk County Press,* 16 October 1886, 12 November 1887.

119. *Polk County Press,* 2 March 1887.

120. *Polk County Press,* 29 October 1887.

121. *Polk County Press,* 19 January 1889.

122. *Polk County Press,* 14 September 1995.

123. Nesbit, *Urbanization and Industrialization,* 143–45.

124. *Polk County Press,* 10 June 1893; Nesbit, *Urbanization and Industrialization,* 145–46.

125. Lampard, *Rise of the Dairy Industry in Wisconsin,* 272, 276.
126. Ibid.
127. Schafer, *History of Agriculture in Wisconsin,* 164; Lampard, *Rise of the Dairy Industry in Wisconsin,* 334–45, 337, 344–99.
128. Sherman E. Johnson, *From the St. Croix to the Potomac—Reflections of a Bureaucrat* (Bozeman: Montana State University, 1974), 19, 33–36.
129. E. Johnson, *An Early Look at Chisago County,* 19–20, 25, 28, 33, 69, 121; History of Rush City, Chisago County, and the St. Croix River, Carl H. Sommer Papers, 1878–1968, Minnesota Historical Society, St. Paul.
130. Ibid.
131. Folsom, *Fifty Years in the Northwest,* 404, 410; Rosenfelt, *Washington,* 146–49; *Stillwater Messenger,* 30 July 1875.
132. Raymond E. Murphy, *The Geography of the Northwestern Pine Barrens of Wisconsin* (Madison: Department of Geography, University of Wisconsin–Madison, 1931), 69, 72.
133. Ed L. Peet, *Burnett County, Wisconsin* (Grantsburg, Wis.: The Journal of Burnett County Print, 1902), 69, 101.
134. Robert Gough, *Farming the Cutover: A Social History of Northern Wisconsin, 1900–1940* (Lawrence, Kansas: University Press of Kansas, 1997), 26; James Fitting et al., *Archeological Survey of the St. Croix National Scenic Riverway: Phase I Report* (Jackson, Mich.: Gilbert/Commonwealth, 1977), 58–60; Peet, *Burnett County, Wisconsin,* 48.
135. Gough, *Farming the Cutover,* 10, 11, 19; *Polk County Press,* 21 September 1895; Eldon M. Marple, *The Visitor Who Came to Stay: Legacy of the Hayward Area* (Hayward, Wis.: Country Print Shop, 1971); Trygg, *Composite Map of United States Land,* nos. 8, 9, 14, 15; James. I. Clark, *Farming the Cutover: The Settlement of Northern Wisconsin* (Madison: State Historical Society of Wisconsin, 1956).
136. Seymour, *Sketches of Minnesota,* 204; *Burnett County Sentinel,* 23 May 1879; *Taylors Falls Journal,* 9 May 1879; Folsom, *Fifty Years in the Northwest,* 230–31; *The Heritage Areas of Burnett County* (Madison: Wisconsin Heritage Areas Program, 1978), 30; *Burnett County Sentinel,* 18 July 1880; *Burnett County Sentinel,* 23 May 1879.
137. Robert F. Fries, *Empire in Pine: The Story of Lumbering in Wisconsin, 1830–1900* (Madison: State Historical Society of Wisconsin, 1951), 169; Peet, *Burnett County, Wisconsin,* 71.
138. *Burnett County Sentinel,* 11 August 1876; Folsom, *Fifty Years in the Northwest,* 230–31; *Burnett County Sentinel,* 29 September 1876, 23 May 1879.
139. Gough, *Farming the Cutover,* 13; Peet, *Burnett County, Wisconsin,* 9.
140. Gough, *Farming the Cutover,* 10, 20–22, 30; Peet, *Burnett County, Wisconsin,* 9, 11, 98; *Burnett County Sentinel,* 16 July 1880, 17 July 1885, 6 August 1886.
141. Peet, *Burnett County, Wisconsin,* 7–9, 99.
142. *Burnett County Sentinel,* 28 September 1883, 20 July 1888, 5 April 1889, 10 April 1889.

143. *Burnett County Sentinel,* 30 May 1890, 17 October 1894; *Rush City Post,* as reported in *Burnett County Sentinel,* 3 October 1895.

144. *Stillwater Messenger,* 21 January 1875.

145. Ibid.

146. *Heritage Areas of Burnett County,* 30–31; *Stillwater Messenger,* 21 January 1875; *Burnett County Sentinel,* 31 August 1877, 10 October 1884; Peet, *Burnett County, Wisconsin,* 48; Nesbit, *Urbanization and Industrialization,* 38.

147. *Burnett County Sentinel,* January 1875.

148. *Burnett County Sentinel,* 10 October 1884, 18 April 1879.

149. *Burnett County Sentinel,* April 1877 and 1887.

150. Folsom, *Fifty Years in the Northwest,* 239–40; *Heritage Areas of Burnett County,* 30–31; *Burnett County Sentinel,* 14 July 1876, 21 July 1876, 20 April 1877, 4 November 1887; Peet, *Burnett County, Wisconsin,* 46, 26–27.

151. Peet, *Burnett County, Wisconsin,* 48, 90, 97; "Burnett County through the Years," n.d., Burnett County Historical Society, Danbury, Wis.

152. Eldon M. Marple, *A History of the Hayward Lakes Region: Through the Eyes of the Visitor Who Came and Stayed* (Hayward, Wis.: Chicago Bay Grafix, 1976).

153. Burnett County Homemakers Clubs, *Pioneer Tales of Burnett County* (Webster, Wis.: Dan-Web Printing, 1976), 11, 22, 101–2.

154. Marple, *History of the Hayward Lakes Region.*

155. *Burnett County Sentinel,* 20 March 1891.

156. Gough, *Farming the Cutover,* 31–33; J. I. Clark, *Farming the Cutover,* 6, 11; Peet, *Burnett County, Wisconsin,* 101; Gough, *Farming the Cutover,* 33.

157. *Polk County Press,* 21 September 1895; Arlan Helgeson, "Nineteenth Century Land Colonization in Northern Wisconsin," *Wisconsin Magazine of History* 36, no. 2 (Winter 1952–1953): 119; William A. Henry, *Northern Wisconsin: A Hand-Book for the Homeseeker* (Madison, Wis.: Democrat, 1896), 162–64.

158. Gough, *Farming the Cutover,* 9.

159. Peet, *Burnett County, Wisconsin,* 69, 101–2; Gough, *Farming the Cutover,* 5; Peet, *Burnett County, Wisconsin,* 5–7, 58–59, 73–74.

160. Peet, *Burnett County, Wisconsin,* 57–58.

161. Marple, *Visitor Who Came to Stay.*

162. Peet, *Burnett County, Wisconsin,* 33, 35–38, 53.

163. Gough, *Farming the Cutover,* 27–28, 34–43; Marple, *Visitor Who Came to Stay.*

164. Marple, "Rocks, Minerals and Folly," in *Visitor Who Came to Stay.*

165. Carl Kuhnly, oral history interview, 1985, Burnett County Historical Society, Danbury, Wis.; Baker interview.

166. Gough, *Farming the Cutover,* 93–94, 68–72, 85.

167. Gough, *Farming the Cutover,* 55–57, 63; Peet, *Burnett County, Wisconsin,* 42.

168. J. I. Clark, *Farming the Cutover,* 11–12, 31, 60.

169. "The 'Far North' Community of Burnett County—Danbury," 1966, Burnett County Historical Society, Danbury, Wis.

170. Gough, *Farming the Cutover,* 2, 49, 51, 54.

171. Ibid., 58–60.

172. Carl Kuhnly, "Homesteading in Wisconsin, Up Hill—Down Hill All The Way," pamphlet, 1985, Burnett County Historical Society, Danbury, Wis.; Gough, *Farming the Cutover,* 59, 63–68.

173. Gough, *Farming the Cutover,* 28, 46.

174. Ibid., 164–65.

175. Ibid., 93–114.

176. Ibid., 150–56.

177. Murphy, *Geography of the Northwestern Pine Barrens,* 70, 79, 84, 106.

178. Gough, *Farming the Cutover,* 117, 123–29, 132–35; Murphy, *Geography of the Northwestern Pine Barrens,* 70, 83, 88–89, 106.

179. Murphy, *Geography of the Northwestern Pine Barrens,* 70.

180. Gough, *Farming the Cutover,* 136–39.

181. Ibid., 146–49.

182. Ibid., 179–81, 218, 225. See also David B. Danbom, *Born in the Country: A History of Rural America* (Baltimore, Md.: Johns Hopkins University Press, 2006).

183. L. G. Sorden, "Some Events in Wisconsin's Forest History," paper delivered to the Forest History Association of Wisconsin, Inc., Wausau, Wis., 28–29 September 1979 (paper located in the Burnett County Historical Society, Danbury, Wis.), 15, 17.

184. Ibid., 15–17; Gough, *Farming the Cutover,* 181–82.

185. Sorden, "Some Events in Wisconsin's Forest History," 17.

186. Gough, *Farming the Cutover,* 221–24.

187. Marple, *Visitor Who Came to Stay;* Gough, *Farming the Cutover,* 221–23; James Kates, *Planning a Wilderness: Regenerating the Great Lakes Cutover Region* (Minneapolis: University of Minnesota Press, 2001), 15–31.

Chapter 4. Up North

1. Leo Marx, *The Machine in the Garden: Technology and the Pastoral Ideal in America* (New York: Oxford University Press, 1964), 43–44.

2. Richard Bardon and Grace Lee Nute, eds., "A Winter in the St. Croix Valley, 1802–1803," *Minnesota History* 28, no. 3 (September 1947): 149.

3. Theodore C. Blegen, "'The Fashionable Tour' on the Upper Mississippi," *Minnesota History* 20, no. 4 (December 1939): 378.

4. Joseph N. Nicollet, *The Journals of Joseph N. Nicollet: A Scientist on the Mississippi Headwaters, with Notes on Indian Life, 1836–37,* ed. Martha Coleman Bray (St. Paul: Minnesota Historical Society, 1970), 142.

5. George Catlin, *Letters and Notes on the Manners, Customs, and Condition of the North American Indians* (Philadelphia, 1857), 590–92.

6. Blegen, "'Fashionable Tour,'" 381–82; Bart Christopher Richardson, "Picturesque Landscape Assessment: The Lower St. Croix," M.A. thesis, University of Minnesota, 1993, 83.

7. Theodore C. Blegen, *Minnesota: A History of the State* (Minneapolis: University of Minnesota Press, 1963), 118, 120, 156, 193; Blegen, "'Fashionable Tour,'" 382–88; William I. Petersen, *Steamboating on the Upper Mississippi* (Iowa City: State Historical Society of Iowa, 1968), 255.

8. E. S. Seymour, *Sketches of Minnesota, the New England of the West* (New York: Harper & Bros., 1850), 217, 205, 212, 216.

9. Daniel Drake, *The Northern Lakes: A Summer Resort for Invalids of the South* (1842; reprint, Cedar Rapids, Iowa: Friends of the Torch Press, 1954), 19–20.

10. Elizabeth Fries Ellet, *Summer Rambles in the West* (New York: Reiker, 1852), 143.

11. *Stillwater Messenger,* 9 August 1859.

12. Willard E. Rosenfelt, *Washington: A History of the Minnesota County* (Stillwater, Minn.: Croixside Press, 1977), 300; *Stillwater Messenger,* 9 August 1859.

13. Willoughby M. Babcock, "The St. Croix Valley as Viewed by Pioneer Editors," *Minnesota History* 17 (1936): 280.

14. *St. Paul Advertiser,* quoted in James Taylor Dunn, *The St. Croix: Midwest Border River* (1965; reprint, St. Paul: Minnesota Historical Society Press, 1979), 204; Anita Albrecht Buck, *Steamboats on the St. Croix* (St. Cloud, Minn.: North Star Press of St. Cloud, 1990), 119–20.

15. Blegen, "'Fashionable Tour,'" 378–79, 386.

16. John W. Bond, quoted in Paul Clifford Larson, *A Place at the Lake* (Afton, Minn.: Afton Historical Society Press, 1998), 15; *St. Paul Pioneer,* 1854; P. C. Larson, *Place at the Lake,* 11–13.

17. Richardson, "Picturesque Landscape Assessment," 110–12.

18. *St. Croix Union,* 10 November 1854.

19. Dunn, *St. Croix: Midwest Border River,* 211–12.

20. *Stillwater Messenger,* 1 December 1863, 8 March 1864.

21. Dunn, *St. Croix: Midwest Border River,* 25; Blegen, *Minnesota,* 237.

22. Buck, *Steamboats on the St. Croix,* 118–20; Dunn, *St. Croix: Midwest Border River,* 197–203 (quotes from *Stillwater Messenger* and poem on page 203).

23. *Stillwater Lumberman,* 29 June 1877; *Burnett County Sentinel,* 22 August 1879.

24. Richardson, "Picturesque Landscape Assessment," 112–15; *Stillwater Lumberman,* 2 July 1875.

25. Blegen, "'Fashionable Tour,'" 398; *Stillwater Messenger,* 13 June 1866; *Stillwater Lumberman,* 18 June 1875; Dunn, *St. Croix: Midwest Border River,* 203–4; William H. Dunne, *The Picturesque St. Croix and Other Northwest Sketches* (Chicago: Poole Brothers, 1881), 16; William H. Dunne, *Captain Jolly on the Picturesque St. Croix: Descriptive and Historical Narrative by a Rambler* (St. Paul: J. W. Cunningham, 1880), 67.

26. Dunne, *Picturesque St. Croix,* 27, 35.

27. Ibid., 27–29; Dunne, *Captain Jolly,* 8.

28. P. C. Larson, *Place at the Lake,* 18–19, 103.

29. Ibid., 98; Dunne, *Picturesque St. Croix,* 45.

30. St. Louis, Minneapolis, & St. Paul Short Line, *The Summer Resorts of Minnesota: Information for Invalids, Tourists and Sportsmen* (Minneapolis: Dimond & Ross, 1882), 35–36.

31. Mark Twain, quoted in P. C. Larson, *Place at the Lake,* 71, 73–74; Dunne, *Picturesque St. Croix,* 49; *Northwest Magazine,* quoted in P. C. Larson, *Place at the Lake,* 71.

32. Dunne, *Picturesque St. Croix,* 49.

33. David L. Fritz, "Historic Resource Study: St. Croix National Scenic Riverway," unpublished draft (Omaha, Neb.: National Park Service, 1989), 141.

34. Rosenfelt, *Washington,* 206; P. C. Larson, *Place at the Lake,* 111; *Stillwater Lumberman,* 24 August 1877.

35. Timothy Bawden, "Escape to Wisconsin: The Early Resort Landscape of Northern Wisconsin, 1890–1920," *Wisconsin Preservation News* 22, no. 4 (July–August 1998): 2–3.

36. Richardson, "Picturesque Landscape Assessment," 116; St. Louis, Minneapolis, & St. Paul Short Line, *Summer Resorts of Minnesota,* 43–45; Raymond H. Merritt, *Creativity, Conflict and Controversy: A History of the St. Paul District U.S. Army Corps of Engineers* (Washington, D.C.: U.S. Government Printing Office, 1979), 270–74.

37. Robert C. Nesbit, *Urbanization and Industrialization* (Madison: State Historical Society of Wisconsin, 1985), 527–28.

38. *Stillwater Lumberman,* 7 May 1875; Dunn, *St. Croix: Midwest Border River,* 203–4.

39. *Polk County Press,* 22 October 1887, 30 June 1888, 7 July 1888, July 1891.

40. Dunn, *St. Croix: Midwest Border River,* 212; *Stillwater Lumberman,* 13 October 1875.

41. *Burnett County Sentinel,* 31 October 1879, 23 August 1878, 15 April 1881, 22 April 1881.

42. Rachel Franklin-Weekley, "Recreation and Tourism along the Saint Croix," unpublished paper (Omaha, Neb.: U.S. Department of the Interior, 1999), 28.

43. *Stillwater Lumberman,* 5 October 1877.

44. *Burnett County Sentinel,* 3 November 1882; Dunne, *Captain Jolly,* 47.

45. *Burnett County Sentinel,* 5 November 1880; *Stillwater Lumberman,* 8 June 1877; *Burnett County Sentinel,* 20 March 1891.

46. *Burnett County Sentinel,* 28 May 1880, 16 February 1883; *Polk County Press,* 20 July 1995.

47. Nesbit, *Urbanization and Industrialization,* 529–30; Peg Meier, *Too Hot, Went to Lake: Seasonal Photos from Minnesota's Past* (Minneapolis: Neighbors Publishing, 1993), 125.

48. John F. Reiger, *American Sportsmen and the Origins of the Conservation Movement* (Norman: University of Oklahoma Press, 1973); Peter J. Schmitt, *Back to Nature: The Arcadian Myth in Urban America* (New York: Oxford University Press, 1969); Franklin-Weekley, "Recreation and Tourism," 31.

49. John Muir, quoted in Franklin-Weekley, "Recreation and Tourism," 35.

50. Ibid., 33; Blegen, *Minnesota,* 500; Harry D. Baker, Baker Land and Title Company, 1879–1958, transcript of interview by W. H. Glover, 17 January 1950, Harry D. Baker Papers, Wisconsin Historical Society, River Falls Area Research Center.

51. George H. Hazzard, *Interstate Park: Lectures, Laws, Papers, Pictures, Pointer* (St. Paul, Minn., 1896), 3.

52. Ward Moberg, "The Dalles: Preserve or Develop?" *The Dalles Visitor* 13 (Summer 1981): 2.

53. Baker interview.

54. Baker interview; Ward Moberg, "The 1890s Fight Created Dalles Parks," *The Dalles Visitor* 14 (Summer 1982): 1, 13.

55. Harry D. Baker, quoted in Franklin-Weekley, "Recreation and Tourism," 40–41.

56. Ibid., 41–42.

57. Dunn, *St. Croix: Midwest Border River,* 98.

58. Franklin-Weekley, "Recreation and Tourism," 42–43; Baker interview; Warren Manning, quoted in Richardson, "Picturesque Landscape Assessment," 116–17.

59. Franklin-Weekley, "Recreation and Tourism," 32.

60. *Dalles Visitor,* Summer 1992.

61. Ibid.

62. Ibid.

63. Merritt, *Creativity, Conflict and Controversy,* 280.

64. *Polk County Press,* quoted in *Dalles Visitor,* Summer 1992.

65. William Blanding, quoted in Merritt, *Creativity, Conflict and Controversy,* 281–83.

66. Ibid., 283–84.

67. Ibid., 284.

68. Ibid., 286; *Dalles Visitor,* Summer 1992.

69. Richardson, "Picturesque Landscape Assessment," 118; *Stillwater Messenger,* 9 October 1909.

70. *Free Press,* 26 February 1980, 25 March 1980; Jay Walljasper, "When the Trolley Was King," *Minneapolis.St.Paul Magazine,* April 1981.

71. *Stillwater Messenger,* 18 July 1908.

72. *Stillwater Evening Gazette,* 13 July 1910; Mary Beth Burkholder and Susan Mary Dahlby, *Willow River, St. Croix County, Wisconsin* (Hudson, Wis., 1963), 31–36; James Taylor Dunn, *Saving the River: The Story of the St. Croix River Association, 1911–1986* (St. Paul: St. Croix River Association, 1986), 2.

73. Hans Huth, quoted in Franklin-Weekley, "Recreation and Tourism," 33.

74. Rosemarie Vezina, *Nevers Dam: The Lumberman's Dam* (St. Croix Falls, Wis.: Standard Press, 1965), 6, 17, 27, 33, 37. The dam was washed away in May 1954 and completely removed in 1955.

75. Dunn, *St. Croix: Midwest Border River,* 97-98.

76. Dunn, *Saving the River,* 3–5. Dunn's book offers an account of the history of the St. Croix River Association.

77. Ibid., 3–5; Merritt, *Creativity, Conflict and Controversy,* 289–90. See also Theodore J. Karamanski, "Saving the St. Croix: An Administrative History of the Saint Croix National Scenic River," Omaha, Neb.: National Park Service, Midwest Region, 1992, 16–27.

78. Ira King, quoted in Dunn, *Saving the River,* 12.

79. Karamanski, "Saving the Saint Croix," 29–30.

80. John Dunn, quoted in Dunn, *Saving the River,* 15.

81. Ibid., 15–16. Writing in 1986, Dunn reported that the Army Corps of Engineers had not done any snagging on the river above Stillwater, declaring it "maintenance free."

82. Bawden, "Escape to Wisconsin," 2–3; Eldon M. Marple, *History of the Hayward Lakes Region: Through the Eyes of the Visitor Who Came and Stayed* (Hayward, Wis.: Chicago Bay Grafix, 1976).

83. Marple, *History of the Hayward Lakes Region;* Franklin-Weekley, "Recreation and Tourism," 58–59.

84. Marple, *History of the Hayward Lakes Region,* esp. "A Tour around Lost Land and Teal Lakes"; Aaron Shapiro, "Up North on Vacation: Tourism and Resorts in Wisconsin's North Woods, 1900–1945," *Wisconsin History Magazine* 29, no. 4 (2006).

85. Edward L. Peet, *Burnett County, Wisconsin* (Grantsburg, Wis.: The Journal of Burnett County Print, 1902), 27, 55, 66.

86. Chicago & North Western Railway, *Lakes and Resorts of the Northwest* (Chicago: Chicago & North Western Railway, 1916), 3.

87. Ibid., 8, 19. In Hayward were listed the Giblin Hotel, Cornick's Spider Lake and Teal Lake Resorts, Grindstone of the Pines, Idelhurst Lodge, Boulder Lodge, Hubbard-Teal Lakes Resort, Clover Leaf Lodge, William's Grindstone Lake Resort, Sand Lake Resort, Round Lake Club, Berger's Resort, Sportsmen's Paradise. In Chisago City were listed the Dahl's House, Island Resort, Gleer's House, and Hotel Chisago.

88. Shapiro, "Up North on Vacation," 5; Marple, "A Tour around Lost Land and Teal Lakes," in *History of the Hayward Lakes Region;* Shapiro, "Up North on Vacation," 4; Paul W. Glad, *War, a New Era, and Depression, 1914–1940* (Madison: State Historical Society of Wisconsin, 1990), 211–14.

89. Franklin-Weekley, "Recreation and Tourism," 49–50.

90. Ibid., 52–53; Raymond E. Murphy, *The Geography of the Northwestern Pine Barrens of Wisconsin* (Madison: Department of Geography, University of Wisconsin–Madison, 1931), 95–96.

91. Murphy, *Geography of the Northwestern Pine Barrens,* 72, 93–95, 107.

92. "Here's How Kilkare Lodge Looks Today!" pamphlet, [1929], Burnett County Historical Society, Danbury, Wis.

93. Burnett County Homemakers Clubs, *Pioneer Tales of Burnett County,*

103–4; Iver Johnson, "Johnson's Web Lake Resort," pamphlet, n.d., Burnett County Historical Society, Danbury, Wis.

94. Murphy, *Geography of the Northwestern Pine Barrens,* 110; Shapiro, "Up North on Vacation," 7.

95. *Burnett County Sentinel,* 5, 12, 19, 20, 26, 27 January 2000; 2, 9, 16, 23 February 2000.

96. Glad, *War, a New Era, and Depression,* 216–19; Warren James Belasco, *Americans on the Road: From Autocamp to Motel, 1910–1945* (Cambridge, Mass.: MIT Press, 1978), 8.

97. Bawden, "Escape to Wisconsin," 3.

98. Blegen, *Minnesota,* 532–35; Minnesota State Board of Health, *State Laws and Regulations Relating to Hotels, Restaurants, Places of Refreshment, Lodging Houses, and Boarding Houses* (St. Paul, Minn., July 1932).

99. Murphy, *Geography of the Northwestern Pine Barrens,* 76; Shapiro, "Up North on Vacation," 12.

100. Franklin-Weekley, "Recreation and Tourism," 44–45.

101. Robert C. Nesbit, *Wisconsin: A History* (Madison: University of Wisconsin Press, 1973), 470–71; Glad, *War, a New Era, and Depression,* 494; Marple, *History of the Hayward Lakes Region.*

102. Franklin-Weekley, "Recreation and Tourism," 43–44.

103. Marple, *History of the Hayward Lakes Region; Rice Lake Chronotype,* 28 November 1979; "Work at Camp Riverside," n.d., Burnett County Historical Society, Danbury, Wis.; *St. Paul Pioneer Press,* 6 April 1941.

104. Marple, *History of the Hayward Lakes Region;* Burnett County Homemakers Clubs, *Pioneer Tales of Burnett County* (Webster, Wis.: Dan-Web Printing, 1976), 99; Carol Ahlgren, "The Civilian Conservation Corps and Wisconsin State Park Development," *Wisconsin Magazine of History* 71, no. 3 (Spring 1988): 184–204.

105. "Work at Camp Riverside," 18. See also William Gray Purcell, *St. Croix Trail Country: Recollections of Wisconsin* (Minneapolis: University of Minnesota Press, 1967).

106. Burnett County Homemakers Clubs, *Pioneer Tales of Burnett County,* 73.

107. Writers' Program of the WPA, *Wisconsin: A Guide to the Badger State* (New York: Viking Press, 1938), 22–23; Murphy, *Geography of the Northwestern Pine Barrens,* 93.

108. Marple, *A History of the Wayward Lakes Region,* 81; Writers' Program of the WPA, *Wisconsin: A Guide to the Badger State,* 23–24.

109. Fred Etcherson, oral history interview by David J. Olson, 30 July 1970, Wisconsin Historical Society, River Falls Area Research Center.

110. *St. Paul Pioneer Press,* 6 April 1941.

111. Franklin-Weekley, "Recreation and Tourism," 68–69.

112. Ibid., 69–70.

113. Ibid.

114. Marple, *History of the Hayward Lakes Region,* 67–68.

115. Writers' Program of the WPA, *Wisconsin: A Guide to the Badger State,* 113, 115, 471–72, 477–78.

116. Franklin-Weekley, "Recreation and Tourism," 71–72.

117. Dunn, *St. Croix: Midwest Border River,* 207–8; *St. Paul Pioneer Press,* 1 July 1945.

118. Dunn, *St. Croix: Midwest Border River,* 72; Moira F. Harris, "Ho-ho-ho! It Bears Repeating: Advertising Characters in the Land of Sky Blue Waters," *Minnesota History* 57, no. 1 (Spring 2000): 30.

119. Popular Science Monthly, *How to Build Cabins, Lodges, and Bungalows: Complete Manual of Constructing, Decorating, and Furnishing Homes for Recreation or Profit* (New York: Grosset & Dunlap, 1934; repr., 1946), 5; Ralph P. Dillon, *Sunset's Cabin Plan Book* (San Francisco: Lane, 1938); Franklin-Weekley, "Recreation and Tourism," 62; National Plan Service, *Summer Camps and Cottages* (Chicago: National Plan Service, [194?]).

120. Marple, *History of the Hayward Lakes Region.*

121. Writers' Program of the WPA, *Minnesota: A State Guide* (New York: Viking Press, 1938), 130–31; Writers' Program of the WPA, *Wisconsin: A Guide to the Badger State,* 117; Current, *Civil War Era,* 30–31; Anthony Wise to Mr. Joe Mercedes, Wisconsin Tourist Board, 3 December 1947, Tony Wise Papers, Wisconsin Historical Society, Madison; Anthony Wise, Report on "Winter Tourist Trade In Wisconsin," to the Hayward Chamber of Commerce [1947], Tony Wise Papers.

122. Wise to Mercedes, Tony Wise Papers; Wise, Report on "Winter Tourist Trade In Wisconsin," Tony Wise Papers.

123. "History of Telemark," Tony Wise Papers, Wisconsin Historical Society, Madison.

124. "A Capsule History of the Telemark Recreational Community," Wisconsin Historical Society, Madison; *Chicago Sun-Times,* 13 July 1969.

125. Janice Ward, *Next Stop Dresser Junction* (Osceola, Wis.: The Osceola Sun, 1976), 108–9.

126. Karamanski, "Saving the St. Croix," 32–34.

127. Ibid., 36–39.

128. Dunn, *Saving the River,* 28.

129. Franklin-Weekley, "Recreation and Tourism," 45–46.

130. James Taylor Dunn, *State Parks of the St. Croix Valley: Wild, Scenic, and Recreational* ([St. Paul]: Minnesota Parks Foundation, 1981), 34–35; Franklin-Weekley, "Recreation and Tourism," 72–73. David Backes, *Wilderness Within: A Life of Sigurd F. Olson* (Minneapolis: University of Minnesota Press, 1997), 145; Marple, *History of the Hayward Lakes Region.*

131. Backes, *Wilderness Within,* 145; Marple, *History of the Hayward Lakes Region.*

132. Franklin-Weekley, "Recreation and Tourism," 74–75; James Taylor Dunn, *Marine on St. Croix: From Lumber Village to Summer Haven* (Marine on St. Croix, Minn.: Marine Historical Society, 1968), 92.

133. Ibid.

134. *Chisago County Press,* 5 September 1957, 3 July 1958, 28 August 1958.

135. Karamanski, "Saving the St. Croix," 44.

136. Franklin-Weekley, "Recreation and Tourism," 79.

137. Ibid., 80–81. Northern States Power Company is now Xcel Energy.

138. Gaylord Nelson, oral history interview by Theodore J. Karamanski, 20 December 1991, St. Croix National Scenic Riverway Historic Files, St. Croix Falls, Wis.

139. Robert Bergman to National Park Service Assistant Director, Cooperative Activities, 8 December 1966, St. Croix National Scenic Riverway Historic Files, St. Croix Falls, Wis.; Stewart L. Udall, Secretary of the Interior, to Assistant Secretaries, 2 December 1966, St. Croix National Scenic Riverway Historic Files, St. Croix Falls, Wis.

140. Merle Anderson, oral history interview by Theodore J. Karamanski, 7 January 1992, St. Croix National Scenic Riverway Historic Files, St. Croix Falls, Wis.; Bruce Miller, personal communication with Theodore J. Karamanski, 21 May 1992; *St. Paul Pioneer Press,* 11 October 1970; *Rush City Post,* 9 October 1970.

141. Lower Saint Croix Study Team, "Scenic River Study of the Lower Saint Croix River," unpublished report, June 1971, 3–5; David Shonk, personal communication with Theodore J. Karamanski, 8 June 1992; *Minneapolis Tribune,* 8 June 1971.

142. *Minneapolis Tribune,* 5 May 1971, 14 June 1971; James Harrison, oral history interview by Theodore J. Karamanski, 8 January 1992, St. Croix National Scenic Riverway Historic Files, St. Croix Falls, Wis.

143. Walter F. Mondale, *Hearing before the Subcommittee on Public Lands of the Committee on Interior and Insular Affairs, U.S. Senate, Ninety-Second Congress, S.1928, a Bill to Amend the Wild and Scenic Rivers Act* (Washington, D.C.: U.S. Government Printing Office, 1972), 3–4.

144. Albert H. Quie, oral history interview by Theodore J. Karamanski, 8 January 1992, St. Croix National Scenic Riverway Historic Files, St. Croix Falls, Wis.; Albert Quie to James Harrison, 30 June 1972, Albert H. Quie Papers, Minnesota Historical Society, St. Paul; *Minneapolis Tribune,* 21 July 1972.

145. Karamanski, "Saving the St. Croix," 136–51; *Minneapolis Tribune,* 12 October 1972, 14 October 1972; Karamanski, "Saving the St. Croix," 169–207.

146. *Pioneer Press,* 31 May 1981.

147. Anthony Anderson, oral history interview by Theodore J. Karamanski, 10 January 1992, St. Croix National Scenic Riverway Historic Files, St. Croix Falls, Wis.; *Stillwater Evening Gazette,* 25 June 1990, 30 July 1990; *Minneapolis Star,* 15 July 1990.

148. Ronald N. Satz, *Chippewa Treaty Rights: The Reserved Rights of Wisconsin's Chippewa Indians in Historical Perspective* (Madison: Wisconsin Academy of Sciences, 1991), 98–100.

149. Northern Initiatives Strategic Planning Workgroup, *Northern Initiatives: A Strategic Plan for the Next Decade* (Madison: Wisconsin Department of Natural Resources, Summer 1994).

150. David S. Boyer, "The St. Croix," *National Geographic,* July 1977, 37.

151. "Conserving State's Natural Heritage Made Strides in 2006," *Milwaukee Journal Sentinel,* 31 December 2006; Friends of the St. Croix Headwaters website, http://www.fotsch.org/.

152. Franklin-Weekley, "Recreation and Tourism," 83–84.

153. Ibid., 36–38; Shapiro, "Up North on Vacation," 13.

Bibliography

Archival Sources, Manuscript Sources, and Interviews

Anderson, Anthony. Oral history interview by Theodore J. Karamanski, 10 January 1992. St. Croix National Scenic Riverway Historic Files, St. Croix Falls, Wis.

Anderson, Merle. Oral history interview by Theodore J. Karamanski, 7 January 1992. St. Croix National Scenic Riverway Historic Files, St. Croix Falls, Wis.

Archives and Manuscripts Division. Wisconsin Historical Society, Madison.

Baker Family File. Minnesota Historical Society, St. Paul.

Baker, Harry D. Papers. Wisconsin Historical Society, River Falls Area Research Center.

Bergman, Robert. Letter to National Park Service Assistant Director. Cooperative Activities. 8 December 1966. St. Croix National Scenic Riverway Historic Files, St. Croix Falls, Wis.

Blanding, William H. Papers. Wisconsin Historical Society, River Falls Area Research Center.

Etcherson, Fred. Oral history interview by David J. Olson, 30 July 1970. Wisconsin Historical Society, River Falls Area Research Center.

Folsom, William H. C. Papers. Wisconsin Historical Society, River Falls Area Research Center.

Hanks, Stephen B. "Memoir of Captain S. B. Hanks." Minnesota Historical Society, St. Paul.

Harrison, James. Oral history interview by Theodore J. Karamanski, 8 January 1992. St. Croix National Scenic Riverway Historic Files, St. Croix Falls, Wis.

Homeseekers Land Company Papers. Minnesota Historical Society, St. Paul.

Johnson, Iver. "Johnson's Web Lake Resort." Pamphlet, n.d. Burnett County Historical Society, Danbury, Wis.

Johnston, James. "Memoir." Wisconsin Historical Society, River Falls Area Research Center.

Kuhnly, Carl. "Homesteading in Wisconsin: Up Hill—Down Hill, All the Way." Pamphlet, 1985. Burnett County Historical Society, Danbury, Wis.

———. Oral history interview, 1985. Burnett County Historical Society, Danbury, Wis.

Logging Clipping File. Burnett County Historical Society, Danbury, Wis.

Lower St. Croix Study Team. "Scenic River Study of the Lower Saint Croix River." Unpublished report, June 1971. St. Croix National Scenic Riverway Historic Files, St. Croix Falls, Wis.

Miller, Bruce. Personal communication with Theodore J. Karamanski, 21 May 1992.

Montman, H. H. Letter to parents, 1858. Wisconsin Historical Society, River Falls Area Research Center.

Muench, Virgil J. Papers. Wisconsin Historical Society, Madison.

Nelson, Gaylord. Oral history interview by Theodore J. Karamanski, 20 December 1991. St. Croix National Scenic Riverway Historic Files, St. Croix Falls, Wis.

Peabody, R. A. Papers, 1941–1954. Wisconsin Historical Society, River Falls Area Research Center.

Peterson, Hjalmar Otto. Papers. Minnesota Historical Society, St. Paul.

Quie, Albert H. Oral history interview by Theodore J. Karamanski, 8 January 1992. St. Croix National Scenic Riverway Historic Files, St. Croix Falls, Wis.

———. Papers. Minnesota Historical Society, St. Paul.

St. Croix County. Register of Deeds Collection. Wisconsin Historical Society, River Falls Area Research Center.

Sawyer County. Plat Book, 1897. Wisconsin Historical Society, Madison.

Shonk, David. Personal communication with Theodore J. Karamanski, 8 June 1992.

Sommer, Carl H. Papers, 1878–1968. Minnesota Historical Society, St. Paul.

Stuntz, Albert C. Diaries, 1858–1882. Wisconsin Historical Society, River Falls Area Research Center.

Thatcher, O. M. "The Mission of Folle Avoine." Yellow Lake Pamphlet File. Burnett County Historical Society, Danbury, Wis.

Udall, Stewart L. Letter, Secretary of the Interior to Assistant Secretaries, 2 December 1966. St. Croix National Scenic Riverway Historic Files, St. Croix Falls, Wisconsin.

Wise, Tony. Papers. Wisconsin Historical Society, Madison.

Wisconsin Historical Collections. Wisconsin Historical Society, Madison.

Yellow Lake Pamphlet File. Burnett County Historical Society, Danbury, Wis.

Newspapers

Burnett County Sentinel (Grantsburg, Wisconsin)

Chisago County Press (Lindstrom, Minnesota)

Dalles Visitor (Taylors Falls, Minnesota, and St. Croix Falls, Wisconsin)

Free Press (Stillwater, Minnesota)

Milwaukee Journal Sentinel

Minneapolis Tribune (now the *Minneapolis Star Tribune*)

Mississippi Valley Lumberman (Minneapolis)

Northwestern Lumberman (Chicago)
Pine County Courier (Sandstone, Minnesota)
Polk County Press (Osceola, Wisconsin)
Rice Lake Chronotype (Rice Lake, Wisconsin)
Rush City Post (Rush City, Minnesota)
Stillwater Evening Gazette (Minnesota)
Stillwater Lumberman (Minnesota)
Stillwater Messenger (Minnesota)
St. Croix Falls (Stillwater, Minnesota)
St. Croix Union (Stillwater, Minnesota)
St. Paul Pioneer Press
Taylors Falls Journal (Minnesota)
Taylors Falls Reporter (Minnesota)
Wausau Record-Herald (Wisconsin)

Books, Reports, Periodicals

Ahlgren, Carol. "The Civilian Conservation Corps and Wisconsin State Park Development." *Wisconsin Magazine of History* 71, no. 3 (Spring 1988): 184–204.

Ahlgren, Clifford E., and Isabel F. Ahlgren. "The Human Impact on the Northern Forest Ecosystems." In *The Great Lakes Forest: An Environmental and Social History,* ed. Susan L. Flader. Minneapolis: University of Minnesota Press, 1983.

Ahlgren, Dorothy Eaton, and Mary Cotter Beeler. *A History of Prescott, Wisconsin: A River City and Farming Community on the St. Croix and Mississippi.* [Prescott, Wis.]: Prescott Area Historical Society, 1996.

Allen, James. "Journal and Letters of Lieutenant James Allen." In Henry Rowe Schoolcraft, *Schoolcraft's Expedition to Lake Itasca: The Discovery of the Sources of the Mississippi,* ed. Philip P. Mason. East Lansing: Michigan State University Press, 1958.

Anderson, Gary Clayton. *Kinsmen of Another Kind: Dakota-White Relations in the Upper Mississippi Valley, 1650–1862.* St. Paul: Minnesota Historical Society Press, 1997.

Babcock, Willoughby M. "The St. Croix Valley as Viewed by Pioneer Editors." *Minnesota History* 17 (1936): 280–95.

Backes, David. *A Wilderness Within: A Life of Sigurd F. Olson.* Minneapolis: University of Minnesota Press, 1997.

Bailey, Kenneth P., ed. and trans. *Journal of Joseph Marin, French Colonial Explorer and Military Commander in the Wisconsin Country, August 7, 1753–June 20, 1754.* N.p.: Privately printed, 1975.

Baker, Ray Stannard [David Grayson, pseud.]. *Adventures in Contentment.* New York: Doubleday, 1907.

Bardon, Richard, and Grace Lee Nute, eds. "A Winter in the St. Croix Valley, 1802–1803." *Minnesota History* 28, no. 3 (September 1947): 140–60.

Bawden, Timothy. "Escape to Wisconsin: The Early Resort Landscape of Northern Wisconsin, 1890–1920." *Wisconsin Preservation News* 22, no. 4 (July–August 1998): 1–10.

Belasco, Warren James. *Americans on the Road: From Autocamp to Motel, 1910–1945.* Cambridge, Mass.: MIT Press, 1978.

Birk, Douglas A. *John Sayer's Snake River Journal, 1804–1805: A Fur Trade Diary from East Central Minnesota.* St. Paul: Institute for Minnesota Archaeology, 1989.

Blair, Emma Helen, ed. *The Indian Tribes of the Upper Mississippi Valley and the Region of the Great Lakes.* Lincoln: University of Nebraska Press, 1996.

Blegen, Theodore C. "'The Fashionable Tour' on the Upper Mississippi." *Minnesota History* 20, no. 4 (December 1939): 377–96.

———. *Minnesota: A History of the State.* Minneapolis: University of Minnesota Press, 1963.

Bowers, William L. *The Country Life Movement in America, 1900–1920.* Port Washington, N.Y.: Kennikat Press, 1974.

Boyer, David S. "The St. Croix." *National Geographic,* July 1977.

Bracklin, James. "A Tragedy of the Wisconsin Pinery." *Wisconsin Magazine of History* 3, no. 1 (September 1919): 43–51.

Bryant, Ralph Clement. *Logging: The Principles and General Methods of Operation in the United States.* New York: John Wiley, 1914.

Buck, Anita Albrecht. *Steamboats on the St. Croix.* St. Cloud, Minn.: North Star Press of St. Cloud, 1990.

Bunting, Robert. *The Pacific Rainforest: Environment and Culture in an American Eden.* Lawrence: University of Kansas Press, 1997.

Burke, Fred C. *Logs on the Menominee: The History of the Menominee River Boom Company.* Marinette, Wis.: Privately printed, 1946.

Burkholder, Mary Beth, and Susan Mary Dahlby. *The Willow River, St. Croix County, Wisconsin.* Hudson, Wis., 1963.

Burnett County Homemakers Clubs. *Pioneer Tales of Burnett County.* Bicentennial ed. Webster, Wis.: Dan-Web Printing, 1976.

Carr, Ethan. *Wilderness by Design: Landscape Architecture and the National Park Service.* Lincoln: University of Nebraska Press, 1998.

Carver, Jonathan. *Travels through the Interior Parts of North America in the Years 1766, 1767, and 1768.* London: J. Walter and S. Crowder, 1778.

Catlin, George. *Letters and Notes on the Manners, Customs, and Conditions of the North American Indians.* Philadelphia, 1857.

Charlevoix, Pierre de. *Journal of a Voyage to North America.* 2 vols. London: R. & J. Dodsley, 1761.

Chicago & North Western Railway. *Lakes and Resorts of the Northwest.* Chicago: Chicago & North Western Railway, 1916.

Clark, Clifford E., Jr., ed. *Minnesota in a Century of Change: The State and Its People since 1900.* St. Paul: Minnesota Historical Society Press, 1989.

Clark, James I. *Farming the Cutover: The Settlement of Northern Wisconsin.* Madison: State Historical Society of Wisconsin, 1956.

Cleland, Charles E. "Preliminary Report of the Ethnohistorical Basis of Hunting, Fishing, and Gathering Rights of the Mille Lacs Chippewa." In *Fish in the Lakes, Wild Rice, and Game in Abundance,* ed. James M. McClurken. East Lansing: Michigan State University Press, 2000.

Clement, Ralph. *Logging: The Principles and General Methods of Operation in the United States.* New York: John Wiley, 1914.

Connor, Melissa, et al. *Archeological Investigations along the St. Croix National Scenic Riverway, 1983.* Lincoln, Neb.: National Park Service, 1985.

Connor, Thomas [pseud. of John Sayer]. "The Diary of Thomas Connor." In *Five Fur Traders of the Northwest,* ed. Charles M. Gates, 243–78. St. Paul: Minnesota Historical Society, 1965.

Copway, George. *The Traditional History and Characteristic Sketches of the Ojibway Nation.* London: Charles Gilpin, 1850.

Cordes, Jim. *Reflections of Amador.* North Branch, Minn.: Review Corporation, 1976.

Curot, Michel. "A Wisconsin Fur-Trader's Journal, 1803–04." *Collections of the State Historical Society of Wisconsin* 20 (1911): 396–471.

Current, Richard N. *The Civil War Era, 1848–1873.* The History of Wisconsin 2. Madison: State Historical Society of Wisconsin, 1979.

Danbom, David B. *Born in the Country: A History of Rural America.* Baltimore: Johns Hopkins University Press, 2006.

Danziger, Edmund. *The Chippewas of Lake Superior.* Norman: University of Oklahoma Press, 1979.

Davidson, J. N. *In Unnamed Wisconsin: Studies in the History of the Region between Lake Michigan and the Mississippi.* Milwaukee: S. Chapman, 1895.

Day, Genevieve Cline. *Hudson in the Early Days.* Hudson, Wis.: Star-Observer, 1963.

Delfeld, Paula. *The Indian Priest: Father Philip B. Gordon, 1885–1948.* Chicago: Franciscan Harold Press, 1977.

Densmore, Frances. *Chippewa Customs.* Washington, D.C.: Smithsonian Institution, 1929.

Dillon, Ralph P. *Sunset's Cabin Plan Book.* 6th ed. San Francisco: Lane, 1938.

Downing, A. J. *Landscape Gardening and Rural Architecture.* New York: Dover, 1991.

Drake, Daniel. *The Northern Lakes: A Summer Resort for Invalids of the South.* 1842. Reprint, Cedar Rapids, Iowa: Friends of the Torch Press, 1954.

Dunn, James Taylor. *Marine on St. Croix: From Lumber Village to Summer Haven.* Marine on St. Croix, Minn.: Marine Historical Society, 1968.

———. *The St. Croix: Midwest Border River.* 1965. Reprint, St. Paul: Minnesota Historical Society Press, 1979.

———. *Saving the River: The Story of the St. Croix River Association, 1911–1986.* St. Paul: St. Croix River Association, 1986.

————. *State Parks of the St. Croix Valley: Wild, Scenic, and Recreational.* [St. Paul]: Minnesota Parks Foundation, 1981.

Dunne, William H. *Captain Jolly on the Picturesque St. Croix: Descriptive and Historical Narrative by a Rambler.* St. Paul: J. W. Cunningham, 1880.

————. *The Picturesque St. Croix and Other Northwest Sketches.* Chicago: Poole Brothers, 1881.

Durant, Edward W. "Lumbering and Steamboating on the St. Croix River." *Collections of the Minnesota Historical Society* 10, pt. 2 (1905): 647–58.

Easton, Augustus B. *History of the Saint Croix Valley.* 2 vols. Chicago: H. C. Cooper, Jr., 1909.

Edmunds, R. David, and Joseph L. Peyser. *The Fox Wars: The Mesquakie Challenge to New France.* Norman: University of Oklahoma Press, 1993.

Ellet, Elizabeth Fries. *Summer Rambles in the West.* New York: Reiker, 1852.

Engberg, George Baker. "Labor in the Lake States Lumber Industry, 1830–1930." Ph.D. diss., University of Minnesota, 1949.

Engquist, Anna. *Scandia—Then and Now.* Stillwater, Minn.: Croixside Press, 1974.

The Far "North" Community of Burnett County—Danbury. Burnett County Historical Society, 1966.

Fitting, James, et al. *An Archeological Survey of the St. Croix National Scenic Riverway: Phase I Report.* Jackson, Mich.: Gilbert/Commonwealth, 1977.

Folsom, W. H. C. *Fifty Years in the Northwest.* Ed. E. E. Edwards. 1888. Reprint, Taylors Falls, Minn.: Taylors Falls Historical Society, 1999.

————. "Lumbering in the St. Croix Valley." *Minnesota Historical Society Collections* 10, pt. 2 (1905): 645–75.

Folwell, William Watts. *A History of Minnesota.* St. Paul: Minnesota Historical Society Press, 1930.

Forsyth, Thomas. "Journal of a Voyage from St. Louis to the Falls of St. Anthony, in 1819." *Wisconsin Historical Collections* 6 (1908): 188–219.

Franklin-Weekley, Rachel. "Recreation and Tourism along the St. Croix." Unpublished paper. Omaha, Neb.: Midwest Regional Office, National Park Service, 1999.

Fries, Robert F. *Empire in Pine: The Story of Lumbering in Wisconsin, 1830–1900.* Madison: State Historical Society of Wisconsin Press, 1951.

Fritz, David L. "Historic Resource Study: St. Croix National Scenic Riverway." Unpublished draft. Omaha, Neb.: National Park Service, July 1989.

Gates, Paul W. "Frontier Land Business in Wisconsin." *Wisconsin Magazine of History* 52, no. 4 (Summer 1969): 310–33.

Gautier de Verville, Charles. "Gautier's Journal of a Visit to the Mississippi, 1777–1778." *Wisconsin Historical Collections* 11 (1888): 102–5.

"Geographical Names in Wisconsin." *Wisconsin Historical Collections* 1 (1849): 101–19.

Gibbon, Guy. *The Sioux: The Dakota and Lakota Nations.* Malden, Mass.: Blackwell, 2003.

Gibbs, Oliver, and C. E. Young. "Sketch of Prescott, and Pierce County." *Wisconsin Historical Collections* 3 (1857): 453–65.

Glad, Paul W. *War, a New Era, and Depression, 1914–1940.* The History of Wisconsin 5. Madison: State Historical Society of Wisconsin, 1990.

Goddard, James Stanley. "Journal of a Voyage, 1766–67," ed. Carolyn Gilman. In *The Journals of Jonathan Carver and Related Documents,* ed. John Parker. St. Paul: Minnesota Historical Society Press, 1976.

Goodman, Nancy, and Robert Goodman. *Joseph R. Brown: Adventurer on the Minnesota Frontier, 1820–1849.* [Rochester, Minn.]: Lone Oak Press, 1996.

Gough, Robert. *Farming the Cutover: A Social History of Northern Wisconsin, 1900–1940.* Lawrence: University Press of Kansas, 1997.

Griffin, Duane. "Wisconsin's Vegetation History and the Balancing of Nature." In *Wisconsin Land and Life,* ed. Robert C. Ostergren and Thomas R. Vale. Madison: University of Wisconsin Press, 1997.

Grignon, Augustin. "Seventy-two Years' Recollections of Wisconsin." *Wisconsin Historical Collections* 3 (1857): 244–88.

Harris, Moira F. "Ho-ho-ho! It Bears Repeating: Advertising Characters in the Land of Sky Blue Waters." *Minnesota History* 57, no. 1 (Spring 2000): 23–35.

Haugen, Nils P. "Pioneer and Political Reminiscences." *Wisconsin Magazine of History* 11, no. 2 (December 1927): 146–53.

Hazzard, George H. *Interstate Park: Lectures, Laws, Papers, Pictures, Pointers.* St. Paul, 1896.

Helgeson, Arlan. "Nineteenth Century Land Colonization in Northern Wisconsin." *Wisconsin Magazine of History* 36, no. 2 (Winter 1952–53).

Hennepin, Louis. *A Description of Louisiana.* Ed. John Gilmary Shea. New York: John G. Shea, 1880.

Henry, Alexander. *Travels and Adventures in Canada and the Indian Territories between the Years 1760 and 1776.* Edmonton, Alberta: M. G. Hurtig, 1969.

Henry, William A. *Northern Wisconsin: A Hand-Book for the Homeseeker.* Madison, Wis.: Democrat, 1896.

"Here's How Kilkare Lodge Looks Today!" Pamphlet. Danbury, Wis.: Burnett County Historical Society, [1929].

The Heritage Areas of Burnett County. Madison: Wisconsin Heritage Areas Program, 1978.

Hickerson, Harold. *The Chippewa and Their Neighbors: A Study in Ethnohistory.* New York: Holt, Rhinehart, 1970.

———. *Ethnohistory of the Chippewa in Central Minnesota.* New York: Garland, 1974.

———. *Mdewakanton Band of Sioux Indians.* New York: Garland, 1974.

Hidy, Ralph W., Frank Ernest Hill, and Alan Nevins. *Timber and Men: The Weyerhaeuser Story.* New York: Macmillan, 1963.

History Network of Washington County. *Minnesota Beginnings: Records of Saint*

Croix County, Wisconsin Territory, 1840–1849. Stillwater, Minn.: Washington County Historical Society, 1999.

Holbrook, Stewart. *Holy Old Mackinaw: A Natural History of the American Lumberjack*. Sausalito, Calif.: Comstock, 1938.

Holmquist, June Drenning, ed. *They Chose Minnesota: A Survey of the State's Ethnic Groups*. St. Paul: Minnesota Historical Society Press, 1981.

Hurst, James Willard. "The Institutional Environment of the Logging Era in Wisconsin." In *The Great Lakes Forest: An Environmental and Social History*, ed. Susan L. Flader. Minneapolis: University of Minnesota Press, 1983.

———. *Law and Economic Growth: The Legal History of the Lumber Industry in Wisconsin, 1836–1915*. Cambridge, Mass.: Belknap Press of Harvard University Press, 1964.

Huth, Hans. *Nature and the American: Three Centuries of Changing Attitudes*. Lincoln: University of Nebraska Press, 1990.

Innis, Harold A. *The Fur Trade in Canada*. Toronto: University of Toronto Press, 1956.

Johnson, C. A., J. Pastor, and R. J. Naiman. "Effects of Beaver and Moose on Boreal Forest Landscapes." In *Landscape Ecology and Geographic Information Systems*, ed. Roy Haines-Young, David R. Green, and Steven. H. Cousins. London: Taylor & Francis, 1994.

Johnson, Emeroy. "Early History of Chisago Lake Reexamined." Minnesota Historical Society Collection, n.d.

———. *An Early Look at Chisago County*. North Branch, Minn.: Chisago County Bicentennial Committee, 1976.

Johnson, Hildegard Binder. "Towards a National Landscape." In *The Making of the American Landscape*, ed. Michael P. Conzen. Boston: Unwin Hyman, 1990.

Johnson, Iver. "Johnson's Web Lake Resort." Pamphlet. Danbury, Wis.: Burnett County Historical Society.

Johnson, Sherman E. *From the St. Croix to the Potomac—Reflections of a Bureaucrat*. Bozeman: Montana State University, 1974.

Kappler, Charles J., ed. *Indian Treaties, 1778–1883*. New York: Interland, 1972.

Karamanski, Theodore J. "Saving the St. Croix: An Administrative History of the Saint Croix National Scenic Riverway." Omaha, Neb.: National Park Service, Midwest Region, 1992.

Kates, James. *Planning a Wilderness: Regenerating the Great Lakes Cutover Region*. Minneapolis: University of Minnesota Press, 2001.

Kay, Jeanne. "The Land of La Baye: The Ecological Impact of the Green Bay Fur Trade, 1634–1836." Ph.D. diss., University of Wisconsin–Madison, 1977.

Kellogg, Louise Phelps. "A Portrait of Wisconsin." *Wisconsin Magazine of History* 25, no. 3 (March 1942): 264–72.

Kohl, Johann George. *Kitchi-Gami: Life among the Lake Superior Ojibway*. Trans. Lascelles Wraxall; intro by Robert E. Bieder. St. Paul: Minnesota Historical Society Press, 1985.

Kohlmeyer, Fred W. *Timber Roots: The Laird, Norton Story, 1855–1905.* Winona, Minn.: Winona County Historical Society, 1972.

Kolehmainen, John Ilmari. "In Praise of the Finish Backwoods Farmer." *Agricultural History* 24 (January 1950): 1–21.

Krech, Shepard, III. *The Ecological Indian: Myth and Reality.* New York: Norton, 1999.

La Harpe, Jean Baptiste Bénard de. "Le Suer's Voyage up the Mississippi." *Wisconsin Historical Collections* 16 (1902): 178–91.

Lampard, Eric E. *The Rise of the Dairy Industry: A Study in Agricultural Change, 1820–1920.* Madison: State Historical Society of Wisconsin, 1963.

Landes, Ruth. *The Mystic Lake Sioux: Sociology of the Mdewakantonwan Santee.* Madison: University of Wisconsin Press, 1968.

———. *The Ojibway Women.* New York: Columbia University Press, 1938.

Larson, Agnes M. *History of the White Pine Industry in Minnesota.* Minneapolis: University of Minnesota Press, 1949.

Larson, Paul Clifford. *A Place at the Lake.* Afton, Minn.: Afton Historical Society Press, 1998.

Lass, William E. *Minnesota: A History.* New York: Norton, 1998.

———. "Minnesota's Separation from Wisconsin: Boundary Making on the Upper Mississippi Frontier." *Minnesota History* 50, no. 8 (Winter 1987): 310–30.

Lawson, Nelson. "Luck Township." In *Recollections of 1876: Polk County's First Written Histories.* Osceola, Wis.: Polk County Press, 1876–78.

Limerick, Patricia Nelson. "What on Earth Is the New Western History?" In *Trails: Toward a New Western History,* ed. Patricia Nelson Limerick, Clyde A. Milner II, and Charles E. Rankin, 81–88. Lawrence: University Press of Kansas, 1991.

Long, Stephen H. *The Northern Expeditions of Stephen H. Long: The Journals of 1817 and 1823 and Related Documents.* Ed. Lucile M. Kane, June Holmquist, and Carolyn Gilman. St. Paul: Minnesota Historical Society Press, 1978.

Malhiot, François Victor. "A Wisconsin Fur-Trader's Journal, 1804–05." *Wisconsin Historical Collections* 19 (1910): 163–233.

Marple, Eldon M. *A History of the Hayward Lakes Region: Through the Eyes of the Visitor Who Came and Stayed.* Hayward, Wis.: Chicago Bay Grafix, 1976.

———. *The Visitor Who Came to Stay: Legacy of the Hayward Area.* Hayward, Wis.: Country Print Shop, 1971.

———. *The Visitor Writes Again.* Hayward, Wis.: County Print Shop, 1984.

Marx, Leo. *The Machine in the Garden: Technology and the Pastoral Ideal in America.* New York: Oxford University Press, 1964.

Meier, Peg. *Too Hot, Went to Lake: Seasonal Photos from Minnesota's Past.* Minneapolis: Neighbors Publishing, 1993.

Merritt, Raymond H. *Creativity, Conflict and Controversy: A History of the St. Paul District U.S. Army Corps of Engineers.* Washington, D.C.: U.S. Government Printing Office, 1979.

————. *History of the Santee Sioux: United States Indian Policy on Trial*. Lincoln: University of Nebraska Press, 1967.

Minnesota State Board of Health. *State Laws and Regulations Relating to Hotels, Restaurants, Places of Refreshment, Lodging Houses, and Boarding Houses*. St. Paul, 1932.

Moberg, Ward, "The Dalles: Preserve or Develop?" *Dalles Visitor* 13 (Summer 1981).

————. "The 1890s Fight Created Dalles Parks." *Dalles Visitor* 14 (Summer 1982).

Mondale, Walter F. *Hearing before the Subcommittee on Public Lands of the Committee on Interior and Insular Affairs, U.S. Senate, Ninety-Second Congress, S 1928, a Bill to Amend the Wild and Scenic Rivers Act*. Washington, D.C.: U.S. Government Printing Office, 1972.

Morton, Arthur S. *A History of the Canadian West to 1870–71*. Ed. Lewis G. Thomas. Toronto: University of Toronto Press, 1973.

Murphy, Raymond E. *The Geography of the Northwestern Pine Barrens of Wisconsin*. Madison: Department of Geography, University of Wisconsin–Madison, 1931.

National Plan Service. *Summer Camps and Cottages*. Chicago: National Plan Service, [194?].

Neill, E. D. "Battle of Lake Pokegama." *Minnesota Historical Society Collections* 1 (1872): 141–44.

Nelson, George. *My First Years in the Fur Trade: The Journals of 1802–1804*. Ed. Laura Peers and Theresa M. Schenck. St. Paul: Minnesota Historical Society Press, 2002.

————. *A Winter in the St. Croix Valley: George Nelson's Reminiscences, 1802–03*. Ed. Richard Bardon and Grace Lee Nute. St. Paul: Minnesota Historical Society Press, 1948.

Nesbit, Robert C. *Urbanization and Industrialization, 1873–1893*. The History of Wisconsin 3. Madison: State Historical Society of Wisconsin, 1985.

————. *Wisconsin: A History*. Madison: University of Wisconsin Press, 1973.

Nicollet, Joseph N. *The Journals of Joseph N. Nicollet: A Scientist on the Mississippi Headwaters, with Notes on Indian Life, 1836–37*. Ed. Martha Coleman Bray. St. Paul: Minnesota Historical Society, 1970.

Northern Initiatives Strategic Planning Workgroup. *Northern Initiatives: A Strategic Plan for the Next Decade*. Madison: Wisconsin Department of Natural Resources, Summer 1994.

Nute, Grace Lee. "Marin versus La Verndrye." *Minnesota History* 32, no. 4 (December 1951): 226–38.

Ostergren, Robert C. *A Community Transplanted: The Trans-Atlantic Experience of a Swedish Immigrant Settlement in the Upper Middle West, 1835–1915*. Madison: University of Wisconsin Press, 1988.

Ostergren, Robert C., and Thomas R. Vale, ed. *Wisconsin Land and Life*. Madison: University of Wisconsin Press, 1997.

Outwater, Alice. *Water: A Natural History.* New York: Basic Books, 1996.

Page, Brian, and Richard Walker. "From Settlement to Fordism: The Agro-Industrial Revolution in the American Midwest." *Economic Geography* (October 1991): 280–93.

Parkman, Francis. *Montcalm and Wolfe: France and England in North America.* Boston: Little, Brown, 1908.

Peet, Edward L. *Burnett County, Wisconsin.* Grantsburg, Wis.: The Journal of Burnett County Print, 1902.

Perrot, Nicholas. "Memoir on the Manners, Customs, and Religion of the Savages of North America." In *The Indian Tribes of the Upper Mississippi Valley and the Region of the Great Lakes,* ed. Emma Helen Blair. Lincoln: University of Nebraska Press, 1996.

Petersen, William J. *Steamboating on the Upper Mississippi.* Iowa City: State Historical Society of Iowa, 1968.

Peterson, Jacqueline, and Jennifer S. H. Brown. *The New Peoples: Being and Becoming Metis in North America.* Lincoln: University of Nebraska Press, 1985.

Pike, Zebulon Montgomery. *An Account of Expeditions to the Sources of the Mississippi: And through Parts of Louisiana, to the Sources of the Arkansaw, Kans, La Platte, and Pierre Jaun Rivers.* Philadelphia: C. & A. Conrad, 1810.

———. *Sources of the Mississippi and the Western Louisiana Territory.* Baltimore: Fielding Lucas, 1810.

Pond, Peter. "Narrative of Peter Pond." In *Five Fur Traders of the Northwest,* ed. Charles M. Gates. St. Paul: Minnesota Historical Society, 1965.

Pond, Samuel W. "The Dakotas or Sioux of Minnesota as They Were in 1834." *Minnesota Historical Society Collections* 12 (1889): 338–49.

———. "Indian Warfare in Minnesota." *Minnesota Historical Society Collections* 3 (1880): 131–33.

Popular Science Monthly. *How to Build Cabins, Lodges, and Bungalows: Complete Manual of Constructing, Decorating, and Furnishing Homes for Recreation or Profit.* New York: Grosset & Dunlap, 1934. Reprint, 1946.

Prescott, Philander. *The Recollections of Philander Prescott: Frontiersman of the Old Northwest, 1819–1862.* Ed. Donald Dean Parker. Lincoln: University of Nebraska Press, 1966.

Pritzker, Barry M. *A Native American Encyclopedia: History, Culture, and Peoples.* New York: Oxford University Press, 2000.

Purcell, William Gray. *St. Croix Trail Country: Recollections of Wisconsin.* Minneapolis: University of Minnesota Press, 1967.

Radisson, Pierre Espirit. *Voyages of Peter Espirit Radisson.* Ed. Gideon Scull. New York: Peter Smith, 1943.

Recollections of 1876: Polk County's First Written Histories. Osceola, Wis.: Polk County Press, 1876–78.

Rector, William G. "The Birth of the St. Croix Octopus." *Wisconsin Magazine of History* 40, no. 3 (Spring 1957): 171–78.

————. *Log Transportation in the Lake States Lumber Industry, 1840–1918.* Glendale, Calif.: Arthur H. Clarke, 1953.

————. "Lumber Barons in Revolt." *Minnesota History* 31, no. 1 (March 1950): 33–39.

Reiger, John. F. *American Sportsmen and the Origins of the Conservation Movement.* Norman: University of Oklahoma Press, 1973.

Reimann, Lewis C. *Hurley—Still No Angel.* Ann Arbor, Mich.: Northwoods, 1954.

Resolution 263. In *Condition of Indian Affairs in Wisconsin: Hearings before the Committee on Indian Affairs, United States Senate.* Washington, D.C.: Government Printing Office, 1910.

Resolution 2264. In *St. Croix Indians of Wisconsin: Hearings before the U.S. House Committee on Indian Affairs, 23 July 1919.* Washington, D.C.: Government Printing Office, 1919.

Rice, John G. "The Swedes." In *They Chose Minnesota: A Survey of the State's Ethnic Groups,* ed. June Drenning Holmquist. St. Paul: Minnesota Historical Society Press, 1981.

Richardson, Bart Christopher. "Picturesque Landscape Assessment: The Lower St. Croix." M.A. thesis, University of Minnesota, 1993.

Ritzenthaler, Robert E. "Southwestern Chippewa." In *Handbook of North American Indians.* Vol. 15, *Northeast,* ed. Bruce G. Trigger, 747–49. Washington, D.C.: Smithsonian Institution, 1978.

Rohe, Randall. "Lumbering, Wisconsin's Northern Urban Frontier." In *Wisconsin Land and Life,* ed. Robert C. Ostergren and Thomas R. Vale, 221–40. Madison: University of Wisconsin Press, 1997.

Rosenfelt, Willard E. *Washington: A History of the Minnesota County.* Stillwater, Minn.: Croixside Press, 1977.

Rosholt, Malcolm. *Trains of Wisconsin.* Rosholt, Wis.: Rosholt House, 1985.

————. *The Wisconsin Logging Book, 1839–1939.* Rosholt, Wis.: Rosholt House, 1980.

Rowlands, Water A. "The Great Lakes Cutover Region." In *Regionalism in America,* ed. Merrill Jensen. Madison: University of Wisconsin Press, 1965.

Ryan, J. C. *Early Loggers in Minnesota.* 3 vols. Duluth: Minnesota Timber Producers Association, 1975–1980.

Satz, Ronald N. *Chippewa Treaty Rights: The Reserved Rights of Wisconsin's Chippewa Indians in Historical Perspective.* Madison: Wisconsin Academy of Sciences, 1991.

Schafer, Joseph. *A History of Agriculture in Wisconsin.* Madison: State Historical Society of Wisconsin, 1922.

Schmeckebier, Laurence Eli. "A Gothic House at Taylors Falls." *Minnesota History* 27, no. 2 (June 1946): 123–34.

Schmitt, Peter J. *Back to Nature: The Arcadian Myth in Urban America.* New York: Oxford University Press, 1969.

Schoolcraft, Henry Rowe. *Narrative Journal of Travels through the Great Chain of American Lakes to the Sources of the Mississippi River in the Year 1820.* Albany: E. & F. Hosford, 1821.

———. *Personal Memoirs of a Residence of Thirty Years with the Indian Tribes on the American Frontiers.* Philadelphia: Lippincott, Grambo, 1851. Reprint, New York: Arno Press, 1975.

———. *Schoolcraft's Expedition to Lake Itasca: The Discovery of the Source of the Mississippi.* Ed. Philip P. Mason. East Lansing: Michigan State University Press, 1958.

———. *Travels through the Northwestern Regions of the United States.* Albany: E. & F. Hosford, 1821.

Seymour, E. S. *Sketches of Minnesota, the New England of the West.* New York: Harper & Bros., 1850.

Shapiro, Aaron. "Up North on Vacation: Tourism and Resorts in Wisconsin's North Woods, 1900–1945." *Wisconsin History Magazine* 89, no. 4 (2006): 2–13.

Smith, Alice E. "Caleb Cushing's Investments in the St. Croix Valley." *Wisconsin Magazine of History* 28, no. 1 (September 1944): 7–19.

———. *From Exploration to Statehood.* The History of Wisconsin 1. Madison: State Historical Society of Wisconsin, 1973.

Sorden, L. G. "Some Events in Wisconsin's Forest History." Paper delivered to the Forest History Association of Wisconsin, 28–29 September 1979, Wausau, Wis. (Paper in the Burnett County Historical Society, Danbury, Wis.)

St. Louis, Minneapolis, & St. Paul Short Line. *The Summer Resorts of Minnesota: Information for Invalids, Tourists and Sportsmen.* Minneapolis: Dimond & Ross, 1882.

Tarr, Sharon. *Spooner: A History to 1930.* Spooner, Wis.: Privately printed, 1976.

Thompson, John Giffin. "The Rise and Decline of the Wheat Growing Industry in Wisconsin: The Use and Abuse of America's Natural Resources." Ph.D. diss., University of Wisconsin–Madison, 1909. Reprint, New York: Arno Press, 1972.

Trygg, J. William. *Composite Map of United States Land Surveyor's Original Plats and Field Notes.* Minnesota Series. Ely, Minn.: J. Wm. Trygg, 1966.

Turner, Frederick Jackson. *The Frontier in American History.* New York: Henry Holt & Co., 1920.

Van Kirk, Sylvia. *Many Tender Ties: Women in Fur-Trade Society, 1670–1870.* Norman: University of Oklahoma Press, 1980.

Vezina, Rosemarie. *Nevers Dam: The Lumberman's Dam.* St. Croix Falls, Wis.: Standard Press, 1965.

Vogel, John N. *Great Lakes Lumber on the Great Plains: The Laird, Norton Lumber Company in South Dakota.* Iowa City: University of Iowa Press, 1992.

Walljasper, Jay. "When the Trolley Was King." *Minneapolis.St.Paul Magazine,* April 1981.

Ward, Janice. *Next Stop Dresser Junction.* Osceola, Wis.: Osceola Sun, 1976.

Warren, William W. *History of the Ojibway People.* St. Paul: Minnesota Historical Society Press, 1984.

Weatherhead, Harold. *Westward to the St. Croix: The Story of St. Croix County, Wisconsin.* Hudson, Wis.: St. Croix Historical Society, 1978.

Wells, Robert W. *"Daylight in the Swamp!" Lumberjacking in the Late Nineteenth Century* Garden City, N.Y.: Doubleday, 1978.

———. *Fire at Peshtigo.* New York: Prentice Hall, 1968.

West, Elliot. *The Contested Plains: Indians, Goldseekers, and the Rush to Colorado.* Lawrence: University of Kansas Press, 1998.

White, Bruce M. *The Fur Trade in Minnesota: An Introductory Guide to Manuscript Sources.* St. Paul: Minnesota Historical Society Press, 1977.

———. "Indian Visit Stereotypes of Minnesota's Native People." *Minnesota History* 53 (Fall 1992): 103–6.

———. "The Regional Context of the Removal Order of 1850." In *Fish in the Lakes, Wild Rice, and Game in Abundance: Testimony on Behalf of Mille Lacs Ojibwe Hunting and Fishing Rights,* ed. James M. McClurken. East Lansing: Michigan State University Press, 2000.

White, Richard. "The Winning of the West: The Expansion of the Western Sioux in the Eighteenth and Nineteenth Centuries." *Journal of American History* 65, no. 12 (1978): 319–43.

Woodward, C. Vann. *The Burden of Southern History.* 3rd ed. Baton Rouge: Louisiana State University Press, 1993.

"Work at Camp Riverside." [Danbury, Wis.].: Burnett County Historical Society, n.d.

Wovcha, Daniel S., Barbara C. Delaney, and Gerda E. Nordquist. *Minnesota's St. Croix River Valley and Anoka Sandplain.* Minneapolis: University of Minnesota Press, 1995.

Writers' Program of the Works Project Administration. *Minnesota: A State Guide.* American Guide Series. New York: Viking Press, 1938.

———. *Minnesota: Facts, Events, Places, Tours.* New York: Bacon & Wieck, 1941.

———. *Wisconsin: A Guide to the Badger State.* American Guide Series. New York: Duell, Sloan, and Pierce, 1938.

———. *Wisconsin: Facts, Events, Places, Tours.* New York: Bacon & Wieck, 1941.

Wyman, Mark. *The Wisconsin Frontier.* Bloomington: Indiana University Press, 1998.

Index

Note: Page numbers in italics refer to figures and maps.

WISCONSIN LAND AND LIFE

ARNOLD ALANEN
Series Editor